SPRINGER
LAB MANUAL

Springer-Verlag Berlin Heidelberg GmbH

Rolf-Dieter Wegner (Ed.)

Diagnostic Cytogenetics

With 81 Figures, 9 in Color

Springer

PROF. DR. ROLF-DIETER WEGNER, PH.D.
Institut für Humangenetik
Charité Campus Virchow-Klinikum
Augustenburger Platz 1
D-13353 Berlin

ISBN 978-3-642-47813-0 ISBN 978-3-642-59918-7 (eBook)
DOI 10.1007/978-3-642-59918-7

Diagnostic cytogenetics / Rolf-Dieter Wegner (ed.).
p. cm. – (Springer lab manual)
Includes bibliographical references and index.

1. Genetic disorders-Diagnosis. 2. Human chromosome abnormalities-Diagnosis. 3. Human cytogenetics.
I. Wegner, Rolf-Dieter, 1949– II. Series.
RB155.6.D53 1999
616'.042 – dc21 99-14928 CIP

Cover design: design & production GmbH, D-69121 Heidelberg
Typesetting: Mitterweger Werksatz GmbH, D-68723 Plankstadt
SPIN 10517978 27/3134 5 4 3 2 1 0 - Printed on acid free paper

To my family

Contents

Preface

In the late 50s, the era of clinical cytogenetics began with technical developments which significantly improved tissue culture and chromosome preparation. The accidental discovery of the blood culture system together with the employment of colchicine and hypotonic treatment led to a breakthrough allowing an easy analysis of the human chromosomes. Subsequently, efforts of cytogeneticists focused on chromosomal banding techniques to allow a better insight into chromosomal structures and to create the basis for a more precise study of chromosomal aberrations in humans. Starting in the late 60s with the description of Q-banding, the essential techniques developed during the early and mid 70s are still applied as routine diagnostic techniques at the present time. These now traditional cytogenetic approaches still provide the basic methodology for everyday clinical cytogenetics. However, in the last few years the introduction of molecular-cytogenetics proved soon its value in supplementing the routine approaches. While fluorescence in situ hybridization (FISH) has become a standard technique, new ones such as comparative genomic hybridization (CGH), spectral karyotyping (SKY) and multi-color FISH (M-FISH) which rely basically on in situ hybridization, have shown their potential for diagnostic purposes. Thus, it is obvious that for comprehensive diagnostic cytogenetics a combination of both traditional and new cytogenetic techniques are required in a number of cases.

This volume **Diagnostic Cytogenetics** of the Springer Lab Manual series should provide the reader with protocols covering the main techniques needed for cytogenetics diagnostics in a clinical setting - but also for research purposes. Thus, basic techniques are described by experts working actively in these fields. In addition to a step by step description of every technique, much emphasis is given on how to overcome technical problems. Thus, this book is aimed at the beginner in cytogenetics, providing protocols helpful at the work-bench as well as to those colleagues already working in this field and looking for some technical hints.

This book is structured to start with basic issues of tissue culture and chromosome staining techniques (Chapters 1 and 2). With respect to the responsibility taken by anyone active in clinical diagnostics it seemed essential to add some information about karyotyping, nomenclature and quality control (Chapters 3 and 4). Subsequently, the particular needs for postnatal diagnostics (Chapters 5-10) are provided, followed by those for prenatal diagnosis (Chapters 11-13). A small part of the book is dedicated to the diagnosis of specific syndromes (Chapters 14 and 15) to cell fusion to reveal genetic heterogeneity (Chapter 16) and to trace back the origin of the supernumerary chromosome in trisomies (Chapter 17). The last two parts feature techniques developed during recent years. First come molecular cytogenetic protocols of techniques already in general use as common FISH (Chapters 18), FISH in human sperm cells (Chapter 19), reverse chromosome painting (Chapter 20), and comparative genomic hybridization (Chapter 21). Finally, recently invented techniques - enrichment of fetal cells in maternal blood (Chapter 22), spectral karyotyping (Chapter 23) and M-FISH (Chapter 24) - should give some insight into future developments of diagnostic cytogenetics.

Berlin, Summer 1999 ROLF-DIETER WEGNER

Part I

Classical Cytogenetics

Chapter 1

Tissue Culture

FRIEDEL WENZEL

Introduction

The history of *in vitro* tissue culture dates from the closing years of the 19th century. Loeb (1897), one of the first pioneers in this field, used organotypic cultures. In the initial period, tissue culture was used primarily for investigations in the field of physiology and embryology; however, medical and genetic questions soon gained an important position. Until the development of specific culture media (Eagle, 1955), the cytogenetic analysis of chromosomes depended on spontaneously dividing cells. In 1956, Tijo and Levan, using cultured embryonic cells, were the first scientists to report the correct number of human chromosomes as 46. Further fundamental events brought the breakthrough of cytogenetics as a clinical science: Moorhead et al (1960) established an *in vitro* culture method for the accumulation of dividing cells by using colchicine to arrest cells at metaphase. Also in 1960 Nowell discovered the mitogenic property of phytohemagglutinin, which resulted in a further improvement of cytogenetic techniques, notably in the use of peripheral blood cells. In 1966 Steele and Breg succeeded in culturing amniotic fluid cells and karyotyping fetal chromosomes; today the basic cell culture technology of amniotic fluid cells can be attributed to Milunsky (1979). *In vitro* culture of chorionic villi was developed through the 1970s, eg by Hahnemann (1974); however it took several years for the culture technique to be improved by Niazi et al (1981) and Brambati and Simoni (1983). Today chorionic cultures are well established and accompany direct cytogenetic preparation.

Before extending its range to bone marrow analysis, cytogenetics in hematology and oncology initially focused on peripheral blood as a specimen

Friedel Wenzel, UKBB-Universitäts-Kinderspital beider Basel, Abt. Med. Genetik, Römergasse 8, Basel, 4058, Switzerland (*phone* +41 61 685 67 92; *fax* +41 61 685 60 11)

because of its ease of sampling. As a result of technical difficulties in processing and culturing solid tumor tissue, cytogenetic analysis of this material made little progress until a really satisfactory collagenase solution for tissue disaggregation of solid tumor material was introduced by Van Hoff et al (1980), allowing a more standardized treatment of solid tumor specimens; enzymatic treatment remains the method of choice today.

In diagnostic cytogenetics, careful attention must always be paid to detail in cell culturing as well as in attending techniques. Basic protocols as well as technical comments are presented in this chapter, which is based on the experience of various laboratories. However, this may not exclude the possible need to adapt these methods to special laboratory conditions.

General requirements for cell culture

Classic cytogenetic analysis depends on cells undergoing mitosis to obtain metaphase chromosome spreads. Therefore cells have to be cultured *in vitro* (partly with artificially induced cell divisions) either as a short-term or long-term culture; increasing culture time bears the risk of *in vitro* chromosomal changes, especially in the case of tumor material (eg Zankl et al, 1979). Successful culturing requires well-balanced and stable environmental conditions especially with regard to basic parameters like pH, temperature and maintaining sterility. General aspects of nutrition and culture vessels are discussed later. For further details relating to local conditions such as the layout of a cell culture laboratory and essential equipment, readers are referred to special cell culture literature (Rooney and Czepulkowski, 1992; Freshney, 1993; Morgan and Darling, 1993; Lindl and Bauer, 1994; Martin, 1994; Jones, 1996).

pH The most favourable range of pH in cell cultures is between 7.2 and 7.4. Values of pH close to or higher than 8 are not withstood by cells for more than a few hours (depending on cell type). A pH of less than 7.0 causes cells to stop dividing. To avoid changes in ionic concentrations through loss of water, an atmospheric humidity of 90 to 95% is also necessary.

Temperature Human cell cultures need an environmental temperature of 37.0°C to 37.5°C, which is normally obtained by special incubators with or without a supplementary gas supply (CO_2, O_2) depending on the utilised buffer system. Fluctuations in temperature should not be larger than 0.25°C. Increasing temperatures may speed up the cell multiplication rate so long as the temperature remains below a lethal limit. A small decrease in temperature depresses cell metabolism, but has no general damaging effects.

Maintaining sterility

Cells in culture have no defensive mechanism like the immune system *in vivo* to overcome infections. Antibiotics are sometimes used prophylactically in order to compensate for this, but it is hazardous to depend on them for the prevention of contamination. The better way is to strictly adhere to an aseptic working technique, performing cell culture work in a biological safety cabinet class II which is equipped with a HEPA (high efficiency particulate air) filter. This protects both technician and culture from airborne infections. However it is essential to know in detail the air movement and possible turbulences in a safety cabinet and to adapt working techniques accordingly. Some basic rules are

- not to move across an opened culture vessel,
- not to bring the specimen or other sterile equipment into contact with non-sterile objects,
- not to speak, sneeze or cough while working with sterile material, and
- always to bear in mind that although the air in a laminar flow hood may be sterile, the surface of the working area is not.

Cell types

Different tissue and cell types are used for cytogenetic studies. In prenatal diagnosis the most common types are amniotic cells, chorionic villi, fetal blood and also solid tissue from aborted fetuses. Postnatal peripheral blood, bone marrow, lymph node material, pleural fluids, solid tumor material, skin biopsies and established cell lines are used for diagnostic cytogenetics. Molecular cytogenetics is also based primarily on these specimens, although varying techniques are used for analysis of cytogenetic changes.

Laboratory safety

Safety considerations and quality control are very important aspects in a cytogenetic laboratory, but are sometimes pushed into the background. A cytogenetic laboratory is associated with various biological hazards and a principal aim should be to protect the examined cell cultures (protection of the objects) as well as the laboratory staff (personal protection) and the environment. Requirements of Good Laboratory Practice (GLP), national regulations on working with human tissue as well as all aspects of disposal have to be considered. Any human or animal specimen (fresh biopsy, fluids, tissue, cell lines) may contain infectious agents and therefore has to be considered as a biological hazard and as potentially infectious, thus requiring corresponding working techniques. A second important point is the handling of various toxic substances (eg xylene, methanol), corrosive solvents (eg glacial acetic acid) and potentially mutagenic chemicals (eg methotrexate) during processing of the specimen for cytogenetic analysis. Therefore written safety regulations have to be established, with special adaptations to the individual situation in a cytogenetic laboratory. These should be supported by regular laboratory safety training to keep in mind the various risks and what to do in case of an accident, whatever it may be. Knutsen (1991) discussed laboratory safety in detail; the main aspects are summarized in the following.

Biosafety

The basic elements of biosafety include safe workplace practice, the use of protective equipment and the observance of general precautions. Some basic aspects are:

- The cell culture laboratory should be an air-particle reduced room; windows must always be kept closed, door movements should be reduced to a minimum. Smoking, eating and drinking must be avoided.
- Laboratory workers should wear laboratory coats and disposable plastic gloves when processing specimens or handling biopsy material, fresh tissue or hazardous substances.
- Excess sample material as well as all disposable plasticware used for processing must be decontaminated according to special instructions.
- Reusable material like scissors, forceps or glassware must be cleaned after use and autoclaved again.
- Working areas and surfaces must be cleaned with 70% ethanol before and after handling specimens (decontamination by wiping with a 10% sodium hypochlorite solution is also possible), especially if several technicians share the same establishment and equipment.
- The laminar flow hood should contain only absolutely necessary consumable materials; ventilation louvres should be kept free and no toxic or radioactive material should be used in the hood since air is returned to the room atmosphere by some biological safety cabinets.
- Enough time must be allowed for the fan to purge the hood of airborne contaminants before introducing a new cell line.

Chemical safety

The risk of injury from chemical hazards can be greatly reduced if the hazards are identified and steps are taken for proper handling: eg unmistakable labelling of all chemicals and reagents, drawing attention to possible hazards, storage and use of chemicals only in the officially recommended manner, being fully conversant with first aid treatments for chemical exposures.

Physical safety

Certain items of laboratory equipment may cause accidents: centrifuges can cause infections (direct contact with leaking or broken tubes, creation of aerosols) or injury by broken glass; compressed gas cylinders require special storage and handling; autoclaves may cause burns when opened incorrectly.

Quality control management

Every cytogenetic laboratory should have a quality control and management system which covers all procedures within a genetic laboratory (eg administration, biosafety, probe handling inclusive all working procedures, data management, report system, internal and external further education, laboratory equipment) in correlation to GLP. Such systems may differ between individual establishments in so far as only few general and mandatory rules exist; however, several nations are working on national guidelines which will also serve as basis for a possible accreditation of a genetic laboratory. The following suggestions are based on Knutsen (1991) who discussed quality control management in detail.

Record keeping

All human material entering a cytogenetic laboratory must first be recorded in a laboratory-specific intake register. Many laboratories prefer a double entry in both the computer and a hard copy log book. A unique lab-specific number will be given, which contains at least the year and consecutive specimen number for the appropriate year; also a short form for the tissue should be included if a laboratory works with different tissues, especially if these are from the same patient. This unique number attends on every step during analysis of the material until sign-out. Further information will be collected and entered in the computer at the beginning as well as during processing of the specimen; however, the extent of this additional information may vary between laboratories. Essential elements of record keeping are summarized in Table 1.

Table 1. Details of record keeping

lab-specific number	eg 96AF0001 (year 1996, amniotic fluid, consecutive number) or TU/0001/96 (tumor specimen / consecutive number / year1996)
patient information:	first and last name as well as complete address, birthday and sex patient's hospital number or social security number referral indication and name and full address of referring physician inclusive additional information if necessary or available
characteristics of the specimen:	type of sample; date and time of sampling date and time of receipt condition and amount of specimen upon receipt: - volume or size, appearance (eg color of amniotic fluid) - microscopic assessment in case of chorion, biopsies or cell lines transport information, especially in case of suboptimal transport procedure
culture set up:	person who set up the culture; date and time of set up number and size of culture vessels number of cells placed in each culture vessel classification of cell types or cell clones (if possible) used medium, additives, buffer, serum, antibiotics number and date of subculturing special comments
further cytogenetic procedure:	details of culturing procedures see special chapters in this manual

This laboratory master record sheet should be as complete as possible to allow follow-up, especially in difficult cases or if problems arise during later procedures. Additionally a separate patient file can be created which contains all the clinical data and later on also preliminary and final reports of cytogenetic analysis. Some general rules in record keeping have to be mentioned:

- Always use a double check: name as well as consecutive specimen number on all paperwork and cultures.
- Complete labelling of each culture vessel with date, name, number, subculture, special supplements to avoid any possibility of confusion.
- Take care that handwritten labels are clearly legible.

Procedure manuals

According to GLP it is recommended to have a detailed collection of laboratory specific protocols. These protocols should be updated at least once a year; each revision has to be dated and initialled by a responsible person(s).

If changes in procedures become necessary during the year do not wait until the yearly up-date before changing the corresponding protocol, but do it immediately and also inform all persons who are working with this protocol.

Monitoring equipment

Laboratory equipment (incubators, freezers, laminar flow hood, waterbath, CO_2 supply, centrifuges, microscopes) have to be periodically inspected (daily, weekly, monthly or once a year). Make daily checks and regular monitoring controls of the CO_2 and temperature indication to avoid incubator failure, especially breakdown of CO_2 supply and overheating. Incubators should be connected to emergency power sources. If possible use at least two incubators which are not dependent on the same CO_2 source for each specimen.

Analysis of culture failure

Accurate and up-to-date records should be made of cell culture failures. A distinction must be made between culture failure as a laboratory problem and specimen failure due to inadequate collected specimens. The main aspects are sterility testing of media, testing for growth potential of media and sera, and the overall success rate of diagnostic specimens.

Reagents and supplies

A wide variety of different reagents, additives and supplements are used during routine work in a cytogenetic laboratory. Correct preparation, labelling, storage and disposal are necessary, and it is recommended that a log book be kept for this special purpose. This will help to find explanations for unexpected events. Sera, media and supplements should also be regularly tested (eg by measuring the cloning efficiency of a standard fibroblast strain or 7-day incubation of reagent aliquots before use); use of at least two different lots of culture flasks and media (either media of the same kind or different media).

Materials

Culture vessels

Many cell types used in cytogenetic diagnosis grow attached to a substrate (exceptions are hematopoietic cells and some tumor cells). Therefore various forms of cell culture treated surfaces (eg FALCON®, PRIMARIA®, BIOCOAT® tissue culture surface, all from Becton Dickinson, Germany) are available; additional after-treatment (eg coating with polylysin, fibronectin, collagen or others) is also possible in order to meet special culture requirements. Today plastic has become generally accepted as the basic material for culture vessels, allowing a large variety of culture vessel forms and sizes. The most common ones used in cytogenetics are culture flasks with screw caps and a growth area of 25 cm^2 (other sizes offer growth areas between about 12 cm^2 and 500 cm^2), Leighton tubes (120 x 16 mm), Petri dishes, multiwell cluster plates with varying growth areas and chamber slides. The latter consist of a glass or plastic slide which is combined with a chamber; this type of vessel allows in situ harvesting and provides the option of up to 16 chamber configurations per chamber slide.

For transport between sampling and culture set up, sterile centrifuge tubes or multi-purpose containers may normally be used. Material is also transported in prepared syringes, but problems may be presented by specific types of specimens (see below) and special requirements may result.

Culture media

The basic key to any cell culture medium is a balanced composition including the whole range of components required by a cell line. Each cell culture medium consists of a base salt solution providing the principal ion requirements, together with carbohydrates as an energy source. However, nutritional requirements for medium compositions vary between cells from different organs or tissues as well as according to the length of the culture period, the purpose of the culture and possible suppression of unwanted cell types. A wide range of different media have therefore been developed in the past by supplementing the basic medium with different additives:

- Serum or dialyzed serum is added in varying concentrations in order to make good unknown deficiencies; cytogenetic laboratories normally use either fetal calf serum, newborn calf serum, calf serum or autologous or homologous human serum with concentrations between almost serum-free and up to about 25%.
- Biologically active substances like vitamins, essential and non-essential amino acids, hormones, enzymes, fatty acids.
- L-glutamine as a non-essential amino acid serves as an essential nutrient (energy source). L-glutamine is one of the most unstable supplements in media. Depending on the storage temperature it hydrolyzes to ammonia and glutamic acid, which changes the pH and leads to a loss of L-glutamine. Therefore L-glutamine has to be added always just before use. Today stable alternatives are also available: eg GlutaMAX™ (Life Technol-

ogies, Scotland) is a stabilized dipeptide which does not degrade during storage and therefore avoids build-up of toxic metabolites.

- Buffer substances: Two different buffering systems for neutralising acidic waste products are in common use: either a bicarbonate buffer ($NaHCO_3$) requiring equilibration with a 5% CO_2 atmosphere or an organic buffer like HEPES (4-[2-hydroxy-ethyl]-1 piperazine-ethanesulfonic acid) or TRICIN (N-(trishydroxymethyl)-methyl-glycine) which do not need a special external gas atmosphere. However, the HEPES system is temperature dependent: HEPES-buffered medium at 15°C with a pH of about 7.55 will decrease its pH to about 7.25 with a rise in temperature to 37°C.
- Various ingredients like peptone, protein hydrolysate, yeast extracts.
- Antimicrobial agents (generally recommended in primary cultures): The most commonly used antibiotics in cell culture are penicillin G and streptomycin. However, further antimicrobial agents (see Table 2) are available to treat culture contamination; for more details see Subprotocol 5, Prophylaxis Against Microbial Contamination.
- Phenol red as a pH indicator is often added in a final concentration of 8 – 20 mg/l. At a pH of about 7.2 to 7.4 its colour is red-orange; the color changes to magenta when a solution becomes alkaline (pH 8.0 and above) or to yellow when it becomes too acidic (pH 6.8 and lower).

Media are available as ready-to-use solutions, as concentrates (2-fold, 5-fold, 10-fold) or as powder. The latter two forms need dilution with cell culture quality water (double or triple distilled) according to the instructions of the customer. Good manufacturers also provide details of media composition, storage conditions, stability date and quality control during production. Ready to use media containing serum have a shelf-life of about 4 weeks at 4°C. An overview of some common liquid synthetic media is presented in Table 3. Finally some basic and very general rules for use of media are given below.

Note: A change of medium should always be done by aspirating the "old" medium using sterile pipettes. Decanting of the medium must be avoided for it produces wet vessel necks which may serve as an ideal "contamination path" for microorganisms.
A pipette which has been introduced into a culture vessel containing cells must never be returned to stock bottles of media, salt solutions, trypsin or another culture vessel with different cells to avoid any possibility of cross contamination.

Table 2. Common antimicrobial agents used in cell culture and their characteristics

antimicrobial agent	stock	storage	suggested working concentration [per L medium]	stable at 37°C for	antimicrobial effect against[b)] gr-(+)	gr-(-)	myco.	yeast	mold	mode of action	customer[d)]
Amphotericin B	powder	2 - 8°C	2.5 mg	3 days				+	+	see note 1)	S
	solution 250 µg/ml	-20°C	10 ml	3 days				+	+	see note 1)	B, C, I, L, S, V
Ampicillin	powder	2 - 8°C	100 mg	3 days	+	+				see note 2)	S
Antibiotic-antimycotic solution (100x) (10.000) U Penicillin, 10 mg Streptomycin, 25 µg Amphotericin per ml)	solution	-20°C	10 ml	3 days	+	+		+	+	see action of single components	C, I, L, S
Anti-PPLO Agent	solution 6000 µg/mg	-20°C	5 - 10 ml	3 days			+			see note 9)	C, L
Biomyc-1 (100x)	solution	-20°C	10 ml	› 3 days			+			based on tiamutin	I
Biomyc-2 (100x)	solution	-20°C	10 ml	› 3 days			+			based on minocycline	I
Biomyc-3 (100x)	solution	-20°C	10 ml	› 3 days			+			based on ciprofloxacin	I
BM Cyclin 1 + 2	powder	2 - 8°C	5 mg (BMC1) 10 mg (BMC2)	4 days			+			see customer instructions	B

Cephalothin	powder	2 - 8°C	100 mg	3 days	+	+		inhibits synthesis of cell wall	S
Ciproxin ®	solution 0.2 g/ 100 ml	RT[a)]	2.5 ml	> 5 days			+	based on cipro-floxacin	Y
Ciprobay®	solution 0.2 g/ 100 ml	RT[a)]	2.5 ml	> 5 days			+	based on cipro-floxacin	
Dihydro-streptomycin	powder	2 - 8°C	100 mg	5 days	+	+		see note 3)	S
Erythromycin	powder	RT[a)]	100 mg	3 days	+	+		see note 4)	S
Gentamycin sulfate	powder 600 µg/mg	2 - 8°C	50 mg	5 days	+	+	+	see note 3)	S
	solution 10 mg/ml	2 - 8°C	5 ml	5 days	+	+	+	see note 3)	B, C*, I, L, S
Kanamycin	powder	RT[a)]	100 mg	5 days	+	+	+	see note 3)	S, V
	solution 10 mg/ml	2 - 8°C	10 ml	5 days	+	+	+	see note 3)	B, C*, L, S
Lincomycin Hydro-chloride	powder ~850 U/mg	2 - 8°C	100 mg	4 days	+			see note 5)	S
	solution 50000 µg/ml	-20°C		4 days	+			see note 5)	L
Minocycline	powder	2 - 8°C	5 mg				+		N
	solution 0.6 mg/ml	-20°C					+		V
Mycoplasma Removal Agent	solution 50 µg/ml	RT[a)]	0.5 mg	> 5 days			+	inhibits mycoplasma DNA-gyrase	N

Table 2. Continued

antimicrobial agent	stock	storage	suggested working con-centration [per L medium]	stable at 37°C for	antimicrobial effect against[b)] gr-(+)	gr-(-)	myco.	yeast	mold	mode of action	customer[d)]
Neomycin sulfate	powder	RT[a)]	50 mg	5 days	+	+				see note 6)	S
	solution 10 mg/ml	2 - 8°C	5 ml	5 days	+	+				see note 6)	C, L, S
Nystatin	suspension[c)] 10.000 U per ml	-20°C	10 ml	3 days				+	+	see note 1)	L, S, V
Paromomycin	powder	RT[a)]	100 mg	5 days	+						S
Penicillin G	powder 1600 U/mg	RT[a)]	100.000 U	3 days	+					see note 2)	S
Penicillin-Streptomycin-solution: 5.000 IU Penicillin 5 mg Streptomycin per ml	solution	-20°C	20 ml	3 days	+	+				see action of single components	B*, C*, H*, I*, L, S
Penicillin-Strepto-mycin-Neomycin-solution: 5.000 IU Penicillin 5 mg Streptomycin 10 mg Neomycin per ml	solution	-20°C	10 ml	3 days	+	+				see action of single components	I*, L, S
Phenoxymethyl-penicillanic acid (Penicillin V)	powder 1500 U/mg	RT[a)]	100.000 U	3 days	+					see note 2)	S

Polymyxin B	powder 6000 U/mg	2 - 8°C	50 mg	5 days		+		see note 7)	S
	solution 10000 U/ml	-20°C	10 ml	5 days		+		see note 7)	C, L, V
Spectinomycin dihydrochloride	powder	2 - 8°C	7.5 - 20 mg		+[e)]	+		see note 4)	S
Streptomycin sulfate	powder	2 - 8°C	100 mg	3 days	+	+		see note 3)	S
Tetracycline hydrochloride	powder	-20°C	10 mg	4 days	+	+		see note 8)	S
	solution 5000 µg/mg	-20°C	2 ml	4 days	+	+		see note 8)	C, L
Tiamutin	solution 1.2 mg/ml	-20°C	10 ml	4 days			+		V
Tylosin Tartrate	powder 900 µg/mg	2 - 8°C	8 mg	3 days	+		+	see note 9)	S
	solution 8 mg/ml	2 - 8°C	1 ml	3 days	+		+	see note 9)	S, V*

a): RT = room temperature
b): gr-(+) = Gram (+) bacteria, gr-(-) = Gram (-) bacteria, myco. = mycoplasma
c): Nystatin is insoluble in water, but it is effective as suspension
d): B = Roche Diagnostics, C = BioConcept (Amimed®), H = HyClone® Laboratories, I = Biological Industries, L = Life Technologies™, N = ICN Biomedical, S = Sigma BioSciences™, V = Serva, Y = Bayer Leverkusen
e): gonococcus
*): supplier offers the same agent, but with a different stock concentration which has to be considered in use
note 1): interferes with the permeability of cell membranes of sensitive fungi by binding steroids
note 2): interferes with the final stage of bacterial cell wall synthesis
note 3): interferes with bacterial protein synthesis by binding to 30S subunit of ribosomes
note 4): inhibits elongation at transpeptidation step
note 5): alters 50S subunit of bacterial ribosomes
note 6): causes miscoding and inhibits initiation and elongation
note 7): binds to and interferes with the permeability to the bacterial cytoplasmic membrane
note 8): inhibits elongation by blocking binding of aminoacyl-tRNA to the A-site on the 30S subunit
note 9): interferes with bacterial protein synthesis by binding to the 50S subunit

Table 3. Common synthetic liquid media and their field of application (many of these media are also available in other forms and with special additives, for detailed information see customer information)

medium	application	used in cytogenetics	supplier[s)]
Ames' Medium	maintaining central nervous system tissue in vitro		S
AmnioMAX™-C100	medium-kit, ready to use for primary culture of human amniotic fluid cells and chorionic villi	+	C, I, L, S
Basal Medium Eagle (BME)	primary mammalian cells, normal and transformed cells, HeLa cells		I, L, S
BGJb-Medium	culture of cartilaginous embryonic bone		S
BIO-AMF-1	human amniotic fluid cells	+	I
Cell Culture Freezing Medium	cryopreservation of various cell types		L
Chang Medium	medium-kit, ready to use for primary culture of human amniotic fluid cells and chorionic villi	+	B
Chromosome Diagnostic Medium (CDM™)	medium kit, ready to use for primary culture of human amniotic fluid cells and chorionic villi	+	S
Chromosome Medium 1A	ready to use medium for peripheral blood lymphocytes	+	L
Chromosome Medium 4	ready to use medium for peripheral blood lymphocytes	+	L
Chromosome Medium A	ready to use medium for peripheral blood lymphocytes with phytohemagglutinin	+	V
Chromosome Medium B	ready to use medium for peripheral blood lymphocytes without phytohemagglutinin	+	V

Table 3. Continued

medium	application	used in cytogenetics	supplier[s)]
CMRL-1066	initially designed for mouse L-cells, with serum-supplement usable for many cell types		C, L, S
CRCM-30 Medium	supports many animal and human cell lines		S
Dulbecco's Modified Eagle Medium (DMEM)	enriched modification of BME, supporting a broad spectrum of mammalian cell lines		C, H, I, L, S
DMEM-Ham's F-12 (1:1)	used as serum-free medium for a variety of cell lines, Leydig cells, Sertoli cells		C, H, L, S
Fischer's Medium	cells from leukemic mice		I, L, S
Glasgow Minimum Essential Medium	culture of baby hamster kidney cells		I, L, S
Ham's F-12 Coon's modification	special hybrid cells		S
High Cell Density Medium	support of higher cell densities of attached cell lines		S
H-Y Medium	hybrid cells		S
Iscove's Modified Dulbecco's Medium (IMDM)	murine B lymphocytes, hemopoietic tissue from bone marrow, B cells stimulated with lipopolysaccharide, T lymphocytes, hybrid cells	+	C, H, I, L, S
KaryoMAX® Peripheral Blood Karyotyping Medium	ready-to-use medium for lymphocytes	+	L
KaryoMAX® Bone Marrow Karyotyping Medium	ready-to-use medium for blood cells from bone marrow	+	L
KaryoMAX® MEM-Alpha	human amniocytes	+	L
KaryoMAX® RPMI-1640	lymphocytes from peripheral blood or bone marrow	+	L

Table 3. Continued

medium	application	used in cytogenetics	supplier[s)]
Leibovitz's L-15 Medium	used in CO_2-free systems, primary explants of embryonic and adult human tissue	+	C, H, I, L, S
Lymphocyte Separation Medium	aqueous solution of Ficoll® and sodium diatrizoate for separation of lymphocytes	+	L
McCoy's 5A Medium	lymphocytes, bone marrow, skin, gingiva, testes, adrenal glands, lung, spleen and other tissues	+	C, H, I, L, S
MCDB Medium	group of low protein or serum-free media, each MCDB medium supports a specific cell type		L, S
Medium 199 (M-199)	when supplemented with serum usable for various cell types	+	C, H, I, L, S
Medium 199 - FA	without folic acid, modified for fragile X chromosome evaluation	+	S
Minimum Essential Medium (MEM)	cultivation of a wide variety of adherent cells, in modification also usable for suspension cells		C, H, I, L, S
MEM Alpha Medium		+	C, I, L, S
MEM Alpha Medium-FA	without folic acid, modified for fragile X chromosome evaluation	+	S
MEM D-Val Medium	supports epithelial cells while selectively inhibiting fibroblast growth		C, L
NCTC 135 Medium	used in serum-free culture of special mammalian cells		L, S
Nutrient Mixtures Ham's F-10	CHO cells, HeLa cells, mouse L-cells, human diploid cells, white blood cells	+	C, H, I, L, S
Nutrient Mixtures Ham's F-12	primary rat hepatocytes, rat prostate epithelial cells		C, H, I, L, S

Table 3. Continued

medium	application	used in cytogenetics	supplier[a)]
PB-Max™ Karyotyping Medium	ready-to-use RPMI-based medium for lymphocytes from peripheral blood	+	L
RPMI-Medium	group of various media		S
e.g. RPMI 1603			L
e.g. RPMI 1630	suspension cultures, mouse leukemic cells		S
e.g. RPMI 1640	human normal and neoplastic leukocytes, suspension culture, monolayer culture, fresh human lymphocytes using PHA stimulation	+	C, H, I, L, S
e.g. RPMI 1640 - FA	without folic acid, modified for fragile X chromosome evaluation	+	S
Waymouth Medium MB 752/1	mouse L929 cells, organ culture, carcinoma cell lines, potentially tumorigenic cells		C, H, I, L, S
Williams' Medium E	long-term culture of adult liver epithelial cells		C, H, L, S

a): B = Bühlmann Laboratories, C = BioConcept (Amimed®), H = HyClone® Laboratories, I = Biological Industries, L = Life Technologies™, S = Sigma BioSciences™, V = Serva

It is also often recommended that each cell line should have its own medium stock bottle. This advice certainly makes sense, but may sometimes not be practicable if there are many different cell lines and restricted working space and storage capacities.

Culture media for diagnostic cytogenetics

In diagnostic cytogenetics many types of defined media are used for cell culture. However, optimal growth will be achieved with specially defined media (see Table 3) like Chang medium, BIO-AMF-1, Chromosome Diagnostic Medium, AmnioMAX™ or the KaryoMAX®-series. Possible alternatives are: Ham's F-10, Ham's F-12, Alpha-MEM, McCoy's 5A, RPMI 1640, MEM, M 199 and others. Addition of fetal calf serum is recommended in concentrations between 10 and 25%. Lower concentrations are usable only if growth factors are added. One exception are the special media kits mentioned above, some of which do not need further serum supplementation.

Another special group of media are the folic acid deficient media which have been specially developed for fragile X-analysis: eg MEM-FA, RPMI-FA, M-199. This special application is based on Sutherland (1977), who described the first human chromosome abnormality whose detection depended on the type of tissue culture media (M-199) used in cell culture.

Subprotocol 1
Sampling, Transport, Storage

In addition to culture set-up techniques, types of culture and further cytogenetic analysis, which will be discussed in other chapters of this manual, special attention will be given to sampling, transport and storage in order to provide the best specimen for cytogenetic analysis. Every tissue or cell sample has to be worked up as fast as possible for cytogenetic diagnosis. Any extremes of temperature and pH have to be avoided. For transport a medium should be used with a buffer system which does not require equilibration by a special atmosphere (eg HEPES); however, exceptions are possible.

Procedure

Amniotic fluid specimens

During normal amniocentesis a sample of 10 to 30 ml of amniotic fluid is withdrawn through a styletted needle. The first one or two milliliters obtained should be discarded to reduce the risk of maternal cell contamination. After aspiration amniotic fluid is stored and transported without any additive at room temperature using approved cell culture tubes or containers from amniocentesis kits in insulated boxes. The transport of amniotic fluid in toxic syringes or tubes remains a serious hazard (Kohn, 1981). Therefore referring physicians should be informed about this possible danger arising from "non-approved" plasticware and should be provided with hazard-free syringes and tubes. Some cytogenetic laboratories recommend splitting each amniotic fluid specimen for set-up by two different technicians; in any case only a single sample should be handled at one time in the hood.

Approximately 20 percent of amniotic fluid samples exhibit turbidity because of an extremely high content of cells and/or debris as well as erythrocytes, which may simulate bacterial contamination. The effective level of such contamination, however, is low because of the bacteriostatic property of the amniotic fluid itself.

Chorionic villi specimens

Sampling of chorionic villi specimens is performed by a gynaecologist using either a transvaginal or transabdominal technique. For sampling itself as well as the following preparation steps, aseptic techniques are absolutely necessary. Chorionic villi specimens are collected in a collection medium which is also suitable for other surgical specimens and tissue biopsies. The source of the chorionic villi influences further *in vitro* growth; material from the chorion laeve has less growth capacity than material from chorion frondosum.

Collection medium for chorionic villi

1. Add 100 ml of cell culture medium (eg RPMI 1640 including 10 to 20% fetal calf serum and 20 mM HEPES as buffering system) to a sterile 100 ml glass flask. For economy, the serum concentration of the collection medium can be reduced, especially when the time between collection and further processing of the specimen is very short.

Note: In a slight modification the addition of serum is completely avoided, however, 1 ml Liquémin (= 5000 IU Heparin, F. Hoffmann-La Roche & Co.AG Basel) is added per 100 ml collection medium.

2. Add an appropriate amount of amphotericin B stock solution and of gentamycin stock solution to make a final concentration of 1.25 µg/ml amphotericin B and 25 µg/ml gentamycin.

Note: The antibiotic concentration represents half the concentration recommended for treating an existing contamination. In case of specimens with a high risk of contamination, however, the antibiotic concentration should be doubled.

3. Gently mix the antibiotic containing medium and place aliquots of about 25 ml in four capped, sterile 50 ml centrifuge tubes or universal containers.
4. Label tubes or containers with the type of medium and date of preparation.
5. Fresh preparation will always be the method of choice, but storage of collection medium at 4°C for up to four weeks is permissible.

Further processing

1. The specimen will be immediately washed by several transfers to sterile Petri dishes, always using fresh collection medium; at the same time the chorionic villi are evaluated by an inverted microscope and separated from decidua, blood and mucus components to reduce contamination with maternal cells.

2. A sufficient sample should include at least 5 to 10 mg of good quality villi (adequate size, buds and branches present).

Note: Newport et al (1986) offer reference photographs to help with visual weight estimates of sampled chorionic villi.

3. Cleaned chorionic villi are transferred to a sterile 50 ml screw-cap centrifuge tube or universal container containing fresh collection medium. This is preferable to using a 20 ml syringe.
4. Transport to the cytogenetic laboratory has to be performed immediately using an insulated transport container to keep the specimen at room temperature.
5. Set-up of cultures from chorionic villi should be started immediately after arrival. Mailing of chorionic material is possible, although the rate of mosaicism due to maternal contamination seems to be higher after mailing than after direct culture set-up (Holzgreve and Miny, 1987). Hansson et al (1995) describe the preservation of chorionic villi in complete culture medium (Ham's F-10, supplemented with 25% fetal calf serum, 2 mmol L-glutamine, 100 IU/ml penicillin, 100 μg/ml streptomycin) at 37°C for up to 7 days before establishing cell culture; they did not find a delaying or inhibiting effect on the growth capacity of the mesenchymal cells.

Note: Depending on the gynaecologist's experience transport often occurs before washing and separation which then will be done by the experienced staff of the cytogenetic Laboratory.

Fetal blood

Fetal blood can be obtained by fetoscopy, placental aspiration, percutaneous umbilical cord sampling or from an aborted fetus. It will be sampled in a sodium heparin tube and treated in a similar way to a normal blood specimen which is used for cytogenetic analysis. However, the obtained amount of fetal blood is often small, thus necessitating microtechniques for culture. A different number of cells per ml has also to be considered.

Peripheral blood

The standard technique to obtain peripheral blood is venipuncture into a syringe coated with preservative-free heparin (alternatively also a finger stick collecting about 6 drops in 5 ml Chromosome Medium 1A). Normally 2 to 10 ml of peripheral blood (in the case of young babies the amount may be even smaller) are aspirated and transferred into a 15 ml centrifuge tube supplemented with 100 IU to 200 IU of sodium heparin as the anticoagulant of choice, mixed by gently inverting the tube and transporting to a cytogenetic laboratory. Alternatively, flush a 2 ml syringe with heparin and keep about 0.1 – 0.2 ml heparin in it, then aspirate blood up to 2 ml. The syringe can be transferred to the laboratory; however, care should be taken to fix the plunger of the syringe during transport. Lithium heparin is sometimes not recommended though we have good results when using it. EDTA in any case has to be avoided, for it seems to be toxic. Powdered heparin without preservative (freshly made up in sterile saline) is preferable but liquid heparin (containing alcohol as bacteriostatic agent) is also commonly used, whereas heparin solution that contains phenol has to be avoided. Specimens are unacceptable in the following cases: clotted blood, unlabelled or mislabelled tubes, incorrect anticoagulant, older than 72 hours, less than 1 ml in volume or unsuitable transport conditions.

Transport must take place at room temperature using insulated containers and keeping transit time to a minimum (eg express mail or overnight delivery); iced or frozen storage or transport may result in culture failure. Prolonged transit frequently results in loss of cell viability. A possible alternative may be to process the blood specimen directly at the site of collection till the fixation step and then to mail the fixed cells.

In a routine short-term culture of about three days the mitotic activity of cultured blood samples reaches a peak after 60 to 75 hours, which is therefore the optimum harvesting point for cytogenetic analysis. The samples should preferably arrive in the cytogenetic laboratory on Monday, Tuesday or Friday. If this is not possible the following deviations from the schedule are suggested (Barch, 1991): In case of arrival on Wednesday (1) either keep for one day, then culture 96 hours or (2) keep for two days, then culture 72 hours or (3) culture immediately for 48 hours. In case of arrival on Thursday (1) keep for one day, then culture 72 hours or (2) culture immediately for 96 hours. Blood can be accepted for culture up to 48 hours post-mortem.

Bone marrow

The aspirated bone marrow specimen is collected, stored and transported either in preservative-free sodium heparin or mixed with medium. The latter is recommended for maintaining cell viability especially in case of prolonged transit. In this case one suitable medium is RPMI 1640 supplemented with 10% heat inactivated fetal calf serum, 10 mmol/l HEPES, 100 IU sodium heparin, 100 IU/ml penicillin G, 100 µg/ml streptomycin.

Transport itself must take place as fast as possible at room temperature. Further processing of the bone marrow specimen should be initiated immediately upon arrival.

Bone marrow biopsy

As an alternative to aspirated bone marrow a bone marrow biopsy may also be used for cytogenetic analysis. In this case the procedure is modified to suit the different consistency of the specimen:

1. After sampling of a bone marrow biopsy, place the biopsy specimen in a sterile Petri dish and transfer to a sterile working area (laminar flow hood).
2. Mince the tissue with a scalpel blade or small surgical scissors using a sterile working technique. Be careful not to squeeze the specimen too much.
3. Resuspend the minced tissue in a little culture medium (RPMI 1640, 10% heat inactivated fetal calf serum, 10 mmol/l HEPES, 100 IU penicillin G, 100 µg/ml streptomycin).
4. Transfer the cell suspension to a sterile centrifuge tube, centrifuge at 800 rpm for 8 minutes and discard the supernatant.
5. Resuspend the cell and minced tissue pellet in 2 ml culture medium. In case of a homogeneous cell suspension, determination of the cell count may be useful in order to provide cultures with an appropriate number of cells. Transport to a cytogenetic laboratory is also possible at this point. In this case the amount of culture medium should be increased to about 5 ml.

Lymph node biopsy

The employed collection and transport medium is the same as in bone marrow biopsy. The lymph node sample (at least 2 to 3 mm^3 in size) must be transported immediately from the surgery center to the cytogenetic laboratory and undergo further processing without delay using a sterile universal container with a complete medium. Tissue preparation and centrifugation is the same as in bone marrow biopsy processing. Possible modifications in case of a non sterile biopsy specimen are: (1) washing the specimen in culture medium containing 500 IU/ml penicillin G, 500 µg/ml streptomycin and 500 µg/ml amphotericin B for several minutes before mincing the tissue and (2) adding a further washing step with the same antibiotic containing fresh medium after centrifugation.

Skin biopsy

The physician must pay special attention to an aseptic collection procedure, for collection often takes place in a room without an aseptic atmosphere. Handling of skin biopsies after collection is less critical compared to other tissues mentioned above, provided sterility is maintained. A complete culture medium based on RPMI 1640 (see bone marrow biopsy), MEM or DMEM should be used in a sterile universal container. This is suitable, however, only for short transit times. A special transport medium in case of longer transit times (much better suited to actual conditions) is based on L-15 medium. It is supplemented with 200 IU/ml penicillin, 200 μg/ml streptomycin, 1.25 μg/ml amphotericin B, all final concentrations. This medium is independent of pH regulation by an external gas phase and exhibits only small fluctuations in pH during transport (Leibovitz, 1986). Transport time is less critical if a complete medium is used at room temperature. Storage of a skin biopsy in a complete medium may be possible at room temperature for several days before further processing occurs. Personally I recommend setting up cultures of every skin specimen irrespective of possible unfavorable conditions during collection, transport and storage, for there is at least a small chance of obtaining outgrowing cells especially in skin biopsies.

Aborted fetal tissue

Solid tissue specimens from an aborted fetus can be transported in a suitable transport medium (see tumor biopsy) using a sterile universal container. It is recommended to already add antibiotics to the transport medium. If, however, no antibiotics are used, the specimen should be transferred into fresh complete medium (eg Alpha-MEM or Ham's F-10) containing 20% fetal calf serum and in addition 1.25 μg/ml amphotericin B and 50 μg/ml gentamycin after arrival in the cytogenetic laboratory. The specimen is stored overnight at room temperature before further processing.

Subprotocol 2 Cryopreservation

Cryopreservation means that living cells can be frozen and stored at low temperatures, maintaining their viability and even their function after thawing. Depending on the temperature, cells can be stored for many years. However, some biological features have to be considered when bringing cells to subzero temperatures. The basic reason for problems in cryopreservation is the occurrence of unbalanced concentrations in the extra- and intracellular solution during freezing and thawing procedures. Freezing first causes ice formation in the extracellular environment, which again induces cell dehydration due to the resulting chemical potential difference between intra- and extracellular water. In the case of a fast freezing rate compared to the rate of exosmosis of water from the cell, the cytoplasm will supercool and intracellular ice formation may occur. This in turn will dramatically damage cellular structures by mechanical forces and will result in a loss of viability. If, on the other hand, freezing occurs too slowly, this will result in large local concentration gradients which will irreversably damage cell membranes.

Both problems can be minimized by using special additives and correct timing during the freezing procedure.

- Penetrating cryoprotectants such as DMSO (dimethylsulphoxide) or glycerol afford protection by their colligative properties. The large number of these protectant molecules in the solution lowers the freezing temperature of the medium so that the proportion of water transformed into ice and also the extent of cellular dehydration are reduced (Rowley and Anderson, 1993). Further non-penetrating additives like dextran, polyvinylpyrrolidone and other basic polymers mainly support membrane stability when used in a final concentration of 7 to 12 percent. The combined use of representatives of both groups is also possible.
- Correct timing during freezing means a defined cooling rate over a defined range of temperature. Normally −1°C per minute up to −80°C is recommended; however, deviations from this rule occur due to the special sensitivity of different cell types. Such a precise cooling rate can be achieved only with controlled-rate freezers, but a manual alternative using special boxes and a −80°C freezer (see below) or a special cooling adapter which is placed in the top of the liquid nitrogen tank may be sufficient in most cases. Cells to be frozen must not be infected with microorganisms; they must be viable and preferably in log phase growth.

The number of cells recommended for the freezing procedure ranges between 10^6 and 10^7 cells per ml. Cells frozen in lower or higher cell concentrations may tend to have poorer viability.

Cryopreservation is possible at various low temperatures (eg -20°C, -80°C, -196°C), which have different biological consequences as regards metabolic activity and viability after thawing. At -20°C metabolic activity is only slowed down; frozen cells keep their viability only for several weeks. Storage at -80°C will maintain viability for about six to nine months. Only storage below -130°C allows indefinite storage as no biological process takes place below this temperature. In a liquid nitrogen container storage may occur either in the vapour phase or in the liquid phase. Liquid phase storage offers a uniform temperature of -196°C, but involves the danger of exploding cryotubes. Vapour phase storage avoids this risk, but offers not a uniform temperature but a gradient between about -150°C under the lid of the container and the temperature of the liquid. The temperature range of the gradient can be reduced by raising the liquid level or by adding conductive materials.

Safety concerns when using liquid nitrogen

Liquid nitrogen is a cryogen with a boiling point of -196°C (-320°F). Therefore some special safety aspects have to be considered when using liquid nitrogen:

- Nitrogen gas is colorless, odorless and tasteless and cannot be detected by the human senses, but it can cause dizziness, unconsciousness and in extreme cases death. Therefore it is necessary not to seal liquid nitrogen or prevent it from venting; positioning of liquid nitrogen tanks in the basement is dangerous because of missing gas drainage.
- Wear protective clothes and special gloves, providing good antifreeze insulation and adequate flexibility when handling small cryotubes, and a face shield or at least safety goggles.
- Use only containers which are designed for low-temperature liquids. Do not overfill them and avoid using hollow rods as measuring sticks.
- During cryopreservation liquid nitrogen may enter cryotubes. When retrieving such a tube there is a risk of explosion because entrapped liquid nitrogen will revert to gas with a manifold expansion in volume (nearly 700x).

- It is extremely important to organize a satisfactory monitoring and replenishment system. Refilling has to take place not later than when the level of liquid nitrogen has dropped down to a minimum of 5 to 10 cm to avoid any unobserved rise of temperature combined with partial thawing, for cells will quickly lose their viability.
- Electronic monitoring systems as well as electronic systems for autofilling are commercially available, but themselves need another control system in case of power failure.
- Special attention must be given to the risk of possible transmission of blood-borne virus infections from contaminated cryopreservation tanks (Tedder et al, 1995). Liquid nitrogen is not sterile but a trap for possible contaminants of different types.

Record keeping

An important point is the durable and complete labelling of the tubes. In combination with a detailed cryo-logbook the necessary information must be written on the cryotubes; the information should include at least the name, individual laboratory number, date of freezing and subculture of a frozen cell line. The labelling should clearly allow the identification of the content of each cryotube. Take care to write clearly, so that other persons are also able to read the labels. Written labels which will be stuck on the cryotubes should not be used, for they will become detached in the course of time. Special cryomarkers in different colours are available from Semadeni AG (Switzerland). Nunc (Denmark) is also working on a special labelling technique using bar codes, but this is still in development. In combination with adequate labelling an appropriate register of the stored cryotubes should also be kept. In the era of the computer it is quite easy to establish a cryo database which contains all the necessary information on the corresponding frozen cell lines. Specific software is also available: **CryoBase** (Microtech, Switzerland as well as Merck, Germany) and **KryoTrak** (Messer Griesheim, Germany); both are specially designed programs based on Windows.

Cryomedium

Numerous variations of cryomedia are proposed in the literature. Most of these cryomedia consist of a complete growth medium (Chang medium is not suitable), supplemented with at least 10 percent fetal calf serum and/or other sera and additional cryoprotectants in a final concentration of about 10 percent. In the case of adherent cells DMEM or Ham's F-12, and in the case of suspension cultures RPMI-1640 are recommended as basic synthetic media. But other synthetic media adapted to special cell types may also be used. The concentration of serum may be increased up to 90 percent to the debit of the growth medium. Addition of antibiotics is also possible. Life Technologies Ltd (Scotland) offer a ready-to-use cryomedium (cat.no.: 11101-011). DMSO, although sometimes classified as toxic to cells, can be used at concentrations and exposure times mentioned in the following protocol; if this is adhered to, cultured cells are not damaged although different cell types may exhibit small variations in DMSO compatibility.

Materials

For freezing without controlled-rate freezers

- cell culture, preferably in log growth phase
- sterile cryomedium at room temperature
- sterile growth medium at room temperature
- sterile cryotubes with internal thread and silicone gasket
- sterile pipettes and pipette aid
- centrifuge
- styrofoam box with a wall thickness between 2 and 4 cm

For thawing

- water bath
- sterile growth medium
- 70% ethanol
- T-25 tissue culture flasks
- sterile pipettes and pipette aid
- centrifuge

Procedure

Cryopreservation without controlled-rate freezers

1. Gently harvest cells: spin down suspension cultures at a speed not exceeding 400 g (better: 200 to 300 g for 20 to 25 minutes); adherent cultures are gently trypsinized (for further details see special protocols), collected in growth medium and centrifuged as for suspension cultures.
2. Remove the supernatant and resuspend centrifuged cells in growth medium; determine the number of viable cells by viability staining (eg trypan-blue staining).
3. Mix cell suspension and cryomedium at room temperature to a final number of viable cells between 10^6 and 10^7 cells per ml and a final cryoprotectant concentration of 10 percent.
4. Aliquot the cells into cryotube vials. In case of small 1 or 2 ml tubes it is advisable to use sterile gloves when handling the tubes in order to minimize the risk of contamination during aliquoting. Also the maximum filling volume of the tubes should not be exceeded.
5. Allow the cryoprotectant to enter the cells. After 20 to 30 minutes at room temperature the cooling procedure should start. This time may be used to check for leakage of the cryotubes by submerging filled cryotubes in 0.05% methylene blue for 20 minutes. This allows detection of leaking ampoules.
6. Transfer the cryotubes to a rack in a styrofoam box and deposit this box in a −80°C freezer. A suitable alternative to the styrofoam box is the "NALGENE® Cryo 1°C Freezing Container" (Semadeni AG, Switzerland). In combination with 100% isopropyl alcohol an approximate cooling rate of 1°C per minute can be achieved.
7. The cells may be stored at −80°C for several months or can be transferred after about 24 hours to a liquid nitrogen container. This transfer has to be done as fast as possible to avoid any thawing of the cryotubes.
8. In case of storing in the liquid phase it is recommended to enclose the cryotubes additionally in a heat sealable tube wrap - especially if the cryotube contains hazardous material - to avoid any transfer during storage due to leakage. However, in this case special racks are necessary for the sealed cryotubes.

It may always be beneficial - especially if a relatively large number of cryotubes have to be frozen at the same time - to use similar racks for preparation of the cells and later storage in the liquid nitrogen container. This allows the subsequent arrangement in the liquid nitrogen container already to be recorded during preparation so that the transfer from −80°C to liquid nitrogen can be performed without delay caused by writing down the exact position of every tube.

Direct cryopreservation in the culture flask (Lindl and Bauer, 1994)

Adherent cells which are growing in culture flasks or multi-well dishes as monolayers can be frozen directly in a −80°C freezer. This is a very fast and easy method which allows storage of frozen cells for up to 6 months. However, if there are many frozen cultures a large storage capacity is necessary. This method is also only suitable for adherent cells and is offen used for limited storage of clones in case of transtection experiments.

1. Aspirate the normal growth medium from the cells.
2. Add cryomedium so that cells are covered (T-25: 1.5 to 2 ml; T-75: about 4 ml).
3. Transfer the culture flask to a styrofoam box and deposit it in a −80°C freezer. Be careful to keep the culture flask in a horizontal position.

Variations of the basic freezing procedure

Numerous variations of the basic procedure mentioned above are cited in the literature: chorionic villi (Endres et al ,1985); bone marrow and peripheral blood stem cells (Gorin, 1986; Kessinger et al, 1990; Mericka et al, 1996); skin biopsies (Gray et al, 1995 and personal communication).

Cryopreservation using controlled-rate freezers

Controlled-rate freezers for precise freezing processes with individually chosen cooling rates are available on the market, eg IceCube 1810 (SY-LAB, Austria), KRYO 10 (Planer Biomed, UK), NICOOL series (AIR LIQUIDE, France). However such equipment is expensive and should be used primarily for very sensitive cells or for large numbers of cryopreservations. **Advantages** of such a programmable freezer are

- programmable fall in temperature to give constant freezing conditions and high reproducibility,

- possibility of varying freezing rates, eg between 0.1°C and 50°C per minute,
- adaptation of the freezing procedure to cell line specific requirements,
- possibility of compensating for heat of fusion produced during crystallization by the change from a liquid into a solid state and
- availability of different freezing programmes adapted to various freezing procedures.

Disadvantages are:

- expensive basic equipment,
- time consuming compared to uncontrolled freezing,
- higher running costs during freezing because of liquid nitrogen.

Because of the complexity as well as of some technical diversities between different systems no protocol is given here.

Thawing of cryopreserved cells

Thawing of cryopreserved cells must be performed as quickly as possible. Cryotubes may explode if liquid nitrogen has entered the cryotubes during storage; therefore it is necessary to wear protective clothes, glasses and gloves. According to the cryoprotectant used, the procedure may vary in some details.

1. Prepare and label the corresponding tissue culture flasks.
2. Remove the selected cryotube(s) from the freezer or liquid nitrogen container and transport them in a covered, insulated styrofoam box to avoid any injury by possible explosion.
3. Incubate the cryotubes in a 37°C water bath and agitate the cryotubes until the cell suspension is completely thawed.
4. Soak the cryotubes after thawing in 70% ethanol to avoid transfer of microorganisms from the freezer or liquid nitrogen container or from the water bath.
5. Transfer the cryotubes to the laminar flow hood and wrap them with sterile gauze.

6. Aspirate the cell suspension from the cryotube using aseptic technique.
7. In case of glycerol as cryoprotectant the cells can be diluted 10 times directly into a tissue culture flask.
8. In case of DMSO the cell suspension has to be transferred to a centrifugation tube containing 10 ml fresh growth medium. Some cells may be very sensitive to the washing step after thawing; in such a case it may be advisable to dilute the DMSO concentration stepwise to minimize the osmotic stress.
9. Centrifuge the cell suspension at 800 rpm for 5 minutes.
10. Aspirate the supernatant and resuspend the cells in fresh growth medium.
11. Transfer the cells into the corresponding tissue culture flask.
12. If desired, take out a sample to perform a viability test.
13. Incubate the cells in an incubator at standard conditions; in case of adherent cells the cells should be undisturbed for at least 16 hours before assessing the result.
14. Add the necessary information about thawing to the corresponding records of the cell line.
15. In all cases the medium should be changed the next day to remove any residual traces of cryoprotectant.

Subprotocol 3
Bacterial and/or fungal contamination

Bacterial or fungal contamination is normally detected during routine monitoring of cell cultures either macroscopically (eg pH dependent color change of the medium and/or turbidity in the culture) or microscopically. Bacteria appear as single organisms, much smaller than cells and characterised by various forms. Fungi appear either as filamentous structures (mycelic form) or in a yeast form showing evidence of budding.

When possible the method of choice is to discard the contaminated culture and to go back either to an uninfected culture of the corresponding cell line or to obtain new cultures from previously frozen aliquots of the same line, for antimicrobial treatment often only suppresses but does not eliminate contamination. However, in the case of material for cytogenetic diagnosis, previous cryopreservation often was not possible or a healthy back-up culture is not available, so that a repeated sampling could be an alternative; this, however, may be difficult in the case of bone marrow, amniotic fluid or chorionic villi samples, which has to be considered.

In the case of bacterial or fungal contamination the control will start with

- urgent measures to keep the culture alive,
- selection of a suitable antimicrobial agent and evaluation of its most effective concentration and its minimum toxic dose according to the individual cell line susceptibility and
- location and disposal of the source of contamination combined with cleaning of all used equipment, especially laminar flow hood and incubator.

Every antimicrobial treatment is time consuming. Furthermore, antimicrobial agents are biologically active, i.e. they may effect the cells directly by changing their metabolism, by reducing viability and by possibly selecting certain subpopulations of cells. One must also consider the production of resistant microorganisms and the general fact that the presence of a contaminated culture always increases the risk of contaminating other cultures. Especially with regard to cytogenetic analysis Hoehn (1992) pointed out that the use of antimicrobial agents may cause increased chromosomal breakage rates and elevated rates of pseudomosaicism. Amphotericin B also influences the viability of cryopreserved tissue (Villalba et al, 1995).

Antimicrobial agents in cell culture are available in crystalline or lyophilized form as well as ready-to-use stock solutions (cell culture tested quality is recommended). Concentrations may vary between suppliers, a fact which has to be considered in use. Reconstitution of crystalline or lyophilized antibiotics will normally be performed by using cell culture quality water or a balanced salt solution. According to the individual antibiotic and also the instructions of the supplier, antibiotics can be stored either as stock solution or as aliquots. When preparing aliquots correct labelling is absolutely essential. This must include the name of the antibiotic (abbreviations only may be used if there is no risk of confusion), concentration, date of preparation and volume. Table 2 presents an overview of commonly used antimicrobial agents in cell culture, their field of activity, their working concentration (may vary according to cell line susceptibility and type of contaminant) and storage conditions as well as half-life periods. The most commonly used antibiotics in cell culture are penicillin G and streptomycin. The prophylactic use of antimicrobial agents in culture media will always give rise to discussion. Antibiotic-free media should normally be preferred, exept in a few cases such as the set-up of primary cultures and specimens with a high risk of contamination. Besides the widespread use of antimycotic agents for rescuing cells contaminated with mold or yeast, Percoll® may be used for removal of the contaminating agent (Kruk and Auersperg, 1991; Overhauser et al, 1990).

Materials

For dose response test

- 24 multi-well plate(s)
- antimicrobial agent
- antimicrobial agent-free culture medium
- hemocytometer chamber

Procedure

Dose response test to determine cytotoxic concentration

Antimicrobial agents at higher concentrations can be toxic to cells in culture; the toxicity level depends on the antimicrobial agent itself as well as on the specific susceptibilities of the cells. In order to treat a special cell line with an antimicrobial agent and to avoid cytotoxicity it is recommended to perform a dose response test to determine the level of cytotoxicity. The difference between toxic concentration and effective concentration may be low.

1. Dissociate adherent growing cells using trypsin according to standard procedures (in case of suspension culture start with the following point).
2. Estimate the number of cells using a hemocytometer chamber.
3. Dilute the cells in antimicrobial agent-free culture medium to the normal concentration used for regular cell passage.
4. Dispense the cells into a multiwell culture plate.
5. Add the antimicrobial agent to be tested to each well in a range of concentrations (eg for amphotericin B: 0.25, 0.50, 1.0, 2.0, 4.0, 8.0 μg/ml).
6. Observe the cells daily for signs of cytotoxicity: sloughing, appearance of vacuoles, decrease in confluence, rounding.

7. For further treatment use an antimicrobial agent concentration which is one or two times lower than the cytotoxic concentration.

General procedure for treatment of infected cell cultures

The described method is well established for contaminated flask cultures containing adherent growing cells. In the case of other culture vessels the corresponding volumes have to be adjusted.

1. Carefully aspirate the medium from the contaminated flask avoiding any contact of contaminated medium with the top and neck of the flask.
2. Add about 5 ml of special antibiotic containing medium to a T-25 or about 10 ml to a T-75.
3. Gently rinse the adherent cells and aspirate the rinsing fluid completely.
4. Again add 5 ml or respectively 10 ml of special antibiotic containing medium; now turn the flask and rinse the top and sides to remove any residual contaminated media.
5. Remove the whole of the fluid from the second rinse and add the normal amount (T-25: 5 ml; T-75: 20 ml) of special antibiotic containing complete culture medium.
6. Incubate the culture vessels in a separate incubator if available, or at least in a separate incubator section apart from uncontaminated cultures.
7. Daily microscopic evaluation is recommended to check for possible ineffectiveness of the antibiotic and therefore recontamination as well as morphological modifications of the cell as sign of a decreasing healthiness.
8. A complete medium change has to be performed at least twice a week. The frequency of medium changes will also be determined by the different half-life periods of various antibiotics.
9. Treatment with antibiotic-supplemented medium should be continued for about two weeks, including possible subculturing.
10. After that, antibiotic-free medium should be used for further culture and subculturing; however, daily controls should be maintained for further days to check for recontamination. Retesting for sterility is also recommended.

Subprotocol 4
Mycoplasma Contamination

Mycoplasmas are the smallest free-living prokaryotic cells (0.3 – 0.5 μm in diameter), derived from ancestral anaerobic bacteria by gene deletion (in earlier years also named PPLO = pleuropneumonia-like organisms). They occur widely as commensales, parasites or pathologic agents on plants, vertebrates and invertebrates, especially on mucous membranes. They are also inportant as contaminants of cell cultures. Robinson et al (1956) described the first observation of mycoplasma infections in cell cultures. In particular, *Mycoplasma orale, M. hyorhinis, M. arginini, M. fermentans, M. hominis, M. salivarium* and *Acheloplasma laidlawii* are responsible for more than 90% of mycoplasma infected cell cultures. Sources of mycoplasma contaminations can be biological additives like serum or trypsin as well as a poor aseptic working technique of the operator. An important cause of further infection is the spreading of mycoplasmas by aerosols. For example, one drop of about 50 μl may contain up to 5 x 10^5 mycoplasmas.

The infection rate varies in different cultures; primary cultures normally are seldom infected by mycoplasmas (0% to about 5%). In contrast, in continuous cultures (adherent as well as suspension) the infection rate may be much higher. Short term cultures which are often used in diagnostic cytogenetics may have a reduced rate of mycoplasma contamination due to a short turn-around time which does not allow a mycoplasma infection to become evident.

The effects of mycoplasmas on the cultured cells are manifold and often substantial (Table 4), resulting from mycoplasma gene products (eg enzymes, toxins), from mycoplasmal utilisation of media components and host cell components, as well as from secondary side effects of mycoplasmal growth. Though many mycoplasmas grow slowly, their presence results in various modifications of cell metabolism, function and growth as well as genetic modifications (Schneider and Stanbridge, 1975; McGarrity, 1987). This is all the more important since the mycoplasma concentration in the supernatant of a contaminated culture may be as high as 10^8 mycoplasmas per ml without being recognized, ie there may be up to 500 mycoplasmas per cell. In most cases a chronic mycoplasma infection means more than 10^4 CFU (colony forming units) per ml.

Table 4. Effects of mycoplasma contamination on cell cultures

metabolic effects:	- pH shift by acid metabolites
	- consumption of amino acids
	- consumption of sterols, fatty acids and lipids
	- consumption of sugar components
genetic effects:	- consumption of nucleic acid precursors resulting in chromatin breaks, multiple translocations, reduced chromosome numbers
	- delivery of nucleases into the medium
other effects:	- mimicry of a virus infection
	- influence of virus propagation
	- influence on interferon production
	- increase of specific mitogen activity in lymphocytes
	- increase or decrease of unspecific mitogen activity in lymphocytes
	- plaque formation
	- cytophatic effects
	- lack of symptoms

Detection of contamination

Unlike bacterial or fungal contamination, mycoplasma infection cannot be detected with the naked eye and does not reveal its presence by macroscopic alterations of the cells or media (pH change or culture turbidity). Indirect indications of the presence of mycoplasmas (but also of viruses) can be delayed growth, changes in viability, morphology and metabolism, black "granules" on the cells and plaque-shaped detachment of cells. Mycoplasma detection should be a routine procedure in every cell culture laboratory. Today a range of various detection techniques are available. However, every assay has its own characteristics and not every assay is usable by every cell culture laboratory. The choice of a suitable detection assay depends not only on the special features of the test system itself, but also on the individual requirements of a laboratory as well as the number of cell lines, the frequency of testing and last but not least on the availability of funds and resources. Gram-staining - one of the classical staining methods in microbiology - is useless because mycoplasmas lack a cell wall. The following part of this chapter describes the fluorescent staining and direct culture technique (recommended by the FDA). Additional detection systems like PCR and commercially available kits are also mentioned.

A possible risk of false negative results depending on culture treatment may exist in case of low level contamination: eg trypsinisation of adherent cells before mycoplasma testing can partly kill and/or enzymatically remove mycoplasmas from the cell surface. A more gentle form of cell collection by 5 mM EDTA-PBS solution (pH 7.4) for several minutes can increase sensitivity. Once mycoplasma contamination is diagnosed, immediate disposal of the contaminated cultures is probably the method of choice. If all cultures from a cell line or specimen are infected, one must decide between a repeat sampling or an attempt to salvage the cultures with antimycoplasmal agents.

Fluorescent staining

The most common staining procedures using either DAPI or Hoechst 33258 will be described in detail. Both fluorescent stains are DNA-binding molecules; the Hoechst 33258 stain reacts with A + T-rich DNA regions. These intercalating dyes fluoresce under UV-light, so that mycoplasmas will appear as bright extranuclear spots throughout the cytoplasm and along the cell boundary, whereas uninfected cells show fluorescing nuclei against a negative background.

Advantages are:

- fluorescence staining is a fast and low cost technique and allows the testing of various cell lines;
- no further special equipment except a fluorescence microscope is necessary;
- sensitivity is quite high: 10^3 – 10^4 cfu/ml;
- detection of mycoplasma strains that cannot yet be cultivated (eg *M. hyorhinis*) is possible.

Disadvantages are:

- confusion may arise from other staining fragments like mitochondrial DNA (exhibits little visible fluorescence), bacteria or fungi as well as nucleus fragments of disintegrating cells (larger and sometimes brighter fluorescence in different forms from mycoplasma);

- interpretation of staining results requires a certain degree of experience and training; the decision about mycoplasma infection is subject to personal attitude;
- correct and complete performance requires the use of a positive control; i.e. cultivation of mycoplasmas or a mycoplasma-positive cell line gives rise to the risk of further contamination.

The presence of antibiotics may hide an infection, therefore the cells to be tested should complete at least two passages in antibiotic free medium before testing. The same applies to cell cultures from frozen tubes for cryoprotectants may also mask infection. Double testing of a suspected cell line is normally recommended: one day and three days after incubation. This requires an adequate number of cells, for cell overgrowth will make interpretation difficult.

Materials

For staining with DAPI

- monolayer cell culture with 50 – 70% confluency using either Petri dishes (60 mm in diameter), cover slip culture or multiplace culture chamber slide.
- DAPI: Serva (Germany), Nr. 18860 or Sigma Chemicals (USA), Nr. D 8417
- DAPI-stock solution (50x): Dilute 50 μg DAPI in 10 ml distilled water; sterilize by filtration. The stock solution can be stored in 0.2 ml portions at −20°C (stability: 12 months).
- PBS = phosphate buffered saline.
- mountant: see Hoechst staining.
- fluorescence microscope: absorption: lambda = 340 nm; emission: lambda = 488 nm. Complete filter sets are available from eg Zeiss (Germany) or Leitz (Germany).

For staining with Hoechst 33258

- Hoechst 33258 stain stock solution
  - dissolve 5 mg Hoechst 33258 from Serva (Germany), Nr. 15090 or Sigma Chemicals (USA), Nr. B 1155 in 100 ml PBS.
  - sterilize the stock solution by filtration through a 0.1 μm membrane.
  - the stock solution (wrapped with aluminium foil) can be stored in the dark at 4°C.

- Hoechst 33258 stain working solution
  - prepare fresh by diluting 0.1 ml Hoechst 33258 stain stock solution with 10 ml sterile PBS.
- fixative:
  - prepare fresh by adding 3 parts methanol to 1 part acetic acid (glacial).
- mountant:
  - mix 22.2 ml 0.1 M (= 2.1%) citric acid (1 H_2O) with 27.8 ml 0.2 M (= 2.8%) disodium-hydrogenphosphate.
  - add 50 ml glycerol, adjust pH to 5.5, filter sterilize and store at 4°C.

For direct culture detection

Mycoplasma Test Medium Kit Part A (can be stored at 2–8°C) (Sigma Chemicals, USA)

- Mycoplasma Test Medium Kit Part B (can be stored at 2 – 8°C) (Sigma Chemicals, USA)
- 60 x 15 mm culture dishes and sterile pipette equipment
- 3 water baths (one at 95°C, one at 56°C and one at 37°C)
- incubator for anaerobic conditions.

Procedure

Staining with DAPI (4-6-diamidino-2-phenylindol-di-hydrochloride) (according to Lindl and Bauer, 1994)

1. Dilute 0.2 ml DAPI stock solution (50x) with 10 ml methanol (final concentration 0.1 µg DAPI/ml staining solution). The staining solution should always be freshly prepared.
2. Aspirate the medium, rinse cells with PBS and wash once with staining solution.
3. Add fresh staining solution so that the cells are well covered and incubate the cells for 15 minutes at 37°C in an incubator.
4. Aspirate the staining solution and cover the cells with a coverslip using a little PBS.
5. Examine the cells with the fluorescence microscope.

Staining with Hoechst 33258

Note: The toxic properties of Hoechst 33258 (bisbenzimide fluorochrome) are unknown, therefore gloves are recommended during handling of the stain.

1. Aspirate the medium from the cells.
2. Fix the cells with two changes of fixative solution (5 minutes each).
3. Wash the cells in deionised water.
4. Incubate the cells for 30 – 60 minutes at 37°C with Hoechst 33258 stain working solution.
5. Rinse with deionised water and cover the cells with a coverslip using a drop of the mountant.
6. Examine the cells with a fluorescence microscope.

Complete mycoplasma staining kits are also commercially available. Sigma Chemicals (USA), for example, offers the MYC-1 kit which contains all the reagents necessary to perform the Hoechst staining procedure; the kit also contains positive and negative control slides for comparison with test slides. ICN Pharmaceuticals (USA) offers a ready-to-use "Mycoplasma Staining Kit" including stain (Hoechst 33258), diluent, mounting media and controls.

Detection of mycoplasmas by direct culture

Detection of mycoplasmas by direct culture requires at least the basic equipment as well as the practical experience of a microbiological laboratory. Additionally Sigma Chemicals (USA) offers a special two component mycoplasma test medium with ready-to-use prepared media which are designed for isolation of mycoplasma in cell culture. Direct culture of *M. hyorhinis* fails.

1. Heat mycoplasma test medium, Part B until the agar melts at 95°C.
2. Allow the agar to cool to 56°C and hold at this temperature in a water bath.
3. Heat mycoplasma test medium, Part A to 37°C in a water bath.

4. Aseptically add an equal amount of Part B to Part A and mix thoroughly.
5. Aseptically dispense the complete mycoplasma test medium into sterile 60 x 15 mm culture dishes and allow to set; poured plates can be stored at 2 - 8°C.
6. Inoculate 0.1 ml of the cell suspension to be tested onto a culture dish containing the mycoplasma test medium.
7. Store the inoculated plates in an inverted position at 37°C for one week under anaerobic conditions.
8. Examine the plates microscopically: locate the colonies (2.5x objective) and examine their morphology (10x or 16x objective). Mycoplasma colonies range in size from 10 to 55 µm and normally exhibit a characteristic "fried egg" appearance. Typical colonies may develop within four days, but all plates should be incubated for 2 to 3 weeks before being considered "negative". However, the interpretation of a direct culture also needs some microbiological experience, especially to distinguish between mycoplasma colonies and possible artefacts with a colony-like appearance (e.g. calcium and magnesium soap crystals).

PCR-methods

PCR technology has also found its way into the detection of mycoplasmas in cell culture. Specific DNA primers have been developed and used for DNA amplification by PCR. This technology is rapid, very sensitive and also specific, for it allows not only the detection of an infection but also the identification of the contaminating species. However, additional PCR equipment and experience are necessary which will not be available in every cytogenetic laboratory. At the moment further validation work is required to optimize the protocols in order to standardize this technique; but without any doubt it will be one of the most important detection systems in the near future. For further details see Johansson et al (1990), Luczak et al (1991), Harasawa et al (1993), Hopert et al (1993), Roulland-Dussoix et al (1994), Dussurget and Roulland-Dussoix (1994), van Kuppeveld et al (1994).

Commercially available mycoplasma detection kits

A wide range of detection kits based on different detection principles are available. Beside the two staining kits from Sigma and ICN already mentioned above, the following ones are worth reciting, although this list may not be complete.

- The Mycoplasma Detection Kit (Roche Diagnostics, Germany) is based on an ELISA system and utilises polyclonal antibodies against *M. arginini, M. hyorhinis, M. orale* and *A. laidlawii*. The test requires overnight incubation and gives a colorimetric result without the need for specialist facilities. However, detection sensitivity is lower than with staining or PCR; also *M. fermentas* (accounts for approximately 15% of mycoplasma infections in cell culture) is not detected.
- The Mycoplasma PCR ELISA (Roche Diagnostics, Germany) combines the features of PCR with that of a standard ELISA. Using only one protocol this kit allows detection of a wide range of mycoplasma species within one day at a high sensitivity (about 10^3 CFU/ml cell culture medium) (Wirth et al, 1995; Kirchhoff and Schmidt, 1995).
- The Mycoplasma Primer Set (Stratagene, USA) can be performed within a few hours. The PCR products are analysed using standard agarose gel electrophoresis.
- The American Type Culture Collection (ATCC, USA) offers a PCR-based mycoplasma detection kit using a set of mixed primers that amplify the spacer region between the 16S and 23S rRNA genes of mycoplasma.
- The American Type Culture Collection (ATCC, USA) additionally offers a mycoplasma detection service based on combined testing by the direct culture method and indirect Hoechst staining for a price of $ 75 per sample (contact address: ATCC, Applied Sciences Laboratory, 12301 Parklawn Drive, Rockville, MD 20852, phone: ++1 301-231-5594, fax: ++1 301-816-4366).
- MycoTest® (Life Technologies, Scotland) is an enzyme based assay using a mycoplasma-free indicator cell line (eg 3T6, ATCC): 6-methylpurine-deoxyriboside (a non-toxic adenosine analogue) is converted by the mycoplasma specific enzyme adenosine phosphorylase into 6-methylpurine and 6-methylpurine riboside, both of which are toxic to mammalian cells. The amount of cell death directly correlates with the degree of mycoplasma contamination.

- Mycocell Probe (International Mycoplasma SA, France) allows detection of many mycoplasmal contaminations by molecular hybridisation. The nucleic acid probe recognises a region of DNA displaying a high degree of homology in many mycoplasmas, but does not interact with eukaryotic DNA.
- RIDASCREEN® Mycoplasma IFA (R-Biopharm GmbH, Germany) uses fluorescent-marked monoclonal antibodies and detects 16 mycoplasma and acheloplasma species.
- The ImmuMark™MycoTest™ (ICN Pharmaceuticals, USA) is based on a fluorochrome-labelled monoclonal antibody which is specific for a broad range of mycoplasma species accounting for more than 96% of cell culture infections.
- Hybricomb Mycoplasma Test Kit (Biological Industries, Israel) works without radioactive probe labels. It uses a mixture of mycoplasma total genome DNA as a probe for the detection of seven varieties which account for most mycoplasma contaminations in cell cultures.

Subprotocol 5
Elimination of Mycoplasmas from Cell Cultures

Similarly to the wide variety of detection assays, numerous methods have also been developed for the selective killing of mycoplasmas. Here too every approach has its own characteristics and differs in terms of efficacy, time and costs. The use of antibiotics is usually the method of choice. Penicillin (interference with bacterial cell wall synthesis) is valueless since the mycoplasmas lack cell walls, but various other antimycoplasmal agents are available (Table 2). Mycoplasmas are susceptible to tetracyclines, quinolones, macrolides and aminoglucosides. However, not all mycoplasma species and strains react in the same way to single antibiotics. Furthermore, different cell lines and cell types exhibit a varying susceptibility (i.e. cytopathic effects, time for recovery) so that no universal recipe for antimycoplasmal treatment is available. The Mycoplasma Removal Agent (ICN Pharmaceuticals, USA) is recommended by the European Collection of Animal Cell Cultures (ECACC) as one of the most effective methods available. Combinations of antimycoplasmal agents are also sometimes used: thus Coronato et al (1994) propose a combined treatment (especially in case of *M. orale*) by addition of tylosine (250 µg/ml) for 12 days followed by addition of minocycline (5 µg/ml) for a further 10 days. Generally it is preferable not to grow cell lines with antimicrobial agents on a permanent basis; they should be used only for curing contaminated cultures or as short-term prophylactic treatment (eg newly acquired cell lines; newly thawed from frozen stock).

Other elimination techniques like the use of hyperimmune sera (Jeansson and Brorson, 1985; Nair, 1985), passaging cells in athymic nude mice (van Diggelen et al, 1977), or use of nucleic acid analogues followed by exposure to light (Marcus et al, 1980; Tarshis et al, 1994) are more specialised techniques and cannot be proposed as standard elimination procedures.

By way of an example the use of BM Cyclin and Ciprofloxacin are described in detail. BM-Cyclin 1 + 2 is a combination of two different antibiotics which will be used in a sequential and cyclic mode. The normal duration of treatment is 21 days. BM-Cyclin 1 is a pleuromutilin derivative and BM-Cylin 2 is a tetracycline derivative.

Materials

For BM-Cyclin

- BM-Cyclin 1 + 2 (Roche Diagnostics, Germany)
- complete cell culture medium

For Ciprofloxacin

- Ciproxin® (0.2 g ciprofloxacin/100 ml) (Bayer Leverkusen, Germany)
- complete cell culture medium

Procedure

Elimination of mycoplasmas using BM-Cyclin (according to Lindl and Bauer, 1994)

1. Dilute 5 mg BM-Cyclin 1 in 1000 ml ready-to-use medium (final concentration 5 µg/ml).
2. Dilute 10 mg BM-Cyclin 2 in 1000 ml ready-to-use medium (final concentration 10 µg/ml).
3. Make portions of both media and store them at −20°C.
4. Day 1: Remove medium from infected cell culture. Add adequate amount of BM-Cyclin 1-medium and incubate for three days.
5. Day 4: Remove BM-Cyclin 1-medium from infected cell culture. Add adequate amount of BM-Cyclin 2-medium and incubate for four days.

6. Continue this medium alternation (3 days BM-Cyclin 1-medium; 4 days BM-Cyclin 2-medium) for a further two weeks.

Various degrees of success with BM-Cyclin 1 + 2 are reported: Branch and Guilbert (1986) described good results in hematopoietic cells whereas Kotani et al (1991) and Somasundaram et al (1992) could not cure all infected cell lines and found varying susceptibilities in different cell lines. Nissen (1995) had to increase the BM-Cyclin concentration recommended by the manufacturer 2.5-fold to achieve successful elimination. Difficulties also arose with the elimination of *M. hyorhinis* and *A. laidlawii*. Use of BM-Cyclin seems to require empirical determination of the optimal concentration for treating BM-Cyclin sensitive cell lines.

Elimination of mycoplasmas using ciprofloxacin

Ciprofloxacin is a fluoroquinolone antibiotic. It acts by inhibiting the enzyme DNA-gyrase (= topoisomerase II) which is essential for the supercoiling of bacterial DNA. The normal duration of treatment is between two and three weeks (Mowles, 1988; Schmitt et al, 1988; Gignac et al, 1991 and Somasundaram et al, 1992).

1. Add 1 ml ciprofloxacin to 200 ml ready-to-use medium (final concentration 10 μg/ml).
2. Incubate contaminated cell cultures with ciprofloxacin medium for 15 to 20 days.
3. Two medium changes per week are recommended.

Note: As an alternative product Ciprobay® (0.2 g ciprofloxacin / 100 ml; Bayer Leverkusen, Germany) may also be used in a similar final concentration.

Prophylaxis against microbial contamination

In every cell culture laboratory the importance of precautionary measures to avoid microbial contamination has to be emphasized. Most of the following tips are intended to avoid microbial contamination, although some are especially designed for avoiding mycoplasmas:

- The most effective method of eliminating microbial contamination is to prevent it from occurring.
- Biological solutions like serum or trypsin should be used only if they have passed through an improved manufacturing and in-factory control system. Additionally they may be sterile filtered with 0.1 µm filters using a low pressure (at pressures above 15 pounds per square inch (1 bar) mycoplasmas will pass through a 0.2 µm filter). Suitable filters should have a high density of pores as well as a narrow distribution of pores; i.e. the pores of a filter do not all have the same size of exactly 0.1 µm. This is only a theoretical mean value; the actual pore diameter is more or less widely distributed around this value.
- Heat inactivation of serum (30 minutes at 56°C) has both advantages (denaturation of complement protein avoids immunological reactions) and disadvantages (possible denaturation of other necessary serum proteins). Care should be taken when using a water bath, for it will always be a source of contamination.
- Avoid any production of aerosols.
- Do not use antimicrobial agents without a specific indication.
- Routinely screen for infections, especially mycoplasma contaminations.
- Follow a rigorously aseptic working technique as already mentioned in the section on general quality control.
- Cryotubes from the liquid nitrogen storage container must be cleaned with 70% ethanol.

References

Barch MJ, Lawce HJ, Arsham MS (1991) Peripheral blood culture. In: Barch MJ (ed) The ACT Cytogenetics Laboratory Manual. 2nd ed. Raven press, New York, pp 17 - 30

Brambati B, Simoni G (1983) Fetal diagnosis of trisomy 21 in the first trimester of pregnancy. Lancet 1:586

Branch DR, Guilbert LJ (1986) Practical in vitro assay systems for the measurements of hematopoietic growth factors. J Tissue Culture Methods 10:101-108

Coronato S, Vullo D, Coto CE (1994) A simple method to eliminate mycoplasma from cell cultures. J Virol Methods 46:85-94

Dussurget O, Roulland-Dussoix D (1994) Rapid, sensitive PCR-based detection of mycoplasmas in simulated samples of animal sera. Appl Environ Microbiol 60:953-959

Eagle H (1955) The specific amino acid requirements of mammalian cells (strain L) in tissue culture. J Biol Chem 214:839-842

Endres M, Dawson G, Wirtz A, Haindl E (1985) Freezing of chorionic villi. In: Fraccaro M, Simoni G, Brambati B (eds) First trimester fetal diagnosis. Springer, Berlin, Heidelberg, New York, pp 201 - 204

Freshney RJ (1993) Culture of animal cells. A manual of basic techniques. 3rd ed. Alan R. Liss Inc., New York

Gignac SM, Brauer S, Häne B et al (1991) Elimination of mycoplasma from infected leukemia cell lines. Leukemia 5:162-165

Gorin NC (1986) Collection, manipulation and freezing of haemopoietic stem cells. Clin Haematol 15:19-48

Gray RGF, Ryan D, Green A (1995) The cryopreservation of skin biopsies - a technique for reducing workload in a cell culture laboratory. Ann Clin Biochem 32:190-192

Hahnemann N (1974) Early prenatal diagnosis. A study of biopsy technique and cell culturing from extraembryonic membranes. Clin Genet 6:294-306

Hansson K, Schuring-Blom GH, Leschot NJ (1995) The preserving of chorionic villi before establishing long-term cell cultures for cytogenetic analysis. Prenatal Diag 15:1067-1069

Harasawa R, Mizusawa H, Nozawa K, Nakagawa T, Asada K, Kato I (1993) Detection and tentative identification of dominant mycoplasma species in cell cultures by restriction analysis of the 16S - 23S rRNA intergenic spacer regions. Res Microbiol 144:489-493

Hoehn H (1992) Amniotic fluid cell culture. In: Milunsky A (ed) Genetic disorders and the fetus. 3rd ed. The Johns Hopkins University Press, New York, London, pp 101-121

Holzgreve W, Miny P (1987) Chorionzottendiagnostic. VCH Publishers, Weinheim

Hopert A, Uphoff CC, Wirth M, Hauser Hj, Drexler HG (1993) Mycoplasma detection by PCR analysis. In Vitro Cell Dev Biol 29A:819-821

Jeansson S, Brorson J-E (1985) Elimination of mycoplasmas from cell cultures utilizing hyperimmune sera. Exp Cell Res 161:181-188

Johansson KE, Johansson I, Göbel UB (1990) Evaluation of different hybridization procedures for the detection of mycoplasma contamination in cell cultures. Mol Cell Probes 4:33-42

Jones GE (1996) Human Cell Culture Protocols. Humana Press, Totowa

Kessinger A, Schmit-Pokorny K, Smith D, Armitage J (1990) Cryopreservation and infusion of autologous peripheral blood steem cells. Bone Marrow Transplant 5 (Suppl. 1):25-27

Kirchhoff H, Schmidt R (1995) Detection of mycoplasma in cell cultures by the mycoplasma PCR ELISA in comparison to the culture method. Biochemica 1:33-35

Knutsen T (1991) Laboratory safety and quality control. In: Barch MJ (ed) The ACT cytogenetics laboratory manual. 2nd ed. Raven Press, New York, pp 563-587

Kohn G (1981) Failure of amniotic fluid cell culture due to syringe toxicity. Prenatal Diag 1:233

Kotani H, Butler G, Heggan D, McGarrity GJ (1991) Elimination of mycoplasmas from cell cultures by a novel soft agar technique. In Vitro Cell Dev Biol 27:509-513

Kruk PA, Auersperg N (1991) Percoll centrifugation eliminates mold contaminants from cell cultures. In Vitro Cell Dev Biol 27A:273-276

Leibovitz A (1986) Development of tumor cell lines. Cancer Genet Cytogenet 19:11-19

Lindl T, Bauer J (1994) Zell- und Gewebekultur. 3rd ed. Gustav Fischer Verlag, Stuttgart, Jena, New York

Loeb L (1897) Ueber die Entstehung von Bindegewebe, Leukozyten und roten Blutkörperchen aus Epithel und über eine Methode, isolierte Gewebsteile zu züchten. M. Stern und Co., Chicago

Luczak J, Knower SA, Cox MS, Dubose J Jr, Harbell JW (1991) Mycoplasma contamination detected in cell lines and their products from 1985 to the present. In Vitro Cell Dev Biol 27:122A

Marcus M, Lavi U, Nattenberg I, Rottem S, Markowitz O (1980) Selective killing of mycoplasmas from contaminated mammalian cells in cell cultures. Nature 285:659-661

Martin BM (1994) Tissue Culture Technique: An introduction. Birkhäuser Verlag, Basel

McGarrity GJ (1987) Mycoplasmas in cell cultures. Isr J Med Sci 23:770-771

Mericka P, Strakova H, Vavra L, Blaha M, Filip S, Jilkova B (1996) Cryopreservation of peripheral blood stem cells. International Workshop on Techniques in Cryopreservation, Brno, Czech Republic, April 16-19, 1996

Milunsky A (1979) Amniotic fluid cell culture. In: Milunsky A (ed) Genetic disorders and the fetus. 1st ed. Plenum Press, New York

Moorhead P, Nowell P, Mellmann W, Battips D, Hugerford DA (1960) Chromosome preparations of leucocytes cultured from human peripheral blood. Exp Cell Res 20:613-616

Morgan SJ, Darling DC (1993) Animal cell culture. BIOS Scientific Publishers Limited, Oxford

Mowles JM (1988) The use of ciprofloxacin for the elimination of mycoplasma from naturally infected cell lines. Cytotech 1:355-358

Nair CN (1985) Elimination of mycoplasma contaminants from cell cultures with animal serum. Proc Soc Exp Biol Med 179:254-258

Newport M, Coleman DV, McPherson K (1986) Estimation of the weight of chorionic villus samples obtained from first trimester pregnancies by transcervical aspiration. Prenatal Diag 6:265-269

Niazi M, Coleman DV, Loeffler FE (1981) Trophoblast sampling in early pregnancy. Culture of rapidly dividing cells from immature placental villi. Br J Obstet Gynaecol 88:1081-1085

Nissen E (1995) Treatment of mycoplasma contamination. In Vitro Cell Dev Biol 31:260

Nowell PC (1960) Phytohemagglutinin: an initiator of mitosis in cultures of normal human leukocytes. Cancer Res 290:462-466

Overhauser J, Chakraborty S, Kelley-Card L (1990) Removal of yeast contamination from lymphoblast cultures. BioTechniques 8:177

Robinson LB, Wichelhausen RB, Roizman B (1956) Contamination of human cell cultures by pleuropneumonia-like organisms. Science 124:1147-1148

Rooney DE, Czepulkowski BH (1992) Human cytogenetics. A practical approach. 2nd ed. IRL Press, Oxford, Washington

Roulland-Dussoix D, Henry A, Lemercier B (1994) Detection of mycoplasmas in cell cultures by PCR: a one year study. J Microbiol Methods 19:127-134

Rowley SD, Anderson GL (1993) Effect of DMSO exposure without cryopreservation on hematopoietic progenitor cells. Bone Marrow Transplant 11:389-393

Schmitt K, Däubener W, Bitter-Suermann D, Hadding U (1988) A safe and efficient method for elimination of cell culture mycoplasmas using ciprofloxacin. J Immunol Methods 109:17-25

Schneider EL, Stanbridge EJ (1975) Mycoplasma contamination of cultured amniotic fluid cells: Potential hazard to prenatal chromosome diagnosis. Science 184:477

Somasundaram C, Nicklas W, Matzku S (1992) Use of ciprofloxacin and BM-Cyclin in mycoplasma decontamination. In Vitro Cell Dev Biol 28A:708-710

Steele MW, Breg WR (1966) Chromosome analysis of human amniotic fluid cells. Lancet 1:383

Sutherland GR (1977) Fragile sites on human chromosomes. Demonstration of their dependence on the type of tissue culture medium. Science 197:265-266

Tarshis M, Katzeel A, Rottem S (1994) Use of Merocyanine 540 and Hoechst 33258 for the selective killing of contaminating mycoplasmas in cell cultures. J Immunol Methods 168:245-252

Tedder RS, Zuckerman MA; Goldstone AH, Hawkins AE, Fielding A, Briggs EM, Irvin D, Blair S, Gorman AM, Patterson KG, Linch DC, Heptonstall J, Brinks NS (1995) Hepatitis B transmission from contaminated cryopreservation tank. Lancet 346:137-140

Tijo JH, Levan A (1956) The chromosome number of man. Hereditas 42:1-16

Van Diggelen OP, Shin S, Phillips DM (1977) Reduction in cellular tumorigenicity after mycoplasma infection and elimination of mycoplasma from infected cultures by passage in nude mice. Cancer Res 37:2680-2687

Van Kuppeveld FJM, Johansson K-E, Galama JMD, Kissing J, Bölske G, Van der Logt JTM, Melchers WJG (1994) Detection of mycoplasma contamination in cell cultures by a mycoplasma group-specific PCR. Appl Environ Microbiol 60:149-152

Villalba R, Alonso P, Villalba JM, Rioja LF, Villagran JLG (1995) The effect of amphotericin B on the viability of cryopreserved human skin. Cryobiology 32:314-317

Van Hoff DD, Casper J, Bradley E, Trent JM, Reichert C, Altman A (1980) Direct cloning of neuroblastoma cells in agar: a measure of response and prognosis. Cancer Res 40:3591-3597

Wirth M, Grashoff M, Schumacher L, Hauser HJ (1995) Mycoplasma detection by the mycoplasma PCR ELISA. Biochemica 3:33-35

Zankl H, Ludwig B, May G, Zang KD (1979) Karyotypic variations in human meningioma cell cultures under different in-vitro conditions. J Cancer Res Clin Oncol 93:165-172

Chapter 2

Chromosome Staining

ANGELIKA KÖHLER

Introduction

The "banding era" in human cytogenetics began in 1970 with the publication of Q-banding of human chromosomes (Caspersson et al 1970). Subsequently, numerous other banding methods have been introduced, many of them designed for very specialized purposes. Some techniques make variations in certain chromosomal segments visible, thus allowing determination of the parental origin of a particular chromosome. Others identify pericentric heterochromatin in all or some autosomes and the distal Y-chromosome, or specific pericentromeric heterochromatin in chromosome 15. A guide to the application of chromosome banding is shown in Figure 1.

An internationally accepted nomenclature for the designation of the various bands and stainings of human chromosomes has been established since 1960 and resulted in the first ISCN (International System for Human Cytogenetic Nomenclature) in 1978 (ISCN 1978). Since then, several adaptations owing to improved or newly introduced techniques and revised nomenclatures have been considered, culminating in the latest update published in 1995 (ISCN 1995).

Chromosomes are made of diverse chromatin stretches composed of DNA and proteins which stain highly specifically, resulting in a pattern that depends on the staining technique applied. "Differential staining" techniques (Verma and Babu 1989) are delineated for general application. G-, R-, and Q-banding produce characteristic patterns along the entire length of chromosomes and allow the unequivocal identification of each chromosome. Moreover, the structure of each chromosome can be analysed. Consequently, variations or alterations of the chromosome structure become

Angelika Köhler, Justus-Liebig-Universität, Institut für Humangenetik, Schlangenzahl 14, Giessen, 35392, Germany (*phone* +49 641 99 41604; *fax* +49 641 99 41609; *e-mail* Angelika.Koehler@Humangenetik.Med.Uni-Giessen.de)

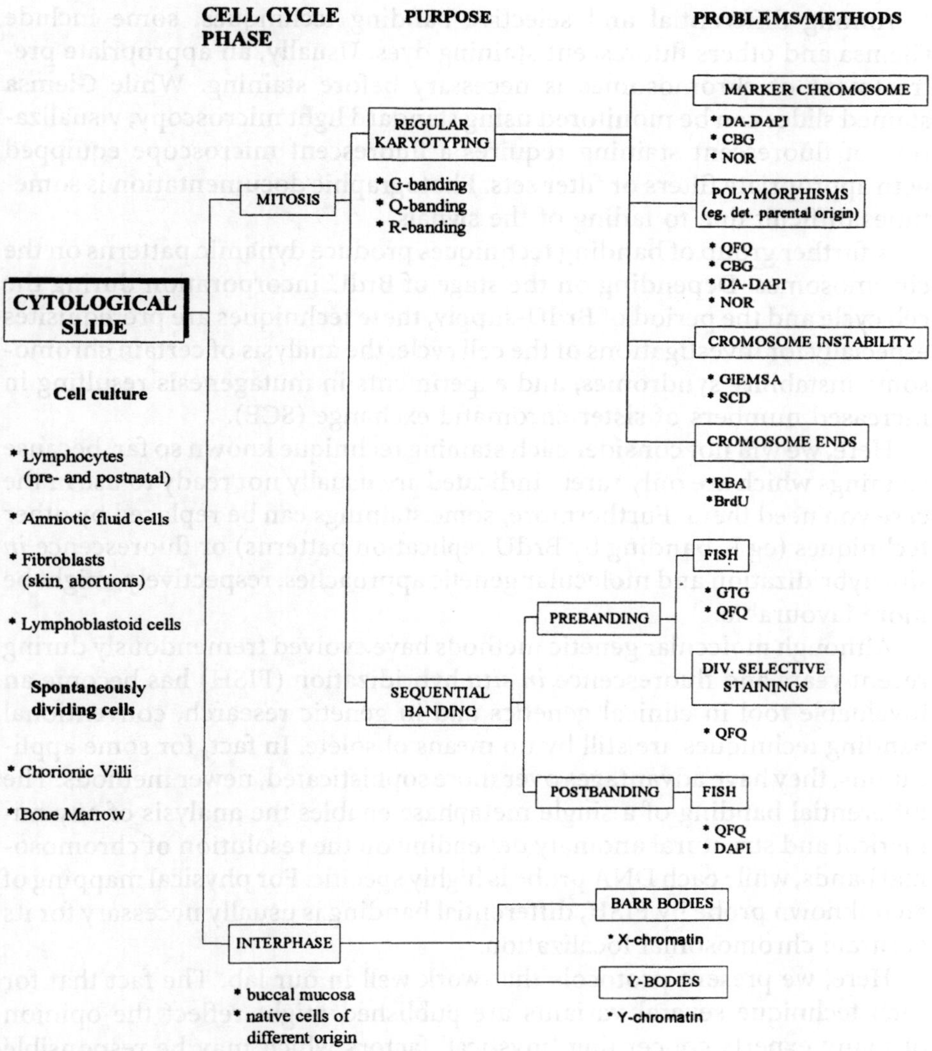

Fig. 1. Guide to application of chromosome banding

obvious, especially when comparing both homologues of a chromosome pair. In contrast, the term "selective banding" (Verma and Babu 1989) indicates that only specific regions of certain chromosomes are labelled. Such methods are usually applied when a specific problem has arisen from a banded karyotype, eg the occurrence of a marker chromosome or an obviously unbalanced chromosomal complement.

Among differential and selective banding techniques, some include Giemsa and others fluorescent staining dyes. Usually, an appropriate pretreatment of chromosomes is necessary before staining. While Giemsa stained slides can be monitored using standard light microscopy, visualization of fluorescent staining requires a fluorescent microscope equipped with appropriate filters or filter sets. Photographic documentation is sometimes difficult due to fading of the signals.

A further group of banding techniques produce dynamic patterns on the chromosomes. Depending on the stage of BrdU incorporation during the cell cycle and the period of BrdU-supply, these techniques are prerequisites especially for investigations of the cell cycle, the analysis of certain chromosome instability syndromes, and experiments in mutagenesis resulting in increased numbers of sister chromatid exchange (SCE).

Here, we will not consider each staining technique known so far, because stainings which are only rarely indicated are usually not ready to use in the case you need them. Furthermore, some stainings can be replaced by other techniques (eg R-banding by BrdU replication patterns) or fluorescence *in situ* hybridization and molecular genetic approaches, respectively, might be more favourable.

Although molecular genetic methods have evolved tremendously during recent years and fluorescence *in situ* hybridization (FISH) has become an invaluable tool in clinical genetics and in genetic research, conventional banding techniques are still by no means obsolete. In fact, for some applications, they have advantages over more sophisticated, newer methods. The differential banding of a single metaphase enables the analysis of any numerical and structural anomaly depending on the resolution of chromosomal bands, while each DNA probe is highly specific. For physical mapping of an unknown probe by FISH, differential banding is usually necessary for its accurate chromosomal localization.

Here, we present protocols that work well in our lab. The fact that for each technique several variants are published might reflect the opinion of many experts concerning 'mystical' factors which may be responsible for success or failure in a particular case.

Subprotocol 1
Giemsa Staining

The term "Giemsa staining" refers to the fact that all chromosomes are entirely and homogeneously painted. Accordingly, only groups of chromosomes can be differentiated by their morphology. Therefore, chromosomes of a karyotype are solely classified with respect to their size and to the position of their centromeres. Giemsa staining is useful when chromatid or chromosome gaps, fragile sites, or breaks have to be scored, as in surveys of induced mutagenesis or for diagnosis of syndromes in which chromosome breakage is enhanced, like Bloom syndrome, Fanconi anemia, ataxia telangiectasia and xeroderma pigmentosum.

Using certain Giemsa stains purchased as ready-for-use mixtures, a slight G-banding might be achieved (Figure 2).

Materials

- Giemsa stain, modified (Sigma cat.no. GS 500): 10% in phosphate buffer
- Phosphate buffer: equal parts of solutions 1 and 2
  - solution 1: 9,073g KH_2PO_4 / 1000ml aqua bidest
  - solution 2: 11,87g $Na_2HPO_4 \cdot 2H_2O$ / 1000ml aqua bidest

Fig. 2. Metaphase stained with Giemsa's solution. A slight G-band-like pattern is produced without further treatment

Procedure

Getting started

Mix 10ml Giemsa solution and 90ml phosphate buffer in a coplin jar.

1. Stain each slide for about 5 minutes.

If chromosomes are too dark reduce the time, if they are too light increase it.

Subprotocol 2
GTG-Banding

GTG means **G**-bands produced by **t**rypsin using **G**iemsa. The treatment with the proteolytic enzyme trypsin and subsequent Giemsa staining generates a pattern of dark (Giemsa positive) and light (Giemsa negative) bands along the entire chromosomes that is characteristic for each individual chromosome (Figure 3) (Seabright 1971; Drets and Shaw 1971). This banding technique can be applied on mitoses of lymphocyte, amnion, fibroblast, and trophoblast cell cultures, as well as on spontaneous mitoses of different origin. Due to the lower quality of spontaneously dividing cells, the quality of GTG-banding is reduced in these cases.

Materials

- trypsin (stock solution, 30mg/ml): Dissolve 3g trypsin 1:250 (Difco, cat.-no. 0152-13-1) in 100ml 0,9% NaCl (dissolves better at 37°C). Make 1ml aliquots and store at -20°C.

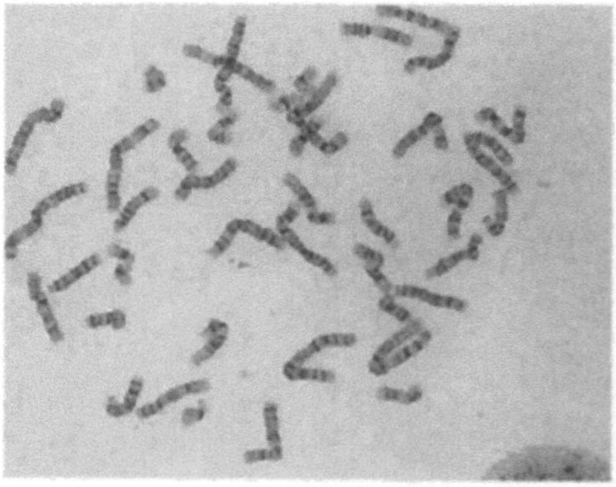

Fig. 3. G-banded metaphase after treatment with trypsin and Giemsa. Resolution is approximately 400 bands per haploid set of chromosomes.

- 0,9% NaCl (sodium chloride); store at 4°C
- Phosphate buffer: equal parts of solutions 1 and 2; store at 4°C
  - solution 1: 9,073g KH_2PO_4 / 1000ml aqua bidest
  - solution 2: 11,87g $Na_2HPO_4 \cdot 2H_2O$ / 1000ml aqua bidest
- Giemsa stain, modified (Sigma cat.no. GS 500): 7-10% in phosphate buffer

Procedure

Getting started

- Thaw one aliquot of trypsin stock solution and make up to 50ml with NaCl in a Coplin jar. Adjust pH to about 7,5-7,8. Usually 1-2 drops of 1M NaOH are sufficient (pH indicator paper).
- Add about 50ml phosphate buffer to a second Coplin jar.
- Prepare another Coplin jar containing a 7-10% solution of Giemsa stain in phosphate buffer.

1. GTG-banding is performed at room temperature and slides should also be at room temperature. All solutions should be renewed every 4-5 hours. The age of slides is not vital but has to be considered. The older a slide, the longer is the trypsin exposure time. In general, the exposure time should be determined first using a test slide. Usually, the optimal trypsin exposure time works for a whole batch of slides afterwards.
   Start with 10 seconds, for example. If the banding is not pronounced enough the exposure time has to be increased, if the chromosomes look fuzzy the time has to be reduced (Figure 4).
2. Rinse slide in phosphate buffer immediately after trypsin exposure.
3. Transfer to the Giemsa solution. Three to five minutes staining should be sufficient.
4. Rinse slide under running tap water and air-dry. If available, compressed air or a fan are useful to speed up drying.

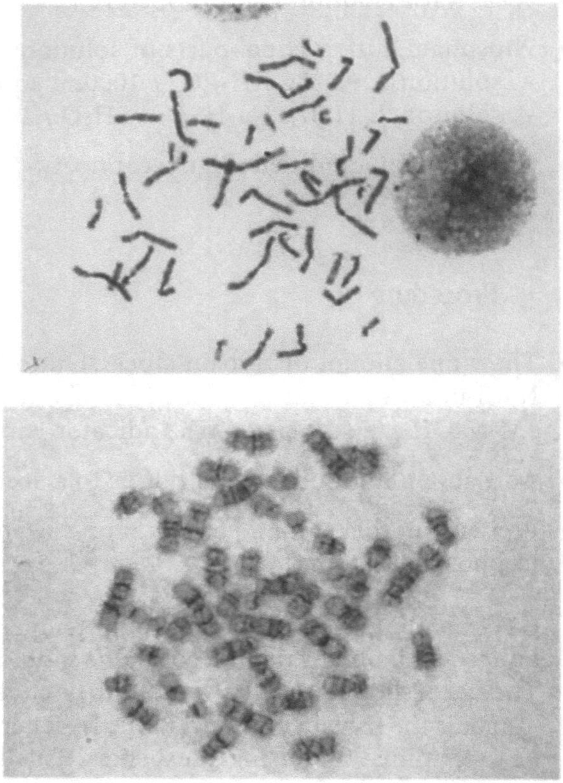

Fig. 4. **a** G-banded metaphase after insufficient exposure to trypsin. Banding is not pronounced enough and evaluation of chromosomal structure is therefore barely possible.

b G-banded metaphase after excessive exposure to trypsin. Chromosomes appear swollen and fuzzy. Evaluation of chromosomal structure is adversely affected.

Troubleshooting

- Quality of cytological preparations is crucial. For making slides, use clean and not greasy slides. If necessary, boil a batch of slides in water containing a detergent. Subsequently, rinse slides several times in deionized water and finally in low concentrated acetic acid (about 1%). Store slides in deionized water at 4°C until use.
- Check mitoses before banding. They should be without cytoplasm and chromosomes should not be too close to each other. It may help to apply ice-cold cell suspension onto cold slides from great height.
- Sometimes GTG-banding can be improved after aging of slides, eg by overnight incubation of slides at 60°C.

Subprotocol 3
QFQ-Banding

Q-banding was not only the first but is also the most simple differential staining procedure (Caspersson et al 1970). It is widely used as an alternative for G- or R-banding in pre- and postnatal cytogenetic diagnosis. Besides the overall banding pattern of the chromosomes, basically a G-band-like pattern, heteromorphisms at pericentromeric regions of chromosomes 3 and 4, pericentromeric regions and satellites of all acrocentrics and the distal portion of the Y chromosome long arm also become visible. The term QFQ implies that Q-bands are produced by fluorescence using quinacrine (Figure 5).

Materials

- Quinacrine · $2H_2O$: Dissolve 20mg quinacrine (Serva, cat.no. 34135) in 50ml McIlvaine's buffer. Store light-protected at 4°C. Solution can be used several times.
- McIlvaine's buffer, pH 5.6: Mix equal parts of solutions 1 and 2 and adjust pH to 5.6 if necessary.
  - solution 1: 0.1M citric acid monohydrate ($C_6H_8O_7 \cdot H_20$). Dissolve 21.01g of citric acid in 1000ml deionized water.
  - solution 2: 0.2M di-sodium hydrogen phosphate dihydrate.
  - ($Na_2HPO_4 \cdot 2H_2O$). Dissolve 35,6g sodium hydrogen phosphate in 1000ml deionized water.

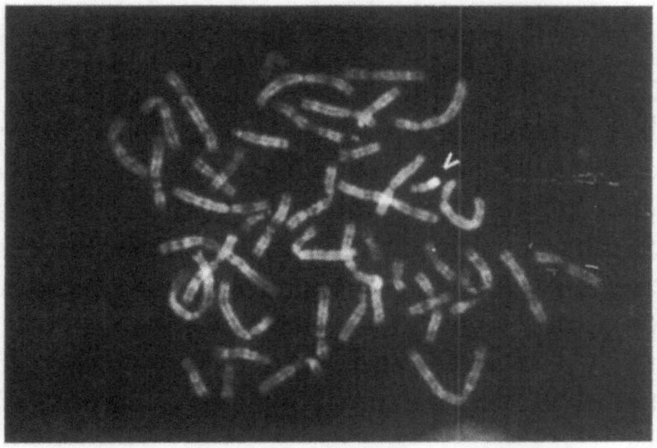

Fig. 5. Q-banded metaphase after treatment with quinacrine. A G-band-like pattern is produced. The bright fluorescence of the Y chromosome is striking (›).

Procedure

Getting started Prepare one Coplin jar with quinacrine solution and two with McIlvaine's buffer.

1. Place slide into a light-protected Coplin jar containing quinacrine solution and stain for 10-20 minutes at room temperature.
2. Rinse slide briefly in McIlvaine's buffer.
3. Immerse slide in a Coplin jar with McIlvaine's buffer until investigation.
4. The slide has to be protected with a coverslip. Excess buffer has to be squeezed out.
5. For analysis, a fluorescence microscope equipped with appropriate filters for quinacrine is necessary. Photographs are taken on Agfa Ortho film

Note: Agfa Ortho films are no longer available. Try Kodak Technical Pan instead. If Q-banding is not performed in an everyday routine it is a good idea to first ascertain exposure times which depend on the brightness of the chromosomes and may vary considerably. Start with 60-80 seconds.

Troubleshooting

- If chromosomes are too bright, change McIlvaine's and/or keep slides for a longer period (eg overnight) in McIlvaine's at 4°C.

Subprotocol 4
C-Banding

The code CBG stands for C-bands by **b**arium hydroxide using Giemsa and describes a method which reveals constitutive heterochromatin (Arringhi and Hsu 1971). Constitutive heterochromatin mainly consists of repetitive DNA that remains condensed during interphase of the cell cycle. In human chromosomes it is located in all centromeric regions, greatly pronounced in chromosomes 1, 9, and 16 and also in the distal part of the Y-chromosome long arm (Figure 6).

The C-banding technique includes a step of denaturation of chromosomal DNA using barium hydroxide or another alkali. This causes loss of DNA

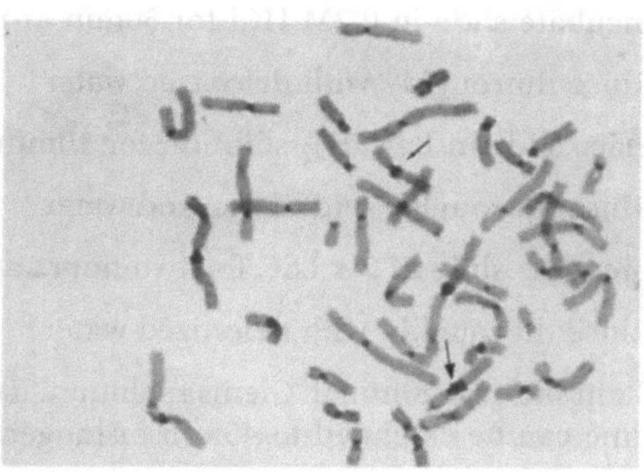

Fig. 6. C-band pattern of chromosomes after treatment with $Ba(OH)_2$ and Giemsa staining. Enlarged and therefore pronounced C-bands in chromosomes 1, 9, and 16 are seen. Note that one chromosome 9 (↓) contains about twice as much heterochromatin as its homologue (↓).

and protein in C-band negative regions. C-band positive regions are better protected against this treatment thus allowing the staining of these chromosomal parts. The underlying mechanism, however, is still unclear.

Materials

- 0.2M HCl: Dilute 10ml HCl with 40ml distilled water
- 0.03M $Ba(OH)_2 \cdot 8H_2O$: Dissolve 0.552g $Ba(OH)_2$ in 50ml distilled water
- 20 x SSC: Dissolve 175.3gNaCl and 88.29g $C_6H_5Na_3O_7 \cdot 2H_2O$ in 1000ml distilled water
- Giemsa stain
- Phosphate buffer: equal parts of solutions 1 and 2
  - solution 1: 9,073g KH_2PO_4 / 1000ml aqua bidest
  - solution 2: 11,87g $Na_2HPO_4 \cdot 2H_2O$ / 1000ml aqua bidest

Procedure

Getting started

- Prepare a Coplin jar with 0.2M HCl
- Filtrate $Ba(OH)_2$ solution into another Coplin jar
- Preheat a Coplin jar containing 2 x SSC to 60°C.
- Prepare Giemsa staining solution (8-10%) in phosphate buffer.

1. Incubate slide in 0.2M HCl for 30min at room temperature
2. Rinse thoroughly with deionized water
3. Place slide in $Ba(OH)_2$ solution for 10min at 37°C
4. Rinse thoroughly with deionized water
5. Incubate slide in 2 x SSC for two hours at 60°C
6. Rinse thoroughly with deionized water
7. Stain slide for 20min in Giemsa solution. If chromosomes are too pale the time can be extended to 45min or longer.

Tip: A shorter incubation time in $Ba(OH)_2$ solution results in a faint G-band-like pattern allowing thus to identify specific chromosomes. This may support the localization of chromosomal breakpoints.

Subprotocol 5
NOR-Staining

Nucleolar organizer regions (NORs) identified by silver staining reveal transcriptionally active genes located in the short arms (satellite stalks) of the acrocentric chromosomes (Goodpasture and Bloom 1975). They contain tandem repeats of rDNA coding for ribosomal RNA. The number of NORs in unselected individuals usually varies from five to ten and reflects a heritable polymorphism. This polymorphism may be helpful for the classification of certain marker chromosomes as well as for the identification of

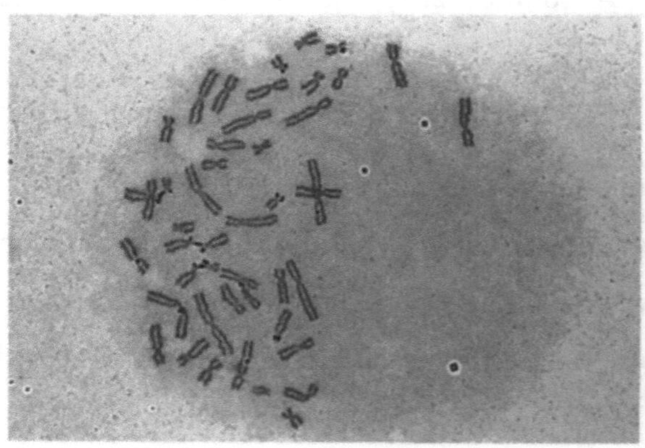

Fig. 7. Active NORs on acrocentric group D and G chromosomes after silver staining. Remarkable patterns of satellite associations stand out.

the parental origin of a certain acrocentric chromosome, for example in trisomic offspring (Figure 7).

Materials

- Colloidal developer: Dissolve 2g granulated gelatine (Fluka, cat.no. 48724) in 100ml distilled water (dissolves better at 37°C) and add 1ml formic acid (HCOOH). Store light protected at 4°C.
- Silver nitrate ($AgNO_3$): Prepare a solution of 4g silver nitrate in 8ml distilled water. Store light protected at 4°C.

Procedure

Getting started

Cover a hot plate with aluminium foil and preheat to 70°C. Because silver nitrate stains heavily it is recommended to wear gloves and a lab coat.

1. First, place two drops of colloidal developer onto slide and subsequently add four drops of silver nitrate solution. Cover with a coverslip. Incubate slide for 2min at 70°C on a hotplate.
2. Remove coverslip and rinse thoroughly with deionized water.

Troubleshooting

- Slides should be not more than a few days old.
- Silver nitrate should be applied immediately after applying the developer.
- If the color of the slide appears too dark (then chromosomes are too dark, too!) reduce the time on hotplate to one minute.
- It might be necessary to slightly increase the temperature of the hotplate.
- In general, working too slowly can result in excessive background of silver deposition on the slides.

Subprotocol 6
DA-DAPI Staining

DA-DAPI is short for subsequent application of distamycin A and 4,6-diamino-2-phenylindole. DAPI is a fluorochrome that produces a slight Q-band-like pattern along the chromosomes when applied alone. When slides are treated with distamycin A before DAPI-staining, fluorescent spots are produced in the pericentromeric regions of chromosomes 1, 9, and 16, the proximal short arm of chromosome 15 and the distal long arm of the Y-chromosome (Schweizer et al 1978). Because of the special chromosome 15 labelling, DA-DAPI staining is often used for characterization of small marker chromosomes. It has been demonstrated that many of these are derivatives of chromosome 15 (Figure 8).

Materials

- Distamycin A (stock: 0.1µg/ml): Dissolve 2mg distamycin A (Sigma, cat.-no. D-6135) in 20ml McIlvaine's buffer, pH 7.0 and store 0.5ml aliquots at -20°C.
- DAPI (4,6-diamino-2-phenylindole; stock: 1µg/ml): Dissolve 0.5mg DAPI (Sigma, cat.no. D-9542) in 500ml McIlvaine's buffer, pH7.0 and store 10ml aliquots at -20°C.
- McIlvaine's buffer, pH7.0: Prepare as described under "QFQ-banding" and adjust pH to 7.0.

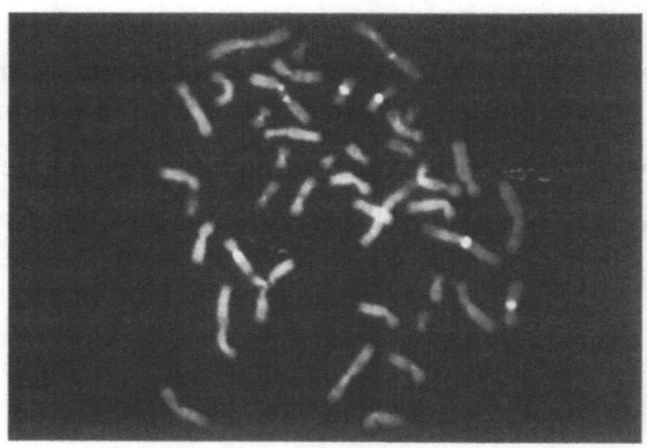

Fig. 8. DA-DAPI staining results in prominent centromeric bands in chromosomes 1, 9, and 16. In addition, pericentromeric regions of chromosomes 15 only are highlighted.

Procedure

Getting started

- Thaw one aliquot (0.5ml) distamycin A and dilute with 0.5ml McIlvaine's buffer (50µg/ml).
- Thaw one aliquot (10ml) DAPI and add 20ml McIlvaine's buffer (0.3µg/ml).
- Prepare a coplin jar containing McIlvaine's buffer.

1. Mount slide with distamycin A for 10-20min at room temperature. Cover with a coverslip.
2. Rinse briefly in McIlvaine's buffer.
3. Stain slide for 8-10min at room temperature in DAPI solution (keep in dark).
4. Rinse briefly in McIlvaine's buffer.
5. If not analyzed immediately, air-dry slide and store in the dark.
6. For viewing, mount slide with McIlvaine's buffer, add a coverslip and squeeze out excess buffer.
7. For visualization a fluorescence microscope equipped with appropriate filters for DAPI is necessary.

Troubleshooting

- If the DA-DAPI signals fade too rapidly, dry slide in the dark at 37°C and apply about 20µl antifading solution (eg ONCOR, cat.no. S1370-5). Cover with a cover slip and squeeze off excess solution.

Subprotocol 7
Replication Pattern (by BrdU-Incorporation)

Depending on the period of BrdU substitution for thymidine during DNA replication, characteristic patterns along the chromosomes are achieved (Latt 1973). A reverse band-like pattern (R-bands) is produced after BrdU application at the end of the S-phase of the cell cycle (Figure 9). Only late replicating chromosomal regions like AT rich DNA stretches, constitutional heterochromatin, and the inactive X chromosome have integrated BrdU and are therefore faintly stained. The resulting bands of the chromosomes are more or less the opposite of G-bands. Whereas after G-banding most chromosome ends are light and consequently unable to be evaluated, following R-banding terminal regions of chromosomes are dark and therefore clearly visible. We describe BrdU incorporation in late replicating DNA only. It should however be mentioned that labelling of DNA during early and middle S-phase results in G-band-like patterns, and in patterns intermediate between G- and R-bands, respectively.

In general, BrdU labelling during various intervals of the S phase is used to study the process of DNA replication. Demonstration of late replicating DNA has widely replaced conventional R-banding. In this case the term RBA is used and refers to **R**-bands prepared by **B**rdU using **a**cridine orange.

Materials

Reagents and buffers

- 5-bromo-2'-deoxyuridine (BrdU; stock 1mg/ml): Dissolve 100mg BrdU (Sigma, cat.no. B-5002) in 100ml Hank's balanced salt solution (HBSS). Store in the dark at 4°C. For the final dilution (20-40mg/ml) add 100-200µl per 10ml media.
- Colchicine (stock 400µg/ml): Dissolve 40mg colchicine in 100ml distilled water. For the final dilution (0,4µg/ml) add 100µl of a 1:10 diluted stock to 10ml media.
- Acridine Orange (stock 0.1%): Dissolve 100mg acridine orange (Fluka, cat.no. 01660) in 100ml distilled water. Store in the dark at 4°C. For the final dilution add 6ml to 100ml phosphate buffer.

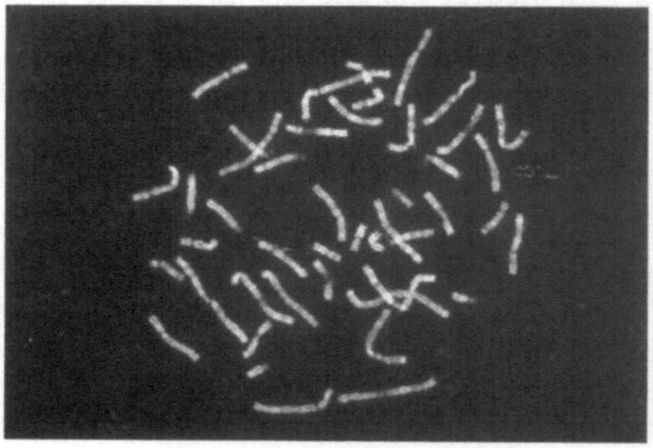

Fig. 9. Staining of late replicated chromosomal segments resulting in an R-band-like pattern. The chromosomal bands are opposite of G-bands. In contrast to G-banded chromosomes the ends of the chromosomes are scoreable.

- Phosphate buffer: equal parts of solutions 1 and 2
  - solution 1: 9,073g KH_2PO_4 / 1000ml aqua bidest
  - solution 2: 11,87g $Na_2HPO_4 \cdot 2H_2O$ / 1000ml aqua bidest
- Ethanol series of 100, 95, 70, and 50%.

Culture of lymphocytes

1. Grow lymphocyte cultures for a total of 72 hours in a tube containing 9ml media (eg RPMI 1640) and 1ml heparinized whole blood as usual.
2. Six hours before harvest add BrdU at a final concentration of about 20-40μg/ml media.
3. Mix carefully by inverting the tube several times.
4. One hour before harvest add colchicine at a final concentration of 0.1μg/ml.
5. Mix carefully by inverting the tube several times.
6. Harvest cells and prepare slides according to the standard protocol.

Procedure

Getting started

- Prepare a Coplin jar with acridine orange solution.
- Prepare a Coplin jar with phosphate buffer.
- Prepare Coplin jars containing decreasing ethanol concentrations.

1. Incubate slide for 10 minutes in acridine orange solution.
2. Wash three times briefly in phosphate buffer.
3. Add a coverslip and squeeze off excess buffer.
4. Analyse under a fluorescence microscope with the appropriate filter sets.
5. Photographs are taken on Agfa Ortho film.

Note: Agfa Ortho films are no longer available. Try Kodak Technical Pan instead. Exposure time has to be determined first. Start with 60 seconds.

Troubleshooting

- Staining is suitable when green and red chromosomal bands are clearly visible and cytoplasm appears reddish. If chromosomes are too green and bands are not pronounced enough, time of staining in acridine orange has to be increased and phosphate buffer rinses should be reduced. If chromosomes are too red, staining time has to be reduced and/or time of washing in phosphate buffer should be increased. During exposure to UV light, staining turns from red to green. Therefore, it is sometimes sufficient to wait for the optimal pattern of the chromosomes.

Subprotocol 8
Sister Chromatid Differentiation (by BrdU Incorporation)

Incorporation of 5-bromodeoxyuridine (BrdU) as a thymidine substitute during the last two successive cell cycles results in sister chromatid differentiation (SCD). As a result, after exposure to UV light, the sister chromatid with BrdU integrated into both DNA strands appears faintly stained while the other chromatid containing only one substituted DNA strand is dark when stained with Giemsa (Figure 10). Consequently, chromosomes of metaphases after only one cell cycle of BrdU incorporation are entirely dark, while metaphases of more than two cycles include entirely light chromosomes, chromosomes with only partial SCD, and chromosomes demonstrating SCE (see below).

Exchanges between sister chromatids result in patterned chromosomes ("harlequin chromosomes"), where the label of one chromatid switches to the sister chromatid (Latt 1974). Basically, this may occur zero to several

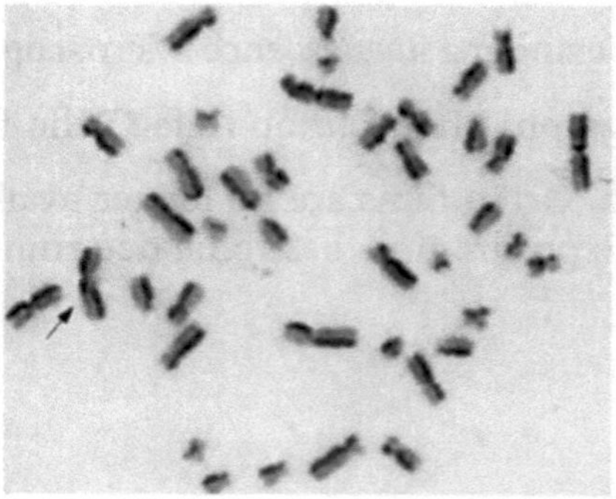

Fig. 10. Harlequin-like chromosomes after exposure to BrdU for the last two cell cycles. On some chromosomes sister chromatin exchanges are visible (↓).

times per chromosome. In normal human metaphases, however, such sister chromatid exchanges (SCEs) occur about five to ten times per metaphase.

The biological relevance of SCE is still unknown. However, the association of particular genetic syndromes with an increased number of SCEs has been noticed. Furthermore, we know that the exposure of cell cultures to certain mutagenic agents may also result in an increasing number of SCEs.

Materials

Reagents and buffers

- 5-bromodeoxyuridine (BrdU) (stock 1mg/ml): Dissolve 100mg BrdU in 100ml Hank's balanced salt solution (HBSS). For the final dilution (10-20µg/ml) add 100-200µl per 10ml media.
- Hoechst 33258 (150µg/ml): Dissolve 15mg Hoechst (bisbenzimid-trihydrochlorid; Sigma, cat.no. B1782) in 100ml distilled water. Store in aliquots at -20°C.
- 2 x SSC: Dissolve 17.5g sodium chloride (NaCl) and 8.8g sodium citrate, dihydrate ($Na_3C_6H_5O_77 \cdot 2H_2O$) in 1000ml distilled water. Adjust pH to 7.0.
- Phosphate buffer: equal parts of solutions 1 and 2
  - solution 1: 9,073g KH_2PO_4 / 1000ml aqua bidest
  - solution 2: 11,87g $Na_2HPO_4 \cdot 2H_2O$ / 1000ml aqua bidest
- Giemsa stain, modified (Sigma cat.no. GS 500): 7-10% in phosphate buffer.
- Deionized water
- UV light source

Culture of lymphocytes

- Grow lymphocyte cultures for a total of 72 hours in a tube containing 9ml media (eg RPMI 1640) and 1ml heparinized whole blood as usual.
- After 24 hours add BrdU at a final concentration of about 10µg/ml media.
- Mix carefully by inverting the tube several times.
- One hour before harvest add 100-200µl colchicine.
- After a total of 72 hours harvest cells and prepare slides as usual.

Procedure

Getting started Preheat 2 x SSC to 60-65°C in a Coplin jar.

1. Mount slide in Hoechst for 15min and keep in dark.
2. Rinse briefly in deionized water.
3. Mount slide horizontally in 2 x SSC under UV light. 2 x SSC should cover slide by about 1cm. Exposure time is 1-2 hours.
4. Rinse briefly in deionized water.
5. Incubate slide in 2 x SSC for 1-2 hours at 60-65°C.
6. Rinse in deionized water and air-dry.
7. Stain slide in Giemsa solution for 5-10min.
8. Rinse under running tap water and air-dry.

Troubleshooting

- Sometimes BrdU-incorporation is ineffective, and differential staining of chromosomes is not seen. The time of exposure to UV light, the optimal temperature of 2 x SSC and the incubation time in 2 x SSC has to be ascertained for each batch of slides. Staining intensity of Giemsa should be monitored under a light microscope without immersion oil. If chromosomes are too faint, slides can be re-incubated in Giemsa for another few minutes.

Subprotocol 9
Sex Chromatin Staining - X Chromatin (Barr Body)

Barr bodies (X chromatin) representing inactive X chromosomes and Y chromatin indicating the presence of a regular Y chromosome can both be identified in interphase nuclei. Analysis of X and Y chromatin is thus the established technique for the rapid determination of a persons chromosomal sex and for the analysis of gonosomal variants. X chromatin-positive cells are found in about 20-60% of buccal mucosa cells of normal females but also in up to 5% of such cells of normal males. Accordingly, false positive and false negative results are possible and, moreover, the identification of gonosomal mosaics is not likely. Moreover, most structural gonosomal aberrations remain undetected in sex chromatin analysis.

In males with an average sized Y chromosome about 70-90% of nuclei exhibit a fluorescent body. However, state of the art in clinical cytogenetics is nowadays a conventional chromosome analysis as from lymphocyte cultures.

In normal female cells an intensively stained body preferentially located at the periphery of the nucleus is observed (Figure 11; Barr and Bertram 1949). The number of Barr bodies found in a single cell is one less than the total number of X chromosomes. Accordingly, a normal female cell exhibits one Barr body while a female donor with a 47,XXX karyotype displays two such bodies. Due to the fact that not all cells exhibit the Barr body, at least 100-200 cells should be analyzed.

Materials

- freshly prepared slides of buccal mucosa from normal female donors or patients indicative of having a variable number of X chromosomes.
- 50, 70, and 100% ethanol
- 1N HCl
- 0.2% diamond fuchsin (Merck, cat.no. 15937.0025) staining solution: 0.2g diamond fuchsin in 100ml distilled water.

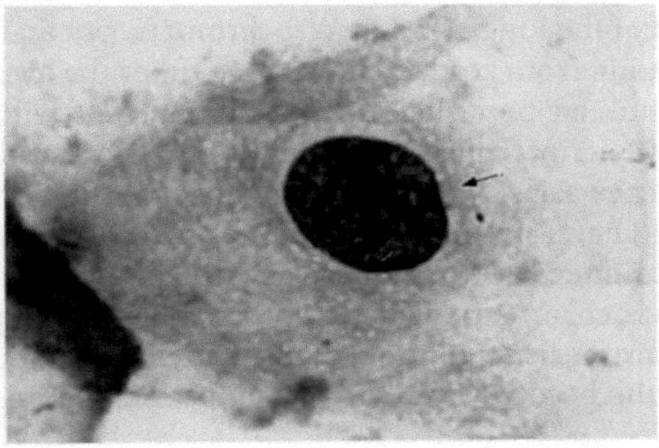

Fig. 11. Buccal mucosa cell of a female with one Barr body (↓) representing the inactive X chromosome.

Procedure

Getting started

- Prepare one Coplin jar each:
  - 100% ethanol
  - 70% ethanol
  - 50% ethanol
  - 1N HCl, prewarmed to 60°C
- Boil diamond fuchsin solution vigorously (add some boiling-stones). Filter the solution and allow to cool.

1. Immediately after preparation, fix slide in 100% ethanol for at least 1hour.
2. Flame-dry slide (optional).
3. Incubate slide in 1N HCl for 10 min at 60°C.
4. Rinse with deionized water and air-dry.
5. Stain in diamond fuchsin solution for 10 min.
6. Rinse with deionized water.
7. Dehydrate in a series of 50, 70, and 100% ethanol, 3 min each at room temperature.
8. Air-dry slide.

Subprotocol 10
Sex Chromatin Staining - Y-Chromatin (Y Body)

Part of the Y chromosome consists of heterochromatin. In many cases about half of the long arm is heterochromatic but an impressive variability in size ranging from zero to about three times the average amount exists. Y-chromatin can be visualized by staining cells with quinacrine (Pearson et al 1970). In interphase nuclei a brightly fluorescent body can be seen, however its size varies according to the size of the heterochromatic segment (Figure 12). Therefore, an analysis based on Y-chromatin staining in interphase nuclei alone may lead to false negative results in case of a very low amount of heterochromatin. On the other hand, marked polymorphisms as for pericentromeric regions of acrocentric chromosomes may lead to false positive results because they can also appear as fluorescent spots in nuclei.

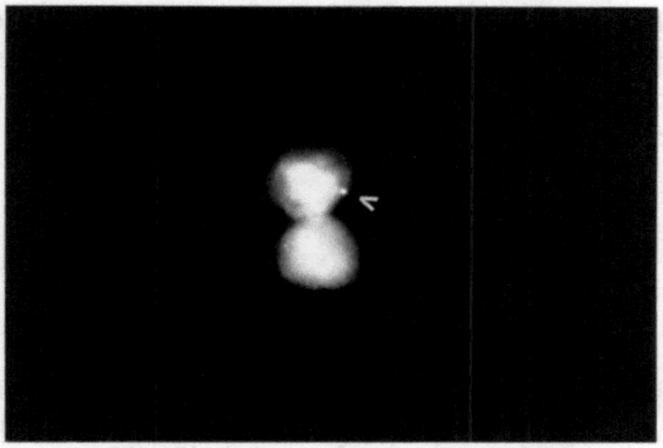

Fig. 12. Male cells after quinacrine staining. The heterochromatin of the Y chromosome appears as a bright fluorescing body in one cell (<).

Materials

- Quinacrine · $2H_2O$: Dissolve 20mg quinacrine in 50ml McIlvaine's buffer. Store light-protected at 4°C. Solution can be used several times.
- McIlvaine's buffer, pH 5.6: Mix equal parts of solutions 1 and 2 and adjust pH to 5.6 if necessary.
  - Solution 1: 0.1M citric acid monohydrate ($C_6H_8O_7 \cdot H_2O$): Dissolve 21.01g of citric acid in 1000ml deionized water.
  - Solution 2: 0.2M di-sodium hydrogen phosphate dihydrate ($Na_2HPO_4 \cdot 2H_2O$): Dissolve 35,6g sodium hydrogen phosphate in 1000ml deionized water.

Procedure

Getting started

Prepare one Coplin jar with quinacrine solution and two with McIlvaine's buffer.

1. Place slide into a light-protected Coplin jar containing quinacrine solution and stain for 15-20 minutes at room temperature.
2. Rinse slide briefly in McIlvaine's buffer.
3. Immerse slide in a Coplin jar with McIlvaine's buffer until investigation.
4. Protect the slide with a coverslip. Squeeze out excess buffer.
5. A fluorescence microscope equipped with appropriate filters for quinacrine is necessary for the analysis. Photographs are taken on Agfa Ortho film.

Note: Agfa Ortho films are no longer available. Try Kodak Technical Pan instead. If Q-banding is not performed as an everyday routine it is a good idea to first ascertain exposure time, which depends on the brightness of the metaphases and may vary considerably. Start with 60-80 seconds.

References

Arringhi FE and Hsu TC (1971): Localization of heterochromatin in human chromosomes. Cytogenetics 10, 81-86

Barr ML and Bertram EG (1949): A morphological distinction between neurones of the male and female, and the behaviour of the nucleolar satellite during accelerated nucleoprotein synthesis. Nature 163, 676-677

Caspersson T, Zech L, and Johansson C (1970): Differential banding of alkylating fluorochromes in human chromosomes. Exp Cell Res 60, 315-319

Drets ME and Shaw MW (1971): Specific banding patterns of human chromosomes. Proc Natl Acad Sci USA 68, 2073-2077

Goodpasture C and Bloom SE (1975): Visualization of nucleolar organizer regions in mammalian chromosomes using silver staining. Chromosoma 53, 37-50

ISCN (1978): An international system for human cytogenetic nomenclature. Birth defects: original article series vol 14, no 8 (National Foundation, New York 1978); also in Cytogenet Cell Genet 21, 309-404

ISCN (1995): An international system for human cytogenetic nomenclature, Mitelman F (ed). S. Karger, Basel

Latt SA (1973): Microfluorometric detection of deoxyribonucleic acid replication in human metaphase chromosomes. Proc Natl Acad Sci USA 70, 3395-3399

Latt SA (1974): Localization of sister chromatid exchanges in human chromosomes. Science 185, 74-76

Pearson PL, Bobrow M, and Vosa CG (1970): Technique for identifying Y chromosomes in human interphase nuclei. Nature 226, 78-79

Schweizer D, Ambros P, and Andrle M (1978): Modification of DAPI banding on human chromosomes by prestaining with a DNA-binding oligopeptide antibiotic, distamycin A. Exp Cell Res 111, 327-332

Seabright M (1971): A rapid banding technique for human chromosomes. Lancet ii, 971-972

Verma RS and Babu A (eds) (1989): Human chromosomes: manual of basic techniques. Pergamon Press, New York

Chapter 3

Karyotyping and Data Interpretation

KARSTEN HELD

Introduction

Research on human cytogenetics began with the work of Arnold (1879) and Fleming (1882) who for the first time examined human mitotic chromosomes. The detailed study of human chromosome morphology began 1956 after Tijo and Levan improved the technique of squash preparation using hypotonic shock and added colchicine to arrest cells in metaphase in order to increase the number of countable cells (Vogel and Motulsky, 1986). In their now classic article they reported that the human chromosome number was 46 and not 48 as was believed from earlier reports. Their work was confirmed in the same year by Ford and Hamerton (1956). The two articles stimulated a renewed interest in human cytogenetics and soon several laboratories were engaged in the study of human chromosomes and a variety of classification and nomenclature systems were proposed. This resulted in confusion in the literature and a need to establish a common system of nomenclature that would improve communication between workers in the field (Harnden, 1985). 1960 "A proposed standard system of nomenclature of human mitotic chromosomes" was reported by a study-group which convened in Denver, Colorado. This report, more usually known as the Denver Conference (1960), has formed the basis for all subsequent nomenclature reports. The present human chromosome nomenclature (ISCN 1995) is based on the results of the following international conferences: Denver 1960, London 1963, Chicago 1966, Paris 1971, Paris 1975, Stockholm 1977, Paris 1981, Memphis 1994. The present ISCN summarizes the current nomenclature, incorporates and supersedes all previous ISCN recommendations.

Karsten Held, Lademannbogen 61-63, Hamburg, 22339, Germany (*phone* 040 538 05800; *fax* 040 538 05821; *e-mail* held@labor-keeser-arndt.de)

The purpose of this chapter is to introduce non-cytogeneticists to the nomenclature of human chromosomes and to aid those who start employing cytogenetic techniques in describing human chromosomes and karyotypes.

The first part of the chapter will focus on the principles applied in describing karyotypes in laboratory reports and literature. The reader should be aware that it is neither intended nor possible to use this chapter as substitute for the ISCN, which in its present form covers 114 pages and which the reader should consult for detailed information.

The second part of the chapter deals with various aspects of data interpretation and quality assessment in clinical cytogenetics.

Normal Karyotype

Explanations and examples on nomenclature given in this chapter are adpoted from ISCN 1995

Non banding techniques

When stained with an appropriate stain, such as Giemsa, several morphological landmarks appear, which can be used to arrange the chromosomes into 7 readily distinguishable groups (A to G) based on descending order of size and of the position of the centromere.

Group A (Chromosomes 1 - 3): Large metacentric or submetacentric chromosomes easily distinguished from each other by size and centromere position.

Group B (Chromosomes 4 - 5): Large submetacentric chromosomes.

Group C (Chromosomes 6 - 12, X): Medium-sized metacentric or submetacentric chromosomes. The X-chromosome resembles the longer chromosomes in this group.

Group D (Chromosomes 13 - 15): Medium-sized acrocentric chromosomes with satellites.

Group E (Chromosomes 16 - 18): Relatively short metacentric or submetacentric chromosomes.

Group F (Chromosomes 19 - 20): Short metacentric chromosomes.

Group G (Chromosomes 21-22, Y): Short acrocentric chromosomes with satellites. The Y-chromosome bears no satellites.

The systematized array of the chromosomes prepared either by drawing, digitized imaging or by photography is called karyotype.

Banding techniques

Numerous technical procedures have been reported that produce banding patterns on metaphase chromosomes. A band is defined as that part of chromosome which is clearly distinguishable from its adjacent segments by appearing darker or lighter with one or more banding techniques. The methods first published for demonstrating bands along the chromosomes were those that used quinacrine mustard or quinacrine dihydrochloride to produce a fluorescent banding pattern. These methods are named Q-staining methods and the resulting bands Q-bands. Techniques which demonstrate an almost identical pattern of dark and light bands along the chromosomes usually use the Giemsa dye mixture as the staining agent. These techniques are generally termed G-staining methods and the resulting bands G-bands. Some banding techniques give patterns that are opposite in staining intensity to those obtained by the G-staining methods, viz, the reverse staining methods, and the resulting bands are called R-bands.

The banding techniques fall into two principle groups: (1) those resulting in bands distributed along the length of the whole chromosome, such as G-, Q-, and R-bands including techniques that demonstrate patterns of DNA replication, and (2) those that stain specific chromosome structures and hence give rise to a restricted number of bands. These include methods which reveal constitutive heterochromatin (C-bands), telomeric bands (T-bands), and nucleolus organizing regions (NORs). For the code to describe banding techniques, see Table 1.

Identification and definition of chromosome bands, landmarks and regions

A band is a part of a chromosome clearly distinguishable from adjacent parts by virtue of its lighter or darker staining intensity. Landmarks are defined as consistent and distinct morphologic features important in identifying chromosomes, e.g., the ends of the chromosome arms, the centromere, and certain bands. A region is defined as an area of a chromosome lying between two adjacent landmarks. The bands and the regions are numbered from the centromere outwards along each chromosome arm. The symbols **p** and **q** are used to designate respectively the short and long arms of each chromosome. The centromere (**cen**) itself is designated 1.0, the two regions adjacent to the centromere are labelled as 1 in each arm, the next more distal regions are 2 and so on.

In designating a particular band ,4 items are required: (1) the chromosome number, (2) the arm symbol, (3) the region number, and (4) the band

Table 1. Examples of the code used to describe banding techniques. In this one-, two-, or three-letter code, the first letter denotes the type of banding, the second letter the general technique, and the third letter the stain.

Q	Q-bands
QF	Q-bands by fluorescence
QFQ	Q-bands by fluorescence using quinacrine
QFH	Q-bands by fluorescence using Hoechst 33258
G	G-bands
GT	G-bands by trypsin
GTG	G-bands by trypsin using Giemsa
GAG	G-bands by acetic saline using Giemsa
C	C-bands
CB	C-bands by barium hydroxide
CBG	C-bands by barium hydroxide using Giemsa
R	R-bands
RF	R-bands by fluorescence
RFA	R-bands by fluorescence using acridine orange
RH	R-bands by heating
RHG	R-bands by heating using Giemsa
RB	R-bands by BrdU
RBG	R-bands by BrdU using Giemsa
RBA	R-bands by BrdU using acridine orange

number within that region. The items are given without spacing or punctuation. For example, 1p31 indicates chromosome 1, short arm, region 3, band 1. Whenever an existing band is subdivided, a decimal point is placed after the original band designation and is followed by the number assigned to each subband, eg 1p31.1, 1p31.2 and so on. If a sub-band is further subdivided, additional digits but no further punctuations are used, eg 1p31.11, 1p31.12, and so on.

In diagnostic cytogenetics a resolution of approximately 400 to 550 bands per haploid set, corresponding to the left and center ideogrammes of the ISCN, will turn out to be sufficient for most practical purposes. High resolution preparations of prophase and prometaphase chromosomes are occasionally required (see Chapter 5) Here one problem in assigning numbers to euchromatic subbands is, that in G-banded preparations, new G-bands appear to arise by subdivision of darkly stained G-bands on less extended chromosomes while in R-staining preparations the R-bands appear to split.

Karyotype designation

The karyotype of any cell is described in three parts, each separated by a comma (,): The first item to be recorded is the total number of chromosomes including the sex chromosomes. The sex chromosome constitution is given next followed by the specific description of unusual chromosomes. Thus the normal human karyotype is designated:

46,XX normal female

46,XY normal male

In the description of chromosome abnormalities sex chromosome aberrations are presented first, followed by abnormalities of the autosomes listed in numerical order, irrespective of aberration type. Each abnormality is separated by a comma.

Numerical aberrations

Sex chromosome aneuploidy is described both numerically and directly in the sex chromosome constitution:

45,X 45 chromosomes, missing a sex chromosome (Turner syndrome).

47,XXY 47 chromosomes, an extra sex chromosome (Klinefelter syndrome).

Autosomal aneuploidy of whole chromosomes is indicated by a plus (+) or minus (-) sign, placed before the chromosome.

47,XX,+21 47 chromosomes, an extra chromosome 21 (Trisomy 21)

45,XX,-22 45 chromosomes, missed a chromosome 22 (Monosomy 22).

Structural aberration

General principles

Letter designation are used to specify rearranged chromosomes. Symbols and abbreviated forms frequently used to designate chromosome abnormalities are listed in Table 2.

Table 2. Symbols and abbreviated terms frequently used in clinical cytogenetics

ace	Acentric fragment
add	Additional material of unknown origin
approximate sign (~)	Denotes intervals and boundaries of a chromosome segment
arrow (→)	From -to, in detailed system
cen	Centromere
colon, single (:)	Break, in detailed system
colon, double (::)	Break and reunion, in detailed system
comma (,)	Separates chromosome numbers, sex chromosomes, and chromosome abnormalities
del	Deletion
de novo	Designates a chromosome abnormality which has not been inherited
der	Derivative chromosome
dic	Dicentric
dir	Direct
fis	Fission, at the centromere
fra	Fragile site
g	Gap
h	Heterochromatin, constitutive
i	Isochromosome
idic	Isodicentric chromosome
ins	Insertion
inv	Inversion or inverted
mar	Marker chromosome
mat	Maternal origin

Table 2. Continuous

min	Minute acentric fragment
mos	Mosaic
pat	Paternal origin
Ph	Philadelphia chromosome
psu	Pseudo-
q	Long arm of chromosome
question mark (?)	Questionable identification of a chromosome or chromosome structure
rep	Reciprocal
rea	Rearrangement
rec	Recombinant chromosome
rob	Robertsonian translocation
s	Satellite
semicolon (;)	separates altered chromosomes and breakpoints in structural rearrangements involving more than one chromosome
slant line (/)	Separates clones
stk	Satellite stalk
t	Translocation
tan	Tandem
tel	Telemore
ter	Terminal (end of chromosome)
trc	Tricentric chromosome
trp	Triplication
upd	Uniparental disomy
v	Variant or variable region

In single chromosome rearrangements the chromosome involved in the change is specified within parentheses ( ) immediately following the symbol identifying the type of rearrangement, eg inv(2) indicates an inversion of chromosome 2. If two or more chromosomes have been altered, a semicolon (;) is used to separate their designations. The + or - signs placed after a chromosome arm symbol (p or q) may be used in text to indicate an increase or decrease of chromosome arm length (eg 4p+, 5q-), but should not be used in the description of karyotype (see below).

Uncertainty in chromosome or band designation may be indicated by a question mark (?) or an approximative sign (). The term **or** is used to indicate alternative interpretations of an aberration.

Specifications of breakpoints:
The location of any given break is specified by the band in which that break has occurred. Since it is not possible at present to define band interfaces accurately, a break suspected at an interface between two bands is assigned arbitrarily to the number of the band more distal to the centromere. If a break can be localized only to a region, the region number may be specified eg 1p1.

Designating structural chromosome aberration:
Two systems for designating structural abnormalities exist. One is the short system, which is commonly used in clinical reports. In this system structurally altered chromosomes are defined only by their breakpoints. The breakpoints are specified within parentheses immediately following the designation of the type of rearrangement and the chromosome(s) involved. The breakpoints are identified by band designations and are listed in the same order as the chromosomes involved. No semicolon is used between breakpoints in single chromosome rearrangements.

The detailed system, besides identifying the type of rearrangement, defines each abnormal chromosome in terms of its band composition. The two systems are not mutually exclusive and can be used to complement each other. In publications it is recommended that the detailed form be given first.

Examples

Abnormal chromosome features

In the following examples of chromosome rearrangements both forms will be given.

Terminal Deletions:

46,XY,del(5)(p13)

46,XY,del(5)(qter→p13:) Terminal deletion with break in band 5p13. The single colon indicates that the segment distal to the 5p13 band is deleted.

Interstitial Deletions:

46,XX,del(5)(q13q33)

46,XX,del(5)(pter→q13::q33→qter)The segment between band 5q13 and 5q33 has been deleted.

Direct Duplication:

46,XX,dup(1)(q22q25)

46,XX,dup(1)(pter→q25::q22→qter)Direct duplication of the segment between band 1q22 and 1q25, which retains the same orientation with respect to the centromere.

Inverted Duplication:

46,XY,dup(1)(q25q22)

46,XY,dup(1)(pter→q25::q25→q22::q25→qter) or

46,XY,dup(1)(pter→q22::q25→q22::q22→qter)Inverted duplication of the segment between bands 1q22 and 1q25. Note that only the detailed system will clarify the location of the duplicated segment.

Insertion within a chromosome:

46,XX,ins(2)(p13q21q31)

46,XX,ins(2)(pter→p13::q31→q21::p13→q21::q31→qter) The nomenclature identifies the recipient region (2p13) first, followed by the segment being inserted (2q21 to 2q31). The original orientation of the inserted segment has been maintained. There is no punctuation when all the segments are from the same chromosome.

Insertion between two chromosomes:

46,XX,ins(5;2)(p14;q22q32)

46,XX,ins(5;2)(5pter→5p14::2q32→2q22::5p14→5qter;
2pter→2q22::2q32→2qter)

The segment of chromosome 2 between bands 2q22 and 2q32 has been inserted directly into band 5p14.

Note that the recipient chromosome is specified first.

Paracentric inversion:

46,XX,inv(3)(q21q26)

46,XX,inv(3)(pter→q21::q26→q21::q26→qter)

The segment of the long arm of chromosome 3 from bands 3q21 to 3q26 has been inverted. The centromere is not involved.

Pericentric inversion:

46,XY,inv(3)(p13q21)

46,XY,inv(3)(pter→p13::q21→p13::q21→qter)

Note that it is apparent from the band designation whether it is a paracentric or pericentric inversion.

Isochromosomes: The symbol **i** is used for isochromosomes and **idic** for isodicentric chromosomes. The breakpoints in isochromosomes are assigned to the centromeric bands p10 and q10 according to the morphology of the isochromosome.

46,X,i(X)(q10)

46,X,i(X)(qter→q10::q10→qter)

One normal X chromosome and an isochromosome for the long arm of one X chromosome.

46,XX,idic(17)(p11)

46,XX,idic(17)(qter→p11::p11→qter)

An isodicentric chromosome composed of the longer arms of chromosomes 17 and the short arm materials between the centromeres and the breakpoints in 17p11.

Reciprocal translocation: In translocations (t) affecting two chromosomes, the sex chromosome or the autosome with the lowest number is always specified first.

Two Break Rearrangements

46,XY,t(2;5)(q21;q31)

46,XY,t(2;5)(2pter→2q21::5q31→5qter; 5pter→5q31::2q21→2qter)

Breakage and reunion have occurred at bands 2q21 and 5q31. The segments distal to these bands have been exchanged. For three breaks and more complex rearrangements the reader is referred to the ISCN (1995).

Whole arm translocations: Whole arm translocations can be adequately described by assigning the breakpoints to the centromeric bands p10 and q10 according to the morphology.

46,XY,t(1;3)(p10;q10)

46,XY,t(1;3)(1pter→1p10::3q10→3qter; 3pter→3p10::1q10→1qter)

Reciprocal whole arm translocation in which the short arm of chromosome 1 has been fused to the centromere with the long arm of chromosome 3 and the long arm of chromosome 1 has been fused with the short arm of chromosome 3.

Robertsonian translocations: The special types of translocations which originate through centric fusion of the long arms of the acrocentric chromosomes 13-15 and 21-22 can be adequately described with the nomenclature for unbalanced whole arm translocations i.e. using the symbol der (see below).

45,XX,der(13;21)(q10;q10)

45,XX,der(13;21)(13qter→13q10::21q10→21qter)

The derivative chromosome has replaced one chromosome 13 and one chromosome 21, there is no need to indicate the missing chromosomes. The resulting net in balance is loss of the short arms of chromosomes 13 and 21.

For historical reasons, the designation **rob** is retained e.g. 45,XX,rob(13;21)(q10;q10). The abbreviation rob should not be used in the description of acquired abnormalities.

Derivative chromosomes: A derivative chromosome **(der)** is a structurally rearranged chromosome generated either by a rearrangement of two or more chromosomes or by multiple rearrangements within a single chromosome. A derivative chromosome resulting from one rearrangement involving two or more chromosomes is specified in parentheses, followed by the type of abnormality.

46,XX,der(1)t(1;3)(p22;q13)

46,XX,der(1)(1qter→1p22::3q13→3qter)

The derivative chromosome 1 has resulted from a translocation of the chromosome 3 segment distal to 3q13 to the short arm of chromosome 1 at band 1p22. It is apparent that the karyotype is unbalanced. The der(1) replaces a normal chromosome 1 and there are two normal chromosomes 3. For more complex and other types of rearrangements the reader is referred to ISCN(1995).

Ring chromosomes: As in other rearrangements affecting a single chromosome, in ring chromosomes derived from one chromosome there is no semicolon between the band designations.

46,XX,r(7)(p22q36)

46,XX,r(7)(::p22→q36::)

Ring chromosome in which breakage and reunion have occurred at bands 7p22 and 7q36. The segments distal to these breakpoints have been deleted.

Ring chromosomes derived from more than one chromosome may contain one or several centromeres. Monocentric ring chromosomes are treated as derivative **(der)** chromosomes, dicentric or tricentric ring chromosomes are designated by the symbol **r** preceded by the triplet **dic** or **trc.** For further details the reader is referred to section 9.2.14 ring chromosomes in the ISCN (1995).

Marker chromosomes: A marker chromosome **(mar)** is a structurally abnormal chromosome in which no part can be identified. Whenever any part of an abnormal chromosome can be recognized it is a derivative chromosome **(der)** and can be adequatly described by the nomenclature for derivative chromosomes. In the description of a karyotype the presence of a mar must be preceded by a + sign.

47,XX,+mar one additional marker chromosome.

Additional material of unknown origin: The symbol **add** should be used to indicate additional material of unknown origin attached to a chromosome region or band. Unknown material that replaces a chromosome segment may, depending on the size of the extra material, result in either increase or decrease in the length of the chromosome arm. Designations such as "1p+" or "1p-" may be used in text to describe such abnormal chromosomes but should not be used in the description of the karyotype.

46,XX,add(19)(p13)

46,XX,add(19)(?::p13→qter) Additional material attached to band 19p13. Neither the origin of the extra segment nor the type of rearrangement is known.

When additional material of unknown origin is attached to both arms of a chromosome and/or replaces more than one segment in a chromosome, the symbol der should be used. Unknown material inserted in a chromosome should be described by the use of the symbols **ins** and **?.**

Normal variable chromosome features

Variation in length: Variation in length of heterochromatic segments (**h**), satellite stalks (**stk**) or satellites (**s**) are distinguished from increases or decreases in arm length as a result of other structural alterations by placing a plus (+) or minus (-) sign after the symbols **h, stk** or **s** following the appropriate chromosome and arm designation.

16qh+ Increase in length of the heterochromatin on the long arm of chromosome 16.

21ps+ Increase in length of the satellite on the short arm of short arm 21.

22pstk+ Increase in length of the satellite stalk on the short arm of chromosome 22.

13cenh+ pat Increase in length of the centromere heterochromatin of the chromosome 13 inherited from the father.

Variation in number and position:

17ps Satellites on the short arm of chromosome 17.

Yqs Satellites on the long arm of the Y-chromosome.

9phqh Heterochromatin in both the short and the long arms of chromosome 9.

Duplicated chromosome structures are indicated by repeating the appropriate designation: 21pss Double satellites on the short arm of chromosome 21.

Fragile sites: Fragile sites (**fra**) associated with specific chromosome bands may occur as normal variants or be associated with a specific disease or phenotype. Several different types of fragile sites may be inducible by culturing cells in media containing different components, all these will be covered by a single nomenclature.

fra(10)(q25.2) A fragile site on chromosome 10 in sub-band 10q25.2.

46,Y,fra(X)(q27.3) A fragile site in sub-band Xq27.3 on the X-chromosome in a male.

In situ hybridisation (ISH)

Techniques utilizing fluorescence in situ hybridisation (FISH) are described [SV1]in Part V Molecular Cytogenetics.

These techniques have provided the cytogeneticist with an increased ability to reveal cryptic abnormalities like the detection of microdeletion syndromes (Table 3) or mosaicism or to correlate chromosome structures with gene locations. These advances have necessitated the development of a specific FISH nomenclature.

In the description of a karyotype the results obtained by conventional cytogenetic analysis are given first, followed by a period (.) followed by the

Table 3. Selected Microdeletion Syndromes

Condition	Chromosomal Localization
Wolf-Hirschhorn syndrome	4p16.3 *
Cri-du-chat syndrome	5p15.2*
Elastin-Williams syndrome	7q11.23 *
Langer-Giedion syndrome	8q24.1
WAGR syndrome	11p13
Prader-Willi/Angelman syndrome	15q11-13*
Rubinstein-Taybi syndrome	16p13.3*
Smith-Magenis syndrome	17p11.2*
Miller-Dieker syndrome	17p13.3*
Alagille syndrome	20p12-11
DiGeorge / Velocardiofacial syndrome	22q11.2*
X-linked Ichthyosis / STS gene	Xp22.3*
Kallman syndrome	Xp22.3*
X-linked Ocular Albinism	Xp22.3*
(* FISH probes available)	

Table 4. List of Symbols and Abbreviations frequently used in FISH nomenclature

-	absent from a specific chromosome
+	present on a specific chromosome
++	duplication on a specific chromosome
x	multiplication sign, precedes the number of signals seen
.	period, separates cytogenetic observations from results of in situ hybridization
;	semicolon, separates probes on different derivative chromosomes
amp	amplified signal
con	connected signals (signals are adjacent)
dim	diminished signal intensity
enh	enhanced signal intensity
fib ish	extended chromatin/DNA fiber in situ hybridization
fish	fluorescence in situ hybridization
ish	in situ hybridization; when used without a prefix applies to chromosomes (usually metaphase or prometaphase) of dividing cells
mv	moved signal (signal moved from original location)
nuc ish	nuclear or interphase in situ hybridization
pcp	partial chromosome paint (hybridization with probe mixtures prepared from partial chromosome scrapings, contigs, etc.)
rev ish	reverse in situ hybridization including comparative genomic in situ hybridization
sep	separated signals (signals are separated)
sp	split signal (single copy probe signal maps to more than one location)
st	stationary signal (signal remaining in original location)
wcp	whole chromosome paint

abbreviation **ish** and the ish results. If a standard cytogenetic observation has not been made, the ish observations only are given. Observation on structural abnormal chromosomes are expressed by the symbol ish followed by the symbol for the structural abnormality followed in separate parentheses by the chromosome(s), the breakpoint(s) and the locus or loci for which probes were used designating according to the genome data base and ordered from pter to qter. The locus designation (or if the locus is not available the probes' name) and the status of each locus is given immediately after the locus designation e.g. present (+) or absent (-).

46,XY.ish del(22)(q11.2q11.2)(D22S75-)

Observation on normal chromosomes are expressed by the symbol ish followed by the chromosome region, band, or sub-band designation of the locus or loci tested followed in parentheses by the locus (loci) tested, a multiplication sign (**x**) and the number of signals seen.

46,XY.ish 22q11.2(D22S75x2)

For further details the reader is referred to ISCN 1995.

Quality Assessment and Quality Assurance in Clinical Cytogenetics

Pre- and postnatal cytogenetic investigations have become routine methods in clincal practice during the last two decades. Keeping in mind the importance and the substantial costs of cytogenetic tests, it is not surprising that there is a great deal of interest in maintaining the confidence of the public in the quality and consistency of these services. In the FRG for example an obligation to quality assurance is stipulated by law (Sozialgesetzbuch V) and the guidelines of the Federal Association of Physicians.

Standards can be improved by the introduction of "Internal Quality Assurance (IQA)" through training, education and agreed "best practice" guidelines and by the introduction of "External Quality Assessment (EQA)" through external checks on quality of chromosomal preparations, staining techniques, and checks on accuracy of cytogenetic testing.

In the European countries IQA and EQA have been developing gradually in the last 10 years. Up to now, there is no generally accepted scheme but some basic principles have emerged. For instance in an EURCOMIC workshop in Leuven, November 8-10,1996, guidelines were drawn up by 24 clinical and laboratory geneticists from 15 countries in Europe which were intended for use as a reference manual by genetic centers in their efforts to achieve and maintain high standards in prenatal diagnosis. These guidelines on minimum quality standards cover cytogenetic analysis procedures as well as various aspects on staffing, medical collaboration, laboratory supervision, work load recommendations and minimum standards of equipment and facilities. Considerable attention was given to quality assurance which included the observance of standards of procedures and protocols, the continuing education of the staff, the following of accepted standard nomenclatures, the issue of reports in the standardized manner, the assessment of internal quality by monitoring laboratory standards and reporting time, the participation in external quality programmes, and the storage and filing of patient's data.

Internal quality assessment

In most countries IQA is promoted by workshops introducing new methods etc., which are organized by the national associations of cytogeneticists and cytologists and accordingly by the establishment of chromosome analysis guidelines. In these guidelines recommendations for appropriate analyses are formulated and the lowest standards for a given reason for referral are defined.

In prenatal diagnosis for example, many laboratories adhere to the recommendations established by the Association of Cytogenetic Technologists Task Force (1990). These guidelines however cover other types of analyses as well.

Table 5 summarizes the recommendations with only slight modifications for prenatal and constitutional studies. For cancer chromosome studies the reader is referred to Chapter 9.

In prenatal diagnosis of chromosomal abnormalities the detection of mosaicism poses a diagnostic problem since an aberrant cell line may be the result of *in vitro* mutation or of mosaicism confined to extraembryonic tissues or of true fetal mosaicism. The reader is referred to Chapter 11 for details (good discussions on this subject have been published by Hook et al.(1977) and Claussen et al. (1984).

Chromosome banding studies should be appropriate to case and type of tissues studied. Especially in prenatal diagnosis, there is no generally accepted minimal standard (good discussions on this subject have been published by Hook et al.(1989), Feeny and Tomkins (1981), Claussen et al.(1992), Jacobs et al.(1992)). The recommendations of the British Association of Clinical Cytogeneticists and the Association of Cytogenetic Technologists (USA) are taken into account in Table 5. There are several reasons why the guidelines are not generally accepted. One reason is i.e. an inconsistency in the recommended number of metaphases to be counted using either the flask method or the *in situ* method. One would have to analysize a much higher number of metaphases from a higher number of clones in the flask method in order to obtain comparable results to the *in situ* method (see Clausen et al. 1984 for discussion).

Table 5. Chromosome analysis guidelines for prenatal and constitutional chromosome studies

Type of analysis	Count§	Analyze	Karyotypeτ	Chromosome band resolution#
Amniotic Fluid a) flask method	15-20 cells, 2 independently established cultures	4 -5 cells	min. 2 cells	minimal 400 bphs for most referral eg. maternal age, ↑AFP, etc
b) *in situ* method	15 cells from a minimum of 10-15 colonies from 2 independently estab. cultures	4-5 cells each from different colonies	min. 2 cells	
CVS a) direct preparation*	15-20 cells if possible	4 -5 cells	min. 2 cells	200 bphs if possible
b) culture	15-20 cells from 2 independently established cultures	4 -5 cells	min. 2 cells	400 bphs
c) combination of direct and culture preparation	15-20 cells	4 -5 cells	min. 2 cells	400 bphs
Constitutional	15-20 cells	4 -5 cells	min. 2 cells	300 bphs: most aneuploidies studies and known structural rearrangements, sex chromosome anomalies 400 bphs: expected small structural rearrangements 500 bphs: recurrent abortions, dysmorphic features 650 bphs: microdeletions, eg Aniridia, Wilm's Tumor

* Because of the high proportion of placenta confined mosaicism analyzing cells from both methods, which assay two different cell types, cytotrophoblasts in the direct, and mesenchymal core in culture, is recommended.

§ Noting any numerical / structural aberrations observed.

τ in cases of mosaicism, karyotype a minimum of 1 cell per cell line.

defines minimal quality for a given reason for referral

External quality assurance

The introduction of EQA in cytogenetics has proved to be difficult for the simple reason that it is not possible to distribute a large number of blood or other tissue samples from a single (informative) patient or proband to participating laboratories.

Two aspects are of paramount importance in cytogenetic analyses. One concerns the quality of chromosome preparation and the second correct data interpretation. The two are of course interrelated to a certain extent, however, though a good slide quality is a prerequisite for a correct diagnosis in many instances it is no guarantee for it.

An elaborate scheme to estimate slide quality has been worked out by the British Association of Clinical Cytogeneticists (1988), which has proved to be very effective in improving standards in cytogenetic services. Recently a simplified scheme, which is based on the scoring system developed in the British quality assessment scheme was introduced in the FRG. In a pilot study, evaluating karyotype and the respective slides, it was demonstrated that scoring banding quality from chosen karyotypes is as efficient and less

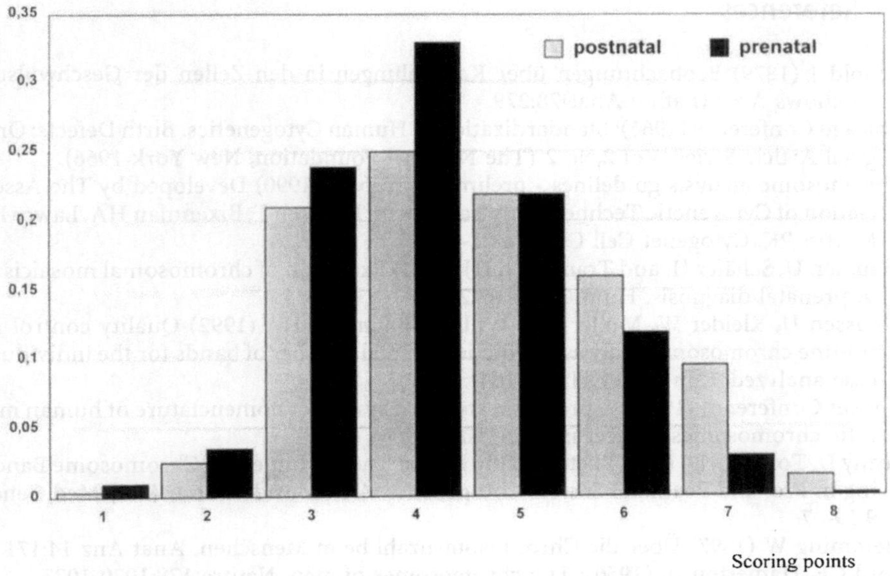

Fig. 1. Distribution of banding quality among 1.450 postnatal and 1.586 prenatal karyotypes from 44 and 42 laboratories (a scoring of 4 points corresponds approximately to the 1971 Paris Convention standard, i.e. 400 bphs and a scoringof 8 points approximately to the right hand set of ISCN 1985, i.e. 850 bphs).

time demanding than scoring from slides. It was further shown, that scoring metaphase availability by external assessment can be omitted, since a reasonable banding score is in general achieved only on slide preparations with a sufficient number of metaphases.

Figure 1 shows the distribution of banding quality among 1.450 postnatal and 1.586 prenatal karyotypes from 44 and 42 participating laboratories respectively. By the second half of 1995 a 400 bands resolution level as minimum goal for most specimens was reached or exceeded by approximately 70% of submitted karyotypes from amniotic cell studies and by 75% in constitutional studies. While this type of EQA is useful to monitor the quality of slide preparation of individual laboratories as well as of the cytogenetic services in general, it does not permit any conclusion as to the correctness of data interpretation. As mentioned above, the only valid test to distribute samples (blood, amniotic fluid, chorionic villi, etc.) from one patient or proband to all participants is not feasible. As yet, no simple and cost effective solution has evolved which would take the various aspects to be considered (eg staining methods, gray scale variation, structural analysis at the microscope, on the computer, or on a screen after projection etc.) into account.

References

Arnold J (1879) Beobachtungen über Kernteilungen in den Zellen der Geschwülste. Virchows Arch (Pathol Anat)78:279

Chicago Conference (1966): Standardization in Human Cytogenetics. Birth Defects: Original Article Series, Vol 2,No 2 (The National Foundation, New York 1966).

Chromosome analysis guidelines - preliminary report (1990) Developed by The Association of Cytogenetic Technologists Task Force: Knutsen T, Bixenman HA, Lawce H, Martin PK. Cytogenet Cell Genet 44:1-4

Claussen U, Schäfer H, and Trampisch HJ (1984) Exclusion of chromosomal mosaicism in prenatal diagnosis. Hum Genet 67:23-28

Claussen U, Kleider W, Müller HG, Wille N, Baumann HA (1992) Quality control in routine chromosome analysis: prediction of total number of bands for the individual case analyzed. Clin Genet 41:100-104

Denver Conference (1960): A proposed standard system of nomenclature of human mitotic chromosomes. Lancet i:1063 – 1065(1960)

Feeny D, Tomkins DJ (1981) Letter to the Editor: The Usefulness of Chromosome Banding in Pre- and Postnatal Service Cytogenetics: A Reconsideration. Am J Med Genet 9:79-87

Flemming W (1897) Über die Chromosomenzahl beim Menschen. Anat Anz 14:171

Ford CE, Hamerton JL (1956): The chromosomes of man. Nature 178:1020-1023

Harnden DG (1985) Historical Introduction. In: ISCN (1985):An International System for Human Cytogenetic Nomenclature, Harnden DG, Klinger HP (eds), Birth Defects: Original Article Series, Vol 21,1 (March of Dimes Birth defects Foundation, New York 1985)

Hook EB (1977) Exclusion of chromosomal mosaicism: Tables of 90%, 95%, and 99% confidence limits and comments on use.Am J Hum Genet 29:94-97

Hook EB, Healy NP, Willey AM (1989) How much difference does chromosome banding make? Adjustment in prevalence and mutation rates of human structural cytogenetic abnormalities. Ann Hum Genet 53:237-242

ISCN (1995): An International System for Human Cytogenetic Nomenclature, Mitelman F (ed); S. Karger, Basel,1995

Jacobs, PA, Browne C, Gregson N, Joyce CH, White H (1992)Estimates of the frequency of chromosome abnormalities detectable in unselected newborns using moderate levels of banding. J Med Genet 29:103-108

London Conference on the Normal Human Karyotype. Cytogenetics 2:264-268(1963)

Paris Conference (1971):Standardization in Human Cytogenetics. Birth Defects: Original Article Series, Vol 8, No 7 (The National Foundation, New York 1972); also in Cytogenetics 11:313-362 (1972)

Tjio JH, Levan A (1956) The chromosome number of man. Hereditas 42:1-16

United Kingdom External Quality Assessment Scheme. Slide Assessment Scoring Guide. Edinburgh. ACC Clin Cytogenet Bull 1988 2:35-26

Vogel F, Motulsky AG (1986) History and Development of Human Cytogenetics. In: Vogel F, Motulsky AG (eds) Human Genetics. Problems and Approaches. Springer-Verlag, Berlin Heidelberg New York Tokyo pp 20-24

Chapter 4

Documentation

PETER MINY AND ROLF-DIETER WEGNER

Introduction

During recent decades cytogenetic analyses evolved from a research instrument available in a limited number of university laboratories to a well established diagnostic tool routinely applied in various medical disciplines. As with other specialized diagnostic methods included in health plans or covered by health insurances, certain requirements defined by legal authorities or professional organisations have to be met regarding the documentation of the outcome and how it was obtained. Obedience to these rules is critical when alleged diagnostic errors lead to law suits, which has not infrequently been the case in the recent past. Most of these instances were related to prenatal chromosome studies. Bearing in mind the limitations and pitfalls of cytogenetic analyses, either pre- or postnatal, a meticulous documentation is obviously at least as important as the ability to produce high quality cytogenetic preparations.

Documentation is also essential for the participation in internal and external quality assessment schemes and has at least two main objectives:

to provide a close as possible account of the course and outcome of an individual cytogenetic examination;

to allow a statistical evaluation of the laboratory performance (by assessing accuracy or duration of testing and other relevant parameters).

There are two main methods of documentation relevant to cytogenetic diagnoses: protocols and image storage. The archiving of image information

Correspondence to Peter Miny, Universitäts-Kinderspital beider Basel, Abt. Medizinische Genetik, Römergasse 8, Basel, 4005, Switzerland (*phone* 0041-61685 62 54; *fax* 0041-61685 60 11; *e-mail* miny@ubaclu.unibas.ch), Rolf-Dieter Wegner, Charité Campus Virchow-Klinikum, Institut für Humangenetik, Augustenburger Platz 1, Berlin, 13353, Germany and use of protocols will be outlined in this chapter after a summary of some pertinent legal requirements and recommendations of professional organisations.

Legal requirements and recommendations of professional organisations

In many countries, appropriate documentation of case history, clinical status, diagnostic procedures and therapy is required by law, usually in more general terms without detailed rules for specialized disciplines such as, for instance, diagnostic cytogenetics. Standards may evolve from litigation trials based on expert opinions, which usually summarize the middle of the road approach for a specific test. In some countries professional organisations have developed guidelines where certain standards in clinical cytogenetics are defined and documentation is briefly addressed (eg American College of Medical Genetics (ACMG), 1993 in the US; The Association of Clinical Cytogeneticists (ACC), 1994 in the UK; The Professional Organisation for Medical Genetics (Berufsverband Medizinische Genetik), 1997 in Germany).

United States

The American College of Medical Genetics started efforts in 1991 to define standards and guidelines for the practices of Clinical Cytogenetics and other medical genetic laboratory disciplines based on several existing standards (Association of Cytogenetic Technologists, ACT), 1991; Great Lakes Regional Genetics Group (GLaRGG), 1992, College of American Pathologists (CAP), 1994; ISCN, 1995 and others (only latest editions cited)). It is explicitly acknowledged that numerous acceptable variations exist in genetic testing methodologies, but also that the accuracy and dependability of all procedures should be documented in each laboratory. The guidelines list the relevant intake information for the patient record (see below) and summarize some general requirements for record keeping such as preservation of their confidentiality and integrity and maintenance which should be for one generation (20 years) if not specified otherwise by laws.

According to the guidelines slides stained with a permanent banding method (G-, C-, or R-banding, NOR) should be kept at least 5 years. For fluorochrome stained preparations individual regulations have to be developed. Residual patient specimen should be kept for one week. Processed patient specimen should be kept until the test has been signed out.

The guidelines also define some terminological standards with relevance for documentation which are briefly summarized:

- **Chromosome count:** Number of centric chromosomes per metaphase cell. Aneuploid metaphases should be characterized for specific gain or loss.
- **Analyzed cells:** Banded metaphase cells in which the individual chromosomes are evaluated in their entirety which includes their structural integrity, either at the microscope or from intact digitized images or photographic prints.
- **Karyotyped cells:** Chromosomes paired after cutting out from a photograph or arranged by the computer from a digitized image.
- **Scored cells:** Cells checked for presence or absence of a specific cytogenetic feature.
- **Abnormal clone:** At least two cells with the same hyperdiploidy or structural abnormality or at least three cells with hypodiploidy for the same chromosome.

In the guidelines the use of one or more objective methods to assess and document the banding resolution is recommended (Stallard et al. 1983; Kao et al. 1990; Josifek et al. 1991)

United Kingdom

The Association of Clinical Cytogeneticists (1994) has published guidelines for clinical cytogenetics which do not specifically address documentation but include a list of points to consider for the final reporting which matches the recommendations of the American College of Medical Genetics (1993) (see below).

Germany

The German Professional Organisation for Medical Genetics has recently issued guidelines for clinical cytogenetics which are an amended version of former guidelines. Documentation is addressed only marginally. Basic general rules are outlined in the legal framework relevant to doctors participating in the health care services.

Protocols and final report

Receipt of sample

A unique labcode which frequently also identifies the type of tissue for every specimen arriving in the cytogenetic lab is tradition and standard. Even today when PC's are present in almost every lab many cytogeneticists keep this code together with the date of arrival, patient's name, first name and birth date preferably in an old-fashioned lab record book even if this is not a must and may be replaced by an electronical database. The form accompanying the specimen, which should be provided by the lab and explicitly ask for details of the proband's clinical status and family history is the first part of the patient's file. Since numerous special investigations (eg fluorescence *in situ* hybridisation (FISH) to exclude specific microdeletions) are being developed, the indication for the cytogenetic examination is an important detail to document.

Culture

For all specimen requiring "long term" culture (eg chorionic villi, amniotic fluid cells, fibroblasts and others) a culture protocol is mandatory. First entries include date (and time) of arrival, amount, quality, appearance (eg blood admixture in amniotic fluid samples) and date of culture set up. For chorionic villi intended to be cultured, the result of an inspection of the sample under an appropriate microscope must be recorded with special reference to maternal cell contamination. For each culture vessel visual inspections, changes of medium, growth, special observations, and date of harvest with subsequent chromosome preparation have to be recorded independently (Figure 1). Unequivocal labelling of culture vessels must be of utmost concern to prevent misidentifications and subsequent diagnostic errors, which may be fatal especially in prenatal diagnosis.

Direct preparation

Amount and appearance of samples for short term cultures (eg blood lymphocytes) or direct preparation (eg bone marrow, chorionic villi) have to be documented.

Chromosome analysis

There is a surprising variety of ways to establish a patient's karyotype. This refers to technical details, banding methods and actual chromosome analysis. A major source of variation is the difference in chromosome condensation (number of bands per haploid set). Protocols are widely used to keep a record of the individual metaphases analysed to establish the karyotype. While in some laboratories all metaphases are recorded photographically or by digital image storage (see below) others rely on protocols alone and keep the slides for documentation additionally. Most cytogeneticists, however, use a combination of both methods (i.e. some metaphases are kept as an

Case-ID: **Name:** **Set up date:**

AFC ( ) **Chorionic villi ( )** **Other:......................** **Operator:** **Page:**

Comments:

Flask a			Flask b			Flask c		
Medium:								
Date	Growth/Comments		Date	Growth/Comments		Date	Growth/Comments	

Fig. 1. Example for a culture protocol

image and the analysis of others is recorded in a protocol). The protocol must specify the banding method (in accordance with ISCN, 1995) and contain a slide identification as well as the nonius for a given metaphase. We have developed a form (Figure 2; Pawlowitzki and Miny, unpublished) which allows the individual recording of all chromosomes for 10 metaphases including an estimate of the banding quality of the individual chromosome. Sketches of a metaphase, another popular type of protocol in the past, record mainly the classification of individual chromosomes and their position in the metaphase. Banding characteristics of individual chromosomes are usually not recorded. The technique is helpful during training where the correct chromosome classification is important. In modern routine clinical cytogenetics, however, advanced banding techniques (eg high resolution banding) produce more complex banding patterns which have to be checked for more subtle changes. Using a form, as illustrated in Figure 2, provides instant information on presence, overlap, and overall banding quality of a given chromosome in a total of 10 metaphases.

The same general rules apply for the documentation of FISH studies in metaphase or interphase. Additionally, the source of probes used for the analysis and, if appropriate, further specifications for their use have to be included.

Case-ID: Name: Diagnosis:

Banding: Operator/Microscope: Page: Date:
Comments:

			A			B		C							D			E			F		G				
	Slide	Non.	1	2	3	4	5	6	7	8	9	10	11	12	13	14	15	16	17	18	19	20	21	22	X	Y	Karyotype
1																											
2																											
3																											
4																											
5																											
6																											
7																											
8																											
9																											
10																											

V = 500 bands or more I = identified by banding pattern X = not identified by banding pattern ? = chromosome questionable
ø = chromosome missing A = chromosome abnormal

Fig. 2. Protocol used to document chromosome analysis under the microscope.

Reporting

In addition to the entry form, culture and examination protocols, photographs or prints of digitized images, and karyotypes, the final report is the last item of the patients file. According to ACMG (1993) and ACC guidelines (1994) the report should include:

- Case identification including name, first name, birth date, lab code, type of specimen, date of receipt of specimen and names of physicians to whom reports are sent.
- Number of cells counted and/or analyzed.
- The ISCN (1995) designation of the karyotype.
- Multiple cells with hypodiploidy, hyperdiploidy or structural rearrangements have to be identified.
- Culture and preparation details and banding methods if significant for the cytogenetic interpretation. It is mandatory to differentiate between direct preparation and cultured cells for chorionic villi sampling.
- Interpretation of result including recommendations for additional studies in the patient or family members or genetic counselling, if appropriate. Interpretation should be clear to a non-geneticist physician.
- Possible inaccuracies and test limitations (if not already addressed in a written consent form signed prior to prenatal diagnosis).

Archiving of Image Information

Establishing a karyotype is image analysis. This can be done by examining a metaphase under the microscope, count the chromosomes, scan the structure of each individual chromosome mainly based on length, centromere index and banding pattern. A major step in differentiating normal variation from structural abnormality is the comparison of both homologues of a given pair. This is most conveniently achieved after arranging the chromosomes to a karyogram. The traditional tool of "capturing" image information is classical photography which up to recently has been used in the majority of laboratories. The rapid development of digital image analysis is changing this at present.

It has to be born in mind, however, that both approaches will lead to a certain loss of optical information depending on the quality of equipment and the experience of the technicians involved. While image archiving is doubtless the optimal solution for documentation, the chromosome structure should not be assessed on photographs or digitized images exclusively as practised in some laboratories. An additional microscopic evaluation of the chromosome structure on the slide should be mandatory.

Storing slides

For a limited period of time slides stained with permanent banding methods (G-, C-, or R-banding, NOR) may be kept (see guidelines above) to meet the obligation of documentation. Even under optimal conditions the stain will fade sooner or later and important information may be lost. In our own experience the quality of slides stored for several years is rather unpredictable. Therefore, additional image storage is highly recommended. Figure 3 summarizes the most popular methods available at present.

Photography Photomicrographs of metaphases, film development, preparation of paper prints and the subsequent cutting out of chromosomes to arrange a karyogram are time consuming and therefore expensive, if the labour involved is properly accounted for. Nevertheless, photographic images of metaphases are still the gold standard for traditional banding methods when a hard copy with the best possible conservation of the original optical information is required. This quality aspect as well as the availability of the necessary hardware and experienced staff in many institutions will help traditional photography to survive for some time. A detailed discussion of all aspects of photomicrography can be found in monographs by Thompson and Bradbury (1987) and Delley (1988).

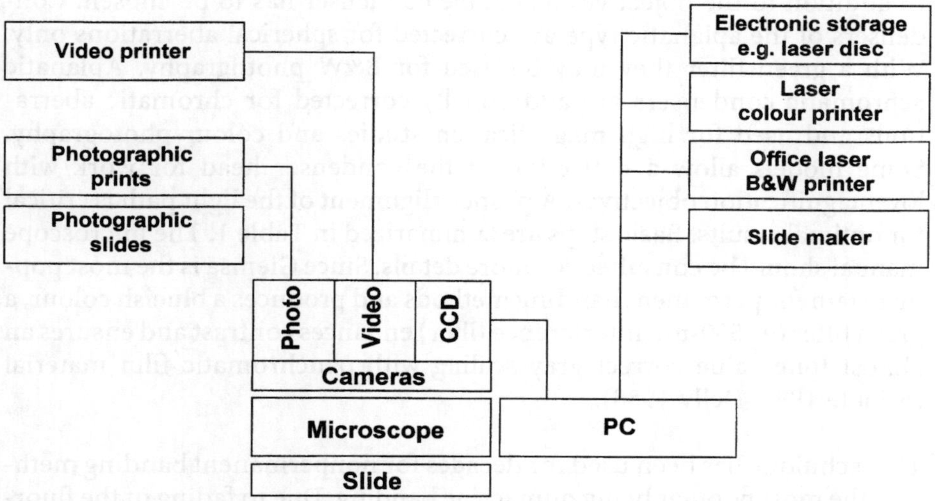

Fig. 3. Methods to store image information for documentation in clinical cytogenetics.

Microscope

The most important prerequisite for optimal results, not only for classical photography but also for digital image processing, is the appropriate equipment of the microscope and well-adjusted parts (Bradbury 1989, Schade 1993, Kapitzka 1994, Kriete et al. 1994). In the past decade a major change in the optical design of the microscope occurred. All major companies now offer new models incorporating infinite tube length optics (eg Delta optics (Leica); ICS infinity colour-corrected system (Zeiss)) which has a number of advantages compared to the traditional equipment. It is important to note that objectives are not interchangable between traditional and new systems. When major investments are planned, the performance of updated old systems should be carefully compared to new ones, something which is best done in the lab.

In our experience two objectives are sufficient for routine clinical cytogenetic examinations including FISH. We always wonder about the numerous objectives attached to some microscopes in routine clinical cytogenetic labs that are never really employed and which useful devices could have been purchased instead. One objective in the 10x to 25x magnification range for metaphase searching and a second one (100x) for chromosome analysis, photography or image capturing are essential. Considering that chromosome analysis requires magnifications close to the resolution limits of light microscopy we recommend to invest in the latter lens. If the microscope is used for diagnostic FISH examinations this objective should be "plan" and apochromatic and have a high numerical aperture (du Manoir et al. 1995).

Bright field microscopy

In addition to the objectives a suitable condenser has to be chosen. Condensers of the aplanatic type are corrected for spherical aberrations only. With a green filter they may be used for B&W photography. Aplanatic achromatic condensers are additionally corrected for chromatic aberrations and used for high magnification studies and colour photography. Some models allow a deflection of the condenser head for work with low magnification objectives. A proper alignment of the light path is critical for optimal results. Basic steps are summarized in Table 1. The microscope manual should be consulted for more details. Since Giemsa is the most popular stain for permanent banding methods and produces a blueish colour, a green filter (eg 550-nm interference filter) enhances contrast and ensures an almost tone value correct gray scaling with panchromatic film material (Schade 1993, Delly 1988).

Epifluorescence microscopy

The technique has been used for decades for nonpermanent banding methods, the most popular being quinacrine banding. Due to fading of the fluorescent dyes the metaphases had to be photographed for chromosome analysis. This inconvenience prompted many labs to introduce permanent banding techniques for routine examinations (mostly GTG-banding). Modern FISH techniques have led to a revival of epifluorescence microscopy in clinical cytogenetics. High pressure mercury lamps are required as a light source. They need to be properly centered (see manual) and have a lifespan of about 200 hours. Burning time should be recorded in a log book. It is

Table 1. Preparing the microscope for bright field examinations (Permanent banding techniques)

- Center lamp filament (refer to manual)
- Adjust eyepieces (also check diopter adjustment)
- Bring condenser to uppermost position
- Close illuminated field iris
- Bring image of the illuminated field iris into focus using condenser height control
- Center this image using adjusting screws of the condenser
- Open illuminated field iris until it just disappears from the field of view
- Set aperture iris to the numerical aperture of the objective in use (engraved on lens body)
- Close aperture iris slowly checking for best resolution by a maximum of 20-40 %
- Check aperture iris by removing an eyepiece and looking into the tube

critical that the selection of suitable filter systems is in accordance with the fluorescence dyes used. Selectable multiple filter sets can be attached to the microscope in various ways depending on the make of the microscope. For image processing systems electrically powered filterwheels controllable by the PC are available. All major companies provide a selection of filter sets for practically all fluorochromes used in clinical cytogenetics. Further details are discussed by Monk (1992) and Becker (no date). More advanced new applications are delt with in detail in chapters 23 and 24.

Photography hardware

Practically all microscopes used in diagnostic cytogenetics can be equipped with tubes to attach a camera. A standard 35mm (reflex) camera with a remote control release to prevent vibrations fitted to this tube and a stop watch to control exposure time represent the low cost end of photomicrography in the cytogenetic lab. Higher investments usually do not improve on image quality but on convenience and efficiency. All major microscope companies offer a range of dedicated photographic equipment with sophisticated automatic systems at the high end of the scale. These include automatic film transportation, adjustable automatic exposure control, interchangeable film cassettes and other useful features. If the microscope is equipped for bright field examinations exclusively, a simple camera mount will be sufficient. Automatic film transportation and exposure control will speed up work. Simultaneous use for epifluorescence microscopy makes interchangeable film cassettes an essential feature. Maximum flexibility is gained by tubes allowing the simultaneous attachment of a CCD and a photographic camera (Figure 3).

A frequent reason for blurred images are vibrations especially during longer exposure times. A firm stand for microscope and camera is essential. Special tables are available but very expensive. Vibrations can be checked for by looking at a metaphase at high magnification and somebody else working on the same table or walking through the room. If present, as is the case in many old buildings, preventive measures must be taken.

Film selection

Based on calculations considering magnification scale of paper prints and image quality (Schade 1993) 35 mm material is more than sufficient for practically all purposes in clinical cytogenetics. This material is widely available and most cost-effective. As a rule B&W material is used for all permanent banding techniques and for those fluorochromes where colour does not matter and hard copies are used to cut out chromosomes for karyotyping. Colour films are mainly used for documentation of FISH examinations in metaphase or interphase, if no digital image processing system is available. CGH is impossible without such a system and a specific software package. Even with modern film technology some old rules apply at least to some extent: The sensitivity or speed of a film is positively correlated to its graininess or resolving power and negatively to its contrast i.e. slow films needing longer exposure times produce high contrast, fine grain and high resolving power. Exposure time is not critical in B&W photomicrography of permanently banded chromosomes but has to be considered in fluorochrome stained preparations due to fading. Modern film and development technology allows variable exposure indices with a given film by modifications of the developing process (eg Kodak Technical Pan, Ilford FP4 Plus). Ilford XP2 films rely on an alternative technology resulting in almost grainfree prints, but need colour development. We used a Kodak Technical Pan film for Giemsa-banded slides developed with the Ilfotec HC procedure and an Ilford FP4 Plus for Quinacrine banding with the Ilford ID 11 procedure (Table 2). In colour photomicrography of FISH preparations slide film should be used exclusively because of problems with the production of paper prints from colour negative films. The "intelligent" printing machin-

Table 2. Examples for film development

Ilford FP4 125 ISO

- ID-11 developer (Ilford)
- 6,5 min in 38 l tank
- 1 min intermediate rinsage
- 15 min fixation in Super Hypam (Ilford)
- 15 min rinsage
- 15 min in drying cupboard

Kodak Technical Pan

- Ilfotec HC developer
- 4,5 min in 38 l tank
- 1 min intermediate rinsage
- 3-4 min fixation in Hypam (Ilford)
- 15 min rinsage
- 15 min in drying cupboard

(use all solutions at 20° C)

ery (Kelly 1988) is not prepared to handle photomicrographs and results are unpredictable. If paper prints are needed they should be copied from the slide. The preferable choice for most of the longer-wave fluorescence stains is a fast daylight colour slide film. We have been using the Kodak Ektachrome 400 for years with good results. For blue fluorochromes (eg DA-DAPI) an artificial light slide film may be considered.

Test strips for calibration

Even with a modern fully automatic exposure system a careful calibration of the exposure time is as important as in the early days of photomicrography in cytogenetics (Davidson 1973). After a meticulous adjustment of the microscope and illumination a test strip with a stepwise increasing exposure time should be produced (in automatic systems by changing ASA settings). It is obviously important to keep all other parameters (eg illumination, filters etc.) constant. A preparation of average quality and contrast should be used for this test. Comparing the different prints should finally result in an optimal setting of the exposure parameters. The darkroom technician has to be familiar with the aspect of chromosomes under the microscope to produce adequate prints.

Videoprinters

A video camera attached to a suitable printer is a low cost alternative to digital image processing systems and offers the advantage of an immediate printout of metaphases. The quality is acceptable although the inconvenience of light sensitive prints, which are difficult to store for documentation remains for the time being. Karyograms may be produced by cutting out chromosomes as from photographic prints. It is also possible to electronically store the video image on suitable media such as magnetic tapes. Cameras and printers of the major companies may be used after some specific adjustment. A company offering complete solutions is for instance AVT-Horn, Aalen, Germany.

Digital image processing

As early as 30 years ago, the use of computers in karyotyping was considered (Ledley 1964). Sophisticated systems were developed on a research basis and later marketed by a small number of companies. Due to the limited capabilities of the hardware in the early years the setup required specific designs with the consequence of high purchase and maintenance costs. Developing and improving software to aid in karyotyping has been a complex task also contributing to the cost. In recent years the computing power in standard off-the-shelf hardware improved dramatically and user-friendly software has been developed. Modern systems run on PC's of the DOS or MAC type and competitive prizing render them an attractive alternative to classical photography. In fact, it is easy to predict that digital image processing will supersede photography as the standard documentation tool in clinical cytogenetics in the near future for cost-benefit reasons.

Basic steps The microscopic image is captured by a CCD camera, then digitized and saved on the computer's hard disk. This digitized metaphase image may be stored for documentation and may subsequently be used for image processing i.e. to arrange chromosomes semi-automatically and interactively to a karyogram which may be further modified with a variety of manipulations including contrast enhancement, straightening of chromosomes and others (Figure 4). Technical details have been addressed in a large number of articles, the following being only a small selection: Graham 1987, Piper and Lundsteen 1987, Lundsteen and Piper 1989, Graham and Piper 1994. For advanced techniques such as CGH, SKY, or M-FISH readers are referred to Chapters 21, 23, and 24.

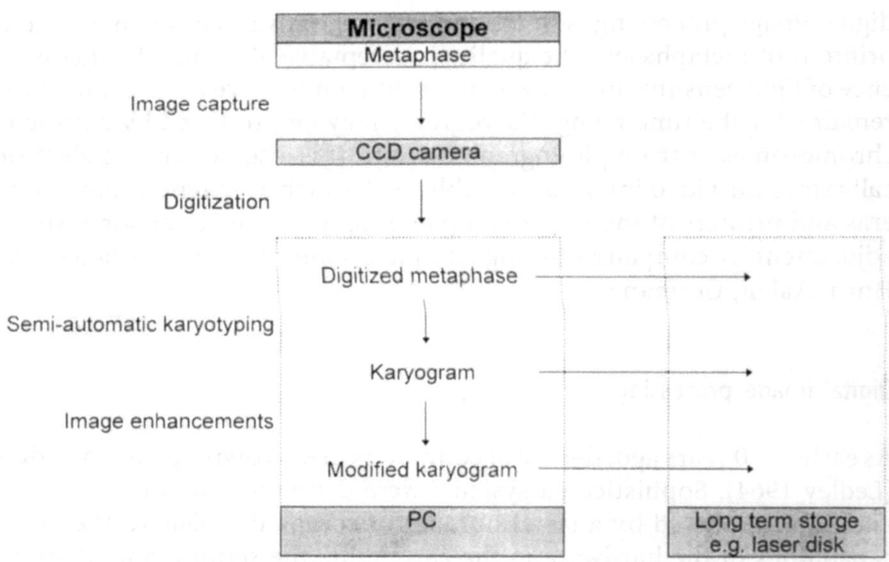

Fig. 4. Basic steps in digital image processing for documentation in clinical cytogenetics.

Hardware

Numerous complete systems are offered commercially. It is beyond the scope of this contribution to compare their performance and costs (Philip and Lundsteen 1985, Lundsteen et al. 1987, Lundsteen and Martin 1989, Korthof and Carothers 1991). We will restrict ourselves to the discussion of some basic issues concerning the use of image analysis hard- and software for documentation in clinical cytogenetics. Only systems running on standard off-the-shelf computer hardware should be chosen for cost reasons. Selecting camera, printer, storage media, and microscope equipment will depend on whether the intended use is restricted to permanent banding techniques (eg GTG-banding) or will include fluorescence *in situ* hybridisation (FISH). Due to the increasing application of FISH techniques it is probably wise to purchase suitable hard- and software from the start.

Camera

The basic choice is between a black and white (B&W) and Colour CCD. Both are equally suitable for standard banding techniques and FISH. The choice is critical for FISH investigations especially if CGH or other advanced examinations are planned. B&W cameras offer the best possible sensitivity which may be needed to demonstrate faint signals with unique sequence probes and in CGH experiments. The disadvantage is the need to capture the image seperately for every colour applied using different filter sets. This procedure, however, has been automatized by software driven electrically powered filterwheels. Multicolour FISH examinations are most conveniently captured in a single step by a colour camera using multiple-bandpass filters, although at the cost of a restricted sensitivity. This issue is dicussed more deeply for instance by Bornfleth et al. (1996) and du Manoir et al (1995). For most routine clinical cytogenetic questions a colour or a lower cost B&W CCD will probably be sufficient.

Printer

Rapid advances in modern laser printer technology and falling prices have provided systems which no longer depend on light sensitive paper. Colour sublimation laser techniques offer close to photographic quality prints of banded metaphase chromosomes as well as metaphase and interphase FISH preparations although at substantial costs. Basic black and white documentation of metaphases and karyotypes used for chromosome analysis under the microscope is available at a low cost by high resolution office laser printers. For a low cost documentation of coloured fluorochrome stained preparations, standard 35 mm slide films may be used (see above). The camera is placed on the second exit of the C-mount and the light is directed either to the CCD or the conventional camera by a switch. We have been working successfully with this setting and kept images of routine FISH studies on a slide film. High quality slides for presentations may also be produced from the digitized and improved image by a slide maker. If this is not possible directly the software should at least be able to save the image in a generally acknowledged file format such as TIFF which can be loaded in standard PC graphics software (Corel Draw, PowerPoint, Harvard Graphics and many others). This allows further modifications eg addition of a headline and can be used for exposure on a slide film.

Storage media

The obligation for documention can be met without immediate preparation of prints or slides altogether, if the images are saved for long-term storage on electronic media. Requirements for the storage of large amounts of data and rapid searching for individual images are probably best met by (re)-writable laser discs. Although such devices have been in use in medical image storaging for some time the experience with long-term data safety is restricted. Therefore, a strict backup scheme should be developed and adhered to. This rule, of course, also applies to patient data recorded electronically.

Table 3. Some major suppliers of hard- and software

Microscopes and Photomicrography	Films
Zeiss	Kodak
Leica	Agfa
Nikon	Ilford
Olympus	Fuji
Digital image processing systems	Video cameras and printers
Applied Imaging	AVT-Horn
PSI	
Oncor	FISH probes
Vysis	Oncor
Metasystems	Vysis
Mikado	Applied Imaging
Karyotec	
Zeiss	
Leica	

For the reader's convenience, some major suppliers of image analysis equipment, eg microscopes, films, cameras, printers and complete digital image processing systems are listed in Table 3. This list represents a subjective selection and is not intended to be complete. There is no ranking involved in list position. Please refer to local or national representatives of the companies or check for Web pages in the Internet.

We sincerely appreciate the support of Mark I. Evans, Detroit, and Rod T. Howell, Bristol by supplying giudelines from their respective countries. We are also indebted to W. Bartosch, Basel, and Mrs. K. Noll, Basel, for giving advice on photomicrography.

References

ACC Association of Clinical Cytogeneticists (1994) Guidelines for Clinical Cytogenetics. ACC, London

ACMG The American College of Medical Genetics (1993) Standards and Guidelines: Clinical Genetics Laboratories. ACMG Inc. Bethesda

ACT Association of Cytogenetic Technologists (1991) Chromosome analysis guidelines. Preliminary report. Am J Med Genet 41:566-569

Becker E (no date) Fluorescence microscopy. Leica, Wetzlar

Berufsverband Medizinische Genetik (1997) Leitlinien zur zytogenetischen Labordiagnostik. Med. Genetik 9:560-561

Bornfleth H, Aldinger K, Hausmann M, Jauch A, Cremer C (1996) Comparative Genomic Hybridisation imaging by the one-chip true-color CCD camera Kappa CF 15 MC. Cytometry 24:1-13

Bradbury S (1989) An Introduction to the optical microscope. Oxford University Press, Oxford CAP College of American Pathologists (1994) Inspection Checklists. CAP, Northfield

Davidson NR (1973) Photographic techniques for recording chromosome banding patterns. J Med Genet 10:122-126

Delly JG (1988) Photography through the microscope. Eastmann Kodak Company, Rochester du Manoir S, Kallioniemi OP, Lichter P, Piper J, Benedetti PA, Carothers AD, Fantes JA, Garcia-Sagredo JM, Gerdes T, Giollant M, Hemery B, Isola J, Maahr J, Morrison H, Perry P, Stark M, Sudar D, van Vliet LJ, Verwoerd N, Vrolijk J (1995) Hardware and software requirements for quantitative analysis of Comparative Genomic Hybridisation. Cytometry 19:4-9

GLaRGG Great Lakes Regional Genetics Group (1992) Quality Assurance for Cytogenetic Laboratories. GLaRGG

Graham J (1987) Automation of routine clinical chromosome analysis. I. Karyotyping by machine. Anal Quant Cytol Histol 9:383-390

Graham J, Piper J (1994) Automatic karyotype analysis. In: Gosden JR (ed) Chromosome analysis protocols. Humana Press, Totowa: 141-219

ISCN (1995) An international system for human cytogenetic nomenclature. Karger, Basel

Josifek K et al (1991) Applied Cytogenet 17:101-105 cited from ACMG (1993)

Kao YS, Kao GA, Walters CS (1990) Banding resolution of amniotic cell chromosome preparations for prenatal diagnosis. Am J Clin Pathol 93:765-770

Kapitzka HG (1994) Microscopy from the very beginning. Carl Zeiss, Oberkochen

Korthof G, Carothers AD (1991) Tests of performance of four automatic metaphase finding and karyotyping systems. Clin Genet 40:441-451

Lacey AJ (ed) (1989) Light microscopy in biology. A practical approach. IRL Press at Oxford University Press, Oxford

Kriete A, Amelinckx S, Reimer L (1994) Microscopy. Ullmann's Encycopledia Of Industrial Chemistry B6:213-278, VCH Publishers, Inc.

Ledley RS (1964) High-speed automatic analysis of biomedical pictures. Science 146:216-223

Lundsteen C, Gerdes T, Maahr J, Philip J (1987) Clinical performance of a system for semi-automated chromosome analysis. Am J Hum Genet 41:493-502

Lundsteen C, Martin AO (1989) On the selection of systems for automated cytogenetic analysis. Am J Med Genet 32:72-80

Lundsteen C, Piper J (ed) (1989) Automation of cytogenetics. Springer, Berlin

Monk AJ (1992) Microscopy, photography, and computerized image analysis systems. In: Rooney DE, Czepulkowski BH (eds) Human cytogenetics. A practical approach. IRL Press at Oxford University Press, Oxford: 223-249

Philip J, Lundsteen C (1985) Semiautomated chromosome analysis. A clinical test. Clin Genet 27:140-146

Piper J, Lundsteen C (1987) Human chromosome analysis by machine. Trends Genet 3:309-313

Schade KH (1993) Lichtmikroskopie: Technologie und Anwendung. Verlag Moderne Industrie AG, Landsberg

Stallard R, Johnson W (1983) Nonsubjective method for estimating the resolution of banded chromosomes. Am J Hum Genet 35:155A

Thompson DJ, Bradbury S (1987) An introduction to photomicrography. Oxford Univeristy Press, Oxford

Part II

Postnatal Diagnosis

Chapter 5

Peripheral Blood

IRIS BARTELS

Introduction

Peripheral blood is the most easily available tissue for postnatal chromosome analysis. Metaphase preparations are finished after two or three days. Though blood does not normally contain spontaneously dividing cells, leucocytes can easily be induced to proliferate by addition of a mitogen. The most commonly used mitogen is phytohaemagglutinin which is isolated from red kidney beans. The cells are cultured in medium supplemented with phytohaemagglutinin for 48 or 72 hours. Cells are then arrested in metaphase by the addition of a spindle inhibitor, eg colcemid, an alkaloid, naturally occurring in *Colchicum* species. After these steps the cultures are ready for harvesting, chomosome preparation and finally for banding or fluorescence *in situ* hybridisation. Hypotonic treatment and fixation are the most critical steps in chromosome preparation. Usually this standard procedure will be adequate to produce well-spread chromosomes satisfactory for most clinical indications.

Elongated chromosomes from early stages of cell division are required for high resolution banding. The proportion of early metaphase and prometaphase cells is low in standard cultures though it can be increased by synchronisation of cultures using chemical blocking agents (eg Methotrexate). Incubation with drugs that partially inhibit chromosome condensation (eg Actinomycin D) also provides a large number of elongated chromosome spreads (Yunis and Lewandowski, 1983).

Iris Bartels, Institut für Humangenetik, Goßlerstr.12 D, Göttingen, 37073, Germany (*phone* +0551-397596; *fax* +0551-399303; *e-mail* ibartel@gwdg.de)

Materials

Equipment

- 37°C incubator
- centrifuge (120 g)
- microscope with phase contrast
- tissue culture flask (30 ml)
- centrifuge tubes (10 to 15 ml)

Culture medium

- 10 ml RPMI 1640 with stable glutamine (Biochrom)
- 1 ml fetal calf serum (Seromed)
- 0.1 ml penicillin (10,000 IU/ml)/streptomycin (10,000 μg/ml)(Biochrom)
- 0.15 ml phytohaemagglutinin HA 15 (Murex)

Note: The complete culture medium can be stored for 5 days in the refrigerator.

Solutions

- Colcemid stock-solution: 10 μg colcemid (Calbiochem)/ml Hanks solution, stored at −20°C
- Hypotonic solution
  - potassium chloride 5.9g/l H_2O (0.079 M)
  - tri-sodiumcitrate x $2H_2O$ 4g/l H_2O (0.0073 M)
  - mix in a ratio of 1:1
- Fixative: mix 3 parts absolute methanol p.a. and 1 part glacial acetic acid (made freshly before use)
- Methotrexate solution: Methotrexate (Lederle) 10^{-5} M
- Thymidine solution: 4 mM Thymidine in PBS
- Actinomycin D solution: 2 mg Actinomycin D (Sigma) dissolved in 2 ml H_2O bidest (store at −20°C).

Preparation of slides

Pretreat glass slides with 80 percent ethanol to remove traces of grease, rinse thoroughly under tap water and store in deionized water in the refrigerator for up to 5 days.

Safety regulations

Lab coat and gloves should be worn.

Procedure

Standard culture

1. Blood samples should be collected in a plastic tube or vacutainer with heparin, eg Liquemin (10 to 20 IU/ml) as an antiagglutinant. Mix gently. Samples can be stored up to 5 days in a refrigerator. 1 ml of heparinized blood from adults and children, respectively and 0.3 ml from newborns is sufficient for 10 ml culture.
2. Dispense 1 ml heparinized blood into a 30 ml plastic vial containing 10 ml culture medium, shake gently, and close the vessels tightly.
3. Incubate for 72 h at 37°C. 48 h incubation usually gives sufficient number of mitoses as well.
4. Add 0.5 ml of colcemid stock-solution (final concentration 0.5 μg/ml) for the final 1.5 h of culture time.

Chromosome preparation

1. Shake the culture gently and transfer the suspension to two 10 ml centrifuge tubes. Spin at 120 g for 10 mins. Remove and discard the supernatant using a pasteur pipette. Leave 0.5 ml of fluid above the pellet and do not remove the upper layer of the pellet.
2. Add 8 ml of prewarmed (37°C) hypotonic solution to each tube and mix by gently drawing the cells up and down using a Pasteur pipette. Incubate for 15 mins at room temperature (20 to 22°C) or for 12 mins at 37°C. Mix gently once during incubation. Spin at 120 g for 10 mins.
3. Remove the supernatant, leaving 0.5 ml above the pellet and resuspend the cells carefully. Hold the vessel in a sloping position and allow 3 drops of freshly prepared ice chilled fixative to run down the wall of the vessel. Add another 6 drops in the same way. Mix gently, but thoroughly using the Pasteur pipette. Add another 5 ml of fixative to each vessel and leave for 10 mins. Centrifuge at 120 g for 10 mins.
4. Remove the supernatant which is coloured red by hemoglobin from disrupted erythrocytes, and add 8 ml fresh cold fixative, and spin. Repeat this step once again.

5. Remove the supernatant except for the last 0.3 ml and resuspend cells and combine the suspension of the two tubes.
6. Drop 2 drops out of a Pasteur pipette onto a wet, cold, grease-free slide from a height of 5 to 10 cm. Excess water should be avoided. Air dry and check the cell density under phase contrast and dilute or concentrate the suspension if necessary.

Preparation of prometaphase chromosomes

Do not process more than 3 cultures for high resolution chromosomes at any one time, as exposure time and temperature may vary too much if more cultures are processed.

Cell synchronisation (MTX method)

1. Set up blood culture as described for standard cultures, step 1 and 2. Incubate at 37°C for 48 to 60 hours.
2. Add 0.2 ml Methotrexate solution 12 to 16 hours before harvesting (in the afternoon of the last but one day).
3. Spin at 120 g for 10 mins after further 12 to 16 hours (in the morning of the last day). Remove the supernatant and resuspend the cells with 10 ml prewarmed culture medium.
4. Centrifuge again and remove the supernatant and resuspend with 10 ml prewarmed culture medium supplemented with 0.1 ml thymidin solution (final concentration 40 μM). Incubate for a further 5 hours at 37°C.
5. Add colcemid solution and incubate for a further 20 mins.
6. Continue with the protocol for chromosome preparation.

Elongated chromosomes without synchronisation

1. Set up culture as described above for standard culture. Steps 1 and 2 should be modified by using 0.1 ml Phytohaemagglutinin and 0.05 ml Pokeweed mitogen (Sigma) instead of 0.15 ml Phytohaemagglutinin. Incubate for 64 to 72 hours.
2. Add 10 μl Actinomycin D solution and incubate for further 30 mins at 37 °C.
3. Add 0.3 ml Colcemid stock solution and incubate for 20 mins at 37°C.
4. Continue with the protocol for chromosome preparation.

Results

Analysis and karyotyping

All staining methods described in chapter 3 can be applied to these preparations. Analyse well spread metaphases under the microscope at 800x to 1000x magnification. The number of cells that should be analysed depends on the indication and the quality of the preparation. Three cells are enough to rule out if a proband is carrier of a familial Robertsonian translocation, but 30 normal cells are necessary to exclude 10 percent mosaicism. For low grade mosaicism 100 cells should be analysed. In most cases 10 cells are adequate. 3 to 5 well banded metaphases ought to be analysed structurally, band by band. For further details see Chapter 3 "Karyotyping and Data Interpretation".

Interpretation of results

- *In vitro* culture can result in the occurrence of chromosomal aberrations. Such culture-induced or spreading-related artifacts include breakages, deletions and rearrangements, as well as trisomies and monosomy X. To distinguish between true mosaicism and chromosomal aberration acquired *in vitro* is one of the ever lasting problems in the interpretation of cytogenetic results. In contrast to long term cultures from amniocytes and fibroblasts more than two lymphocyte metaphases with the same aberration indicate true mosaicism, since one cell divides not more than two times during 72h culture. However, one single aberrant cell among 50 does not exclude constitutional mosaicism. Therefore in some cases the repeat of the lymphocyte culture, harvesting after 48 and 72 hours as well as the examination of another tissue (eg fibroblasts) should prove indicative. Example: Patients with Pallister Killian syndrome exhibit mosaicism of an extra isochromosome 12p, preferably in fibroblasts. Therefore a single cell with tetrasomy 12p in lymphocytes is strongly indicative for chromosome analysis from fibroblasts.
- In healthy probands age dependent losses of X and Y chromosomes occur with increasing age. The clinical meaning of this observation is unknown.
- MTX supplementation for prometaphase preparation may induce chromosome breakages at fragile sites. If fra(X)(q27) is observed in one or more metaphases, DNA analysis for expanded CGG repeat in the FMR1 gene is indicated.

Troubleshooting

- Poorly spread metaphases embedded in a halo of cytoplasm
  The following causes may be underlying:
  - The hypotonic treatment was too short.
  - Clumps were produced during the first fixation step.
  - Fixative or methanol were not free from water. Note that water from the air can get in the bottle, use only p.a. grade methanol. Splashing on very cold wet slides from 20 cm may improve spreading in those situations. Storing of the cell suspension in a fridge for two to 12 h and subsequent replacement of the fixative also may help. Flame drying of slides is not recommended, because subsequent banding is reduced.
- Widely spread, incomplete metaphases
  Hypotonic treatment was too long. Apply the cell suspension from only one cm height to clean dry slides.
- Short, lumpy chromosomes
  The degree of condensation of metaphase chromosomes depends on the colcemid incubation time and concentration. The treatment with colcemid may have to be short in order to obtain less contracted chromosomes. However, the shorter the incubation time, the lower the number of metaphases. The optimal time for this step should be varied.
- Analysis during pregnancy
  Very few metaphases are sometimes obtained from pregnant women or from samples obtained shortly after delivery or abortion. To improve cell growth wash the cells twice in culture medium before culturing.
- Metaphase preparations for FISH analysis
  Successful hybridization requires properly spread metaphases without cytoplasm. Precoating of slides is not necessary and may even impair spreading. In order to save hybridization solution produce slides with an area rich in metaphases. Slides can be used immediately or stored desiccated at −20°C or −70°C for an extended period (see also Chapter 18).

References

Galloway S, Buckton K (1978) Aneuploidy and aging: chromosome studies on a random sample of the population using G-Banding. Cytogenetic Cell Genet 20:78-95

Yunis JJ, Lewandowski RC (1983) High-Resolution Cytogenetics. Birth Defects 19:11-37

Chapter 6

Establishment of Permanent Growing Lymphoblastoid Cell Lines

HEIDEMARIE NEITZEL

Introduction

The development of human cytogenetics demonstrates the dependence of scientific progress on new methodological approaches. Thus, the real beginning of clinical cytogenetics was marked by the introduction of the lymphocyte culture technique as an easily accessible source of mitotic cells (Nowell, 1960). However, due to the limited life-span of these cells new blood sampling is necessary in cases of re-examination. A simple procedure for routine use is available which allows efficient transformation of peripheral B lymphocytes by Epstein-Barr virus (EBV) and thus the establishment of permanent growing lymphoblastoid cell lines (LCL)(Neitzel, 1986) (Figure 1). Compared to other methods of long-term cultivation, ie tissue culture of skin fibroblasts, these cell lines have a number of advantages:

- The material can easily be obtained from any patient.
- EBV-transformed lines exhibit chromosomal stability up to high passages in contrast to SV 40 transformed fibroblasts.
- The growth of LCL in suspension and their minimal pretension concerning medium allows the cultivation up to high cell density without much expenditure of work and costs.
- LCL are the ideal source for molecular studies in humans as repeated DNA and RNA preparations can be obtained without great effort.
- LCL can be used in cell hybridization experiments giving rise to inter- and intraspecific somatic cell hybrids (see Chapter 15).

Heidemarie Neitzel, Charité Campus Virchow-Klinikum, Institut für Humangenetik, Augustenburger Platz 1, Berlin, 13353, Germany (*phone* 030-450-66411; *fax* 030-450-66933; *e-mail* heidemarie.neitzel@charite.de)

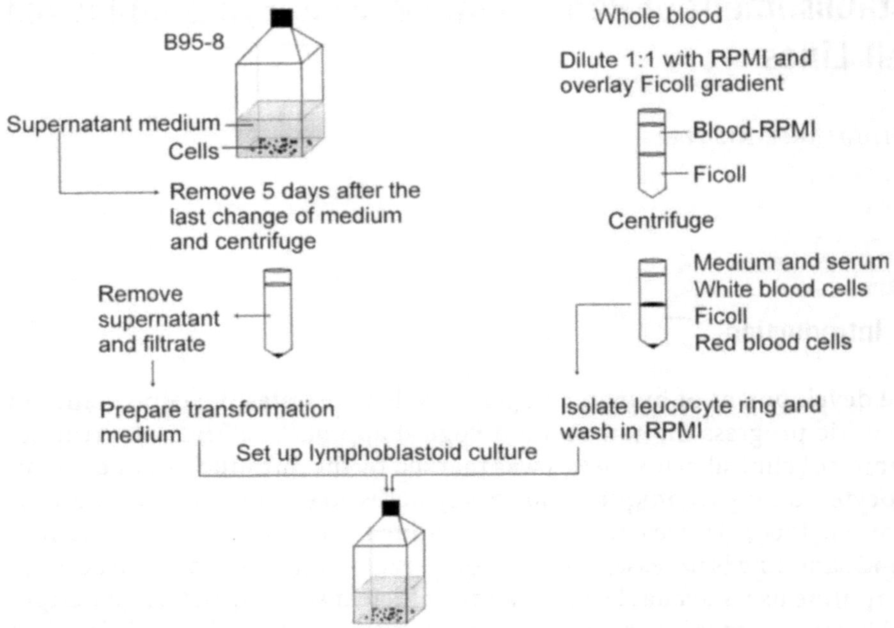

Fig. 1. Flow sheet demonstrating the establishment of lymphoblastoid cell lines.

Biology of Epstein-Barr virus

Epstein-Barr virus is a human lymphotropic herpes-virus which is endemic in all human populations. Most people are infected in early childhood without any apparent clinical features but leading to a positive antibody titer. If the initial infection is delayed, infectious mononucleosis, a benign lymphoproliferative disease, frequently results. In these patients, as well as in a surprisingly high proportion of normal individuals, the EB viral genome persists latently for prolonged periods in the epithelial cells of the saliva and in the parotid duct cells (Miller, 1984; Rickinson, 1984). Epstein-Barr virus has a double-stranded DNA genome consisting of about 172.000 base pairs and exists as a circular episome inside the nucleus of infected cells (Baer et al., 1984).

Transformation of B lymphocytes with Epstein-Barr Virus in vitro

In vitro, Epstein-Barr virus can convert selectively B lymphocytes of humans and old-world primates to continously dividing efficiently immortalized lymphoblastoid cells (Miller and Lipman, 1973). Transforming virus can be easily obtained from the lymphoblastoid marmoset cell line B95-8. Most of the B95-8 cells are considered to be latently infected, while a few support virus production spontaneously and release virus particles into the culture medium.

EBV binds selectively to B cells, presumably only to a subpopulation of B cells, via specific cell surface receptors (Bird et al., 1981; Pattengale et al., 1973). Following exposure to EBV, B cells undergo lymphoblastoid transformation and express subsequently viral antigens, eg EBNA (Epstein-Barr nuclear antigen). The first cellular DNA synthesis in infected B cell cultures can be detected only in EBNA-positive cells approximately 20 h after the appearance of EBNA. The first cellular division occurs 12 - 24h later (Einhorn and Ernberg, 1978).

Prevention of regression in LCL

During the initial stage i e 1 - 2 weeks post-infection, cells expressing EBNA appear in increasing numbers and aggregate in foci of proliferative lymphoblast cells. However, thereafter increasing cell death and complete degeneration of proliferative foci can be observed in many cultures. This phenomenon has been exclusively observed in cultures of seropositive but not of seronegative donors and proved to be dependent upon the presence of T cells in the cultures. These results suggest that the phenomenon of regression in EBV-transformed LCL is a consequence of long-term T-cell-mediated immunity to B cells presenting EBNA antigen to autologous T cells (Issekutz et al., 1982). The regression can be prevented by treatment with Cyclosporin A (Cy A), a fungal metabolite, which has the potential of a selective immunosuppressive agent (Neitzel, 1986).

Subprotocol 1
Transformation of B Lymphocytes with B95-8 EB Virus

Materials

Equipment

- Tissue culture flasks (50 ml) [Falcon # 3013]
- Single-use syringes (20 ml, Luer) [Braun-Melsungen # 17093 F 0101]

- Sterile filters: Millex HA [Millipore # SLHA 025 BS]
- Snap cap tubes (14 ml) [Falcon # 2001]
- Pipettes (10 ml)[Falcon # 7531]
- Centrifuge [Heraeus Sepatech, Rotor # 3360/BS4402/A]

Solutions

- RPMI 1640 (1x with L-Glutamine)[PAA # E15-840]
- Fetal calf serum, FCS [Gibco # 10270-106](heat-inactivated)
- Streptomycin [Grünenthal #757753B]
- Penicillin [Grünenthal # E744114]
- Cyclosporin A: Sandimmun (1ml=50mg) [Sandoz # PZN-2702663]
- Dissolve Streptomycin (1g) in 5 ml aqua dest. (sterile).
- Dissolve Penicillin (1 000 000 U) in 5 ml aqua dest. (sterile).
- Prepare culture medium:
  - 500 ml medium RPMI 1640
  - 50 ml FCS
  - 0.25 ml Streptomycin
  - 0.25 ml Penicillin
- Dilute Cyclosporin A to a concentration of 5000 µg/ml:
  - 1 ml Sandimmun
  - 9 ml RPMI 1640
- Prepare transformation medium:
  - 50 ml filtered EBV-containing B95-8 supernatant
  - 40 ml RPMI 1640
  - 10 ml FCS
  - 50 µl Streptomycin
  - 50 µl Penicillin
  - 20 µl Sandimmun (end concentration 1µg/ml medium)

Procedure

Cultivation of the lymphoid starter cell line B95-8 and purification of EBV

1. Cultivate Mycoplasma-free B95-8 at a concentration of about 5 x 10^5 cells/ml in culture medium.

2. Remove EBV-containing supernatant medium after 5 days of cultivation.
3. Add new culture medium to B95-8.
4. Centrifuge supernatant at 1200 rpm for 10 min to remove marmoset cells.
5. Draw supernatant into a syringe and place the membrane filter onto it.
6. Pass the supernatant through a 0,45 µm membrane filter to remove all viable cells.
7. Repeat the filtration step of the supernatant.
8. Dilute the filtrate 1:1 with fresh RPMI 1640, supplemented with 20 % heat-inactivated fetal calf serum, 2 mM L-glutamine, antibiotics and Cyclosporin A.
9. Filtered virus pools or transformation medium can be stored at 4° C for 1-2 months.

Subprotocol 2
Ficoll Separation of Unfractionated Mononuclear Leukocytes Obtained from Whole Blood

Materials

Equipment

- Glass pipettes (150 mm) [Recker 9411816]
- Snap cap tubes (14 ml) [Falcon # 2001]
- Pipettes (10 ml)[Falcon # 7531]
- Centrifuge [Heraeus Sepatech, Rotor # 3360/BS4402/A]

Solutions

- Liquemin 25 000 (400 IE /ml blood)[Hoffmann-La Roche # PZN-3441331]
- Ficoll separating solution, Density 1.077 [Seromed # L6115]
- RPMI 1640

Procedure

1. Pipette 5 ml Ficoll separating solution into the tubes.
2. Dilute the heparinized blood sample 1:1 with RPMI 1640 (without serum).
3. Overlay the Ficoll separating solution slowly and gently with 5 ml of the blood-RPMI-mixture.
4. Centrifuge the blood-RPMI-mixture over the Ficoll-Hypaque gradient for 40 min at 1000 rpm.
5. Remove the leukocyte ring from the gradient and transfer to another tube.
6. Add 10 ml RPMI 1640 and resuspend cells.
7. Centrifuge for 10 min at 1000 rpm.
8. Discard supernatant and repeat the washing step another two times.

Subprotocol 3
Establishment of Cultures

Materials

Equipment

- Tissue culture flasks (50 ml) [Falcon # 3013]
- Pipettes (10 ml)[Falcon # 7531]

Solutions

- Transformation medium (see Subprotocol 1)

Procedure

1. Resuspend separated mononuclear leukocytes in transformation medium.
2. Establish 3 ml-cultures in tissue culture flasks.
3. Adjust the pH to 6.8 by addition of CO_2 and incubate flasks upright at 37° C.

4. Thereafter change medium once a week by removing half of the supernatant and replacing it by fresh medium containing 1 µg/ml Cyclosporin A.

5. The first subcultivation after starting the cultures can usually be carried out after 2 - 3 weeks.

6. The amount of fetal calf serum in the medium can be decreased to 10 % 3 - 4 weeks after starting the cultures, further treatment with Cyclosporin A is not necessary.

Criteria indicating transformation

There are two distinct morphologic features that can be observed usually as early as approximately 24 h post-infection with EBV indicating successful transformation of B cells:

- Blastogenesis becomes evident resulting in enlargement of the lymphocytes.
- There is increasing development of cell aggregates of proliferative lymphoblast cells.

Subprotocol 4
Freezing of Lymphoblastoid Cells

Materials

Equipment

- Cryotubes [Greiner # 121278]
- Snap cap tubes (14 ml) [Falcon # 2001]
- Pipettes (10 ml)[Falcon # 7531]
- Centrifuge [Heraeus Sepatech, Rotor # 3360/BS4402/A]

Solutions

- DMSO (Dimethylsulfoxide) [Merck # 9678.0100]
- Culture medium (see Subprotocol 1)1
- Prepare freezing medium:
  - 9 ml culture medium
  - 1 ml DMSO

Procedure

1. Change medium 48 h before freezing by removing half of it and replacing it with fresh medium.
2. Centrifuge cell suspension for 10 min at 1000 rpm in snap cap tubes.
3. Resuspend cell pellet at a concentration as high as 5-10x10^6/ml in ice-cold culture medium containing 10 % DMSO (a cell concentration of less than 5 x 10^6 per tube results in a delay of growth after thawing).
4. Prepare aliquots of 1 ml per cryotube.
5. Freeze at a rate of 1° C per min to -40° C and then transfer directly into liquid nitrogen.

Subprotocol 5
Thawing of Lymphoblastoid Cells

Materials

Equipment

- Snap cap tubes (14 ml) [Falcon # 2001]
- Pipettes (10 ml)[Falcon # 7531]
- Tissue culture flasks (50 ml) [Falcon # 3013]
- Centrifuge [Heraeus Sepatech, Rotor # 3360/BS4402/A]

Solution

- Culture medium

Procedure

1. Remove tubes from liquid nitrogen and thaw in a waterbath at 37° C.
2. Add 10 ml cold medium immediately after complete thawing to decrease the DMSO-concentration.
3. Centrifuge for 10 min at 1000 rpm and discard DMSO-containing medium.
4. Resuspend cell pellet in 10 ml fresh medium and incubate for 24 h at 37° C, pH 6.8 to allow recovery of cells.
5. Add further medium or subcultivate.

Subprotocol 6
Chromosome Preparations from Lymphoblastoid Cells

Materials

Equipment

- Snap cap tube (13 ml) [Greiner # 172101]
- Glass slides [Menzel # 507894]
- Centrifuge [Heraeus Sepatech, Rotor # 3360/BS4402/A]

Solutions

- Colcemid (10μg/ml) [Gibco/BRL # 15210-016]
- Hypotonic solution (KCl 0.075M)[Merck # 4936.1000]
- Cold fixative (3:1 (v/v) methanol/acetic acid)

Procedure

1. Dilute cell culture 48 h before harvesting to a concentration of 3-5x10^5.
2. Add Colcemid (0.2 μg/ml) 1-2 hours before harvesting.
3. Transfer culture to a snap cap tube and spin for 10 min at 1000 rpm.
4. Discard supernatant and resuspend sediment gently.
5. Add 5 ml pre-warmed (37°C) hypotonic solution, resuspend the pellet gently and incubate 12 min at room temperature.
6. Spin 10 min at 1000 rpm, discard supernatant and resuspend sediment.
7. Add 5 ml fixative (-20°C) dropwise (slowly) to the cells and resuspend gently.
8. Spin 10 min at 1000 rpm.
9. Remove supernatant, resuspend sediment and add 5 ml fresh fixative.
10. Spin, discard supernatant, resuspend pellet and add fresh fixative (5 ml).

11. Centrifuge, remove supernatant and adjust density of cell suspension with fresh fixative to get ready for preparing slides.

12. Drop cell suspension on cooled (-20°C) ethanol-cleaned and moistened slides. Air-dry the slides.

Safety instructions

To avoid EBV infection of persons handling EBV-tranformed cell lines the following safety instructions should be practised:

- All persons handling lymphoblastoid cell lines should be tested for anti-VCA-titers. Only seropositive persons are allowed to handle these cells.
- Use a clean bench with vertical flow for all manipulations.
- All material which comes into contact with lymphoblastoid cells or their medium should be autoclaved before it is discarded.

There is no risk of infecting other cell types with the virus, eg amnion cells or fibroblasts because of the high cell specificity of EBV for B lymphocytes.

Personal experience

Following this protocol we have successfully established more than 1000 lymphoblastoid cell lines from patients with chromosomal aberrations, DNA repair deficiencies, and other inherited diseases, such as cystic fibrosis, growth hormone deficiency type IA, insulin receptor defects, atypical muscular dystrophy, and juvenile type of epilepsy. In addition, LCL were set up from lymphocytes of different primates.

We observed no difference in the transformation rate depending on the age of the blood donor. For routine use it is desirable to take 5 ml of whole blood to start an LCL, however, even less than 1 ml might be sufficient for effective transformation. If heparinized blood has to be sent by post no specific handling is necessary, LCL could still be set up several days (5 to 7, in one case 10 days) after blood sampling.

Care should be taken at two important steps during the transformation. Firstly, ensure removal of all cells of the starter line B95-8 from the filtrate used for transformation. Otherwise the newly established human line will be contaminated with marmoset cells. The latter also have a modal chromosome number of 46 but can be clearly identified cytogenetically. Secondly, do not freeze the virus-containing supernatant because this will cause loss of transformation efficiency.

References

Bird AG, Britten S, Ernberg I, Nilsson K (1981) Characteristics of Epstein-Barr virus activation of human B lymphocytes. J Exp Med 154:832-839

Einhorn L, Ernberg I (1978) Induction of EBNA precedes the first cellular S-phase after EBV infection of human lymphocytes. Int J Cancer 21:157-160

Issekutz T, Chu E, Geha RS (1982) Antigen presentation by human B cells: T cell proliferation induced by Epstein-Barr virus B lymphblastoid cells. J Immunol 129:1446-1450

Miller G (1984) Epstein-Barr virus - immortalization and replication. N Engl J Med 310:1255-1256

Miller G, Lipman M (1973) Comparison of the yield of infectious virus from clones of human and simian lymphoblastoid lines transformes by Epstein-Barr virus. J Exp Med 138:1398-1412

Neitzel H (1986) A routine method for the establishment of permanent growing lymphoblastoid cell lines. Hum Genet 73: 320-326

Nowell PC (1960) Phytohemagglutinin: An indicator of mitosis in cultures of normal human leukocytes. Cancer Res 20:462-466

Pattengale PK, Smith RW, Gerber P (1973) Selective transformation of B lymphocytes by EB virus. Lancet II:93-94

Rickenson AB, Moss DJ, Allen DJ, Wallace LE, Rowe M, Epstein MA (1981) Reactivation of Epstein-Barr virus-specific cytotoxic T cells by in vitro stimulation with autologous lymphoblastoid cell line. Int J Cancer 27:593-601

Chapter 7

Solid Tissues

REGINE SCHUBERT AND GESA SCHWANITZ

Introduction

Tissue culture procedures for human cells are described which can be used for several purposes, namely biochemical, molecular or cytogenetic investigations. With respect to the aim of this manual the protocols will focus specifically on applications which result in high quality chromosome preparations. Investigations of solid tissues in clinical cytogenetics are helpful, for example to analyse embryonic or fetal tissues after abortion or to search for mosaics in individuals with phenotype/karyotype discrepancies after karyotyping lymphocytes. For long-term cultures of solid tissues biopsies of numerous organs or somatic cell systems can ,in principle, be used. However, in clinical cytogenetics the analysis is preferably done on tissues which are easily accessible and which show a well-known high growth rate under in vitro conditions. In cases of abortions these tissues are skin, achilles tendon, placenta (CVS long-term culture see Chapter 13), amniotic membrane, or the umbilical cord. For postnatal diagnosis gonadal tissue is examined after biopsy or gonadectomy in cases of gonadal anomalies.

Fibroblasts are the most frequently analysed cell type. In the following protocols, culture procedures and preparations of these cells are presented.

Fibroblasts - or fibrocytes- are differentiated cells which under common physiological conditions have ceased cell division and stay in the G_0-phase of the cell cycle. Thus, when obtained for long-term culture they need some time to adapt to the in vitro conditions before the cells resume proliferation and enter the cell cycle. A cell line is established in a long-term culture in three phases (Hayflick and Moorhead, 1961). First, cells attach to the surface

Correspondence to Regine Schubert, Institut für Humangenetik, Wilhelmstr. 31, Bonn, 53111, Germany (*phone* 0228-287-2183; *fax* 0228-287-2380; *e-mail* cytogen@humgen.-uni-bonn.de), Gesa Schwanitz, Institut für Humangenetik, Wilhemstr. 31, Bonn, 53111, Germany of the culture vessel and alter to a spindle-like morphology indicating their return to the cell cycle. In such primary culture the growth speed is influenced by the cell density, e.g. by the number of seeded viable cells. The next phase is characterized by the potential of rapid growth and regular cell division. At the beginning of each mitosis the cells alter their shape. They round up to a sphere which is only attached to the surface of the culture vessel by a small area. At the end of mitosis both daughter cells return to a spindle shape. The terminal phase of growth in a culture is characterized by a broadening and flattening of cells. They start showing vacuoles in the cytoplasm and finally cease to grow. Skin fibroblast cultures possess the potential of approximately 50 population doublings.

Fibroblasts grow in a monolayer. If the number of cells in a culture vessel is too high, mitotic activity ceases (contact inhibition). To keep cells active, the number of cells in a culture vessel has to remain within an optimal range of cell density. This can be obtained by subculturing, i.e. detachment of the cells with enzymatic treatment and reduction of the cell number by passaging aliquots of the suspension in new culture flasks. It is important to note the number of passages to estimate the final life span of the cell line.

Materials

Media, reagents and tissue culture utensils

Different media for long-term cultures are available. Good results are obtained using medium F10. We do not use Chang medium because of the induction of a very high rate of secondary single cell aberrations. The advantage of the latter is rapid cell growth. However it must be taken into account that the costs are much higher as compared to medium F 10.

Procedure

Storage of biopsies

From the time tissue biopsies are taken until the cultures are set up, keep tissue in transport medium or if not available at least in isotonic salt solution. The samples should be not older than 3 days and stored at 4°C.

Set up of a primary culture

In a primary culture the cells grow out of a tissue specimen. The sample has to be minced into very small pieces. Cutting cells releases growth factors, thus inducing cell division of intact cells. Primary cultures can be set up e.g. in flasks, in Leighton tubes or petri dishes. For chromosome preparation cells must either be detached from the surface of the culture vessel and processed according to the air-dry technique for lymphocytes or they

Table 1. Reagents and culture vessels

Material	Concentration	Volume (ml)	Company	Order No.
a. Transport medium				
Medium F 10		1000	Boehringer Mannheim	209 864
Penicillin/	5.000 IU/ml/			
Streptomycin	5.000 μg/ml	20	ICN Biomedicals	1-800-8540530
Fungizone (Amphotericin B Solution)	250 μg/ml	2,80	ICN Biomedicals	16-723-46
b. Culture medium				
Transport medium see above		100	see above	
Fetal calf serum (FCS)		20	Boehringer Mannheim	210 463
BM-Condimed		10	Boehringer Mannheim	663 573
c. Culture reagents				
Trypsin	0,05% in 0,02% EDTA		Life	45 300-019
Colcemid	1 %		Boehringer Mannheim	259 892
NaCl	0,9 %		Fresenius	
Fixative (methanol-acetic acid 3:1)				
Methanol			Riedel de Haen	32 213
Acetic acid			Merck	1.00063
d. Vessels	Size			
Culture flask (T25)	25 cm^2 (75ml)		Sarstedt	831 810 001
Leighton tube	13x52 mm coverslip		Belco	1903-19095
Quadriperm	76x26 mm slide		Heraeus	2613 6907
Petri dish	100x15 mm		Falcon	041029

should be transferred onto slides or coverslips before in situ harvesting. The latter method will be presented here because we obtaine good results with it. An in situ preparation of primary cultures which are cultivated on slides or coverslips is also possible with the disadvantage of loosing a back-up.

Note: All work has to be done under sterile conditions! Work under the hood.

1. Place the tissue biopsy in a petri dish.
2. Mince the biopsy into fragments of at least 0,5 mm^3 using a sterile scalpel.
3. Transfer at least 6 – 8 fragments in each of two sterile culture flasks (T 75) using a pipette. Distribute tissue on the growth surface of the flask.
4. Fragments have to adhere to the bottom of the flasks for about 2 min before adding carefully 3 ml culture medium (room temperature) avoiding detachment of the adherent fragments from the bottom of the flask. Recap the flask.
5. Gently transfer to the appropriate incubator (37°C, 5% CO_2, 97% humidity).
6. Check cultures for growth and possible contamination after 3 – 4 days using an inverted microscope. Leave undisturbed until this time.
7. When the cells begin to grow out of the tissue fragments the medium has to be changed (Figure 1). This point in time is very variable depending on the vitality and the origin of the tissue and takes from 3 – 14 days. The old medium can be decanted or removed with a pipette. Add 3 ml fresh medium.
8. In the further course of cultivation, medium has to be changed twice a week until 2/3 of the bottom of the flasks are covered with cells. This may take up to 3 weeks and again depends on the origin of the tissue and its vitality. With every change of medium the culture is checked microscopically for vitality and contamination.

Subcultivation

The cells are subcultivated by trypsinization and aliquoting into a required number of vessels. A trypsin/EDTA solution induces detachment of the cells from the growth surface. The proteolytic enzyme trypsin breaks down the adhesive proteins of the cells and EDTA binds the 2$^+$ ions. Because serum proteins of the medium would inhibit the effectiveness of trypsin the medium has to be washed out with PBS buffer at the beginning of the procedure. The incubation time must not be too long because trypsin is toxic to

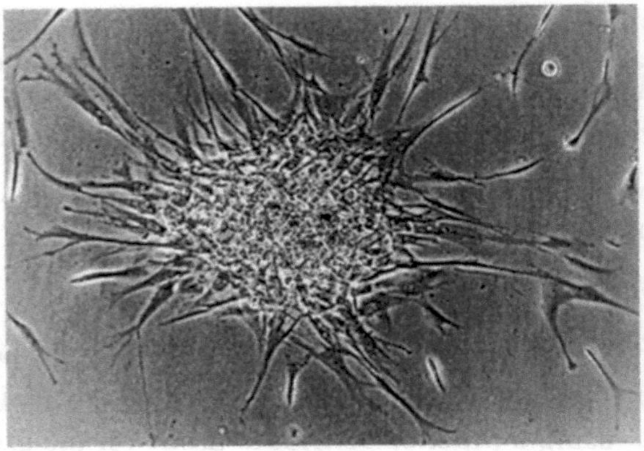

Fig. 1. Tissue biopsy with outgrowing cells.

cells. The action of trypsin is stopped by the addition of serum-supplemented medium. Subcultures for cytogenetic analyses can grow on slides in Quadriperm dishes or glass tubes or on coverslips in Leighton tubes. It is necessary to mark the side of the slide or coverslip on which the cells grow. We break off one corner of the coverslips and place them in the Leighton tube so that the broken edge is located in the right upper corner. Place clean and grease-free slides or coverslips in the vessels which will be used for subcultivation. Close the opening with aluminum caps (home made) and sterilize before use.

1. Remove medium.
2. Add 3 ml PBS and wash for 20 sec gently shaking. Replace with 2 ml of trypsin/EDTA solution. Incubate cultures for 5 – 8 min at 37 °C.
3. In the meantime add the required volume of medium to the vessels which will be used for the subculture (3ml/T25). Avoid air bubbles by dripping medium in one corner and tilting the flask.
4. After 5 – 8 min control detachment of the cells in the invertoscope. The cells round off their edges, detach from the surface of the flask and float in the medium. Gently shaking or tapping can facilitate the trypsinization effect.
5. When the majority of cells are free floating disperse 6 – 8 drops of the cell suspension with a pipette in the prepared vessel. Keep about 0,5 ml of the cell suspension in the primary flask and add 3 ml medium. Incubate the cultures at 37°C, 5% CO_2 and 97% humidity. Use original culture as back up in case a second subculture should be required.

6. Change the medium to eliminate trypsin and damaged or dead cells after the cells are attached to the surface (overnight/ earliest after 3-6 h).

7. Every two days the medium has to be changed. Generally a confluent cell layer will grow within 1 – 4 days.

In situ harvesting

When the cells are in logarithmic growth and the culture shows a high mitotic activity the cells can be harvested (Figure 2). Cell density should not be too high because this will inhibit the spreading of the chromosomes.

1. Add 0,1 ml Colcemid to each vessel, gently mix and incubate at 37°C for 2 – 4 h.

2. Remove medium with Colcemid and replace it carefully with the same volume of prewarmed (37°C) 0,09% NaCl solution and incubate at 37°C for 20 min.

Note: it is important to replace all reagents carefully because the mitoses are easily detached from the surface.

3. Add 3 drops of fixative (3:1= methanol: acetic acid), remove hypotonic/ fixative mixture immediately and replace gently with fresh fixative; leave for 15 min at room temperature (Fixative must be made up freshly every day).
4. Replace with fresh fixative and leave for 15 min.

5. Take out slide or coverslip and air dry. Be sure to label each slide on the upper side.

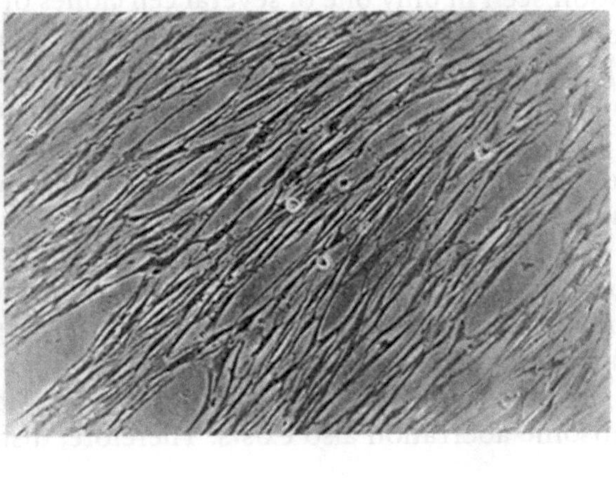

Fig. 2. Cell streams of spindle shaped fibroblasts in an intensively growing cell culture. Mitotic cells appear spheric.

Chromosome preparation

For numerical and structural analysis of the chromosomes specific banding techniques are employed (see Chapter 2).

Other investigations such as testing for mutagenicity by determination of the polyploidy rate, secondary structural chromosome aberrations or mitotic index are performed using homogeneous staining. For in situ hybridization of chromosomes from different tissues after long-term cultures, a specific pretreatment is recommended (see Chapter 18).

Chromosome analysis and karyotyping

Procedures are described in Chapters 3 and 4.

Interpretation

Usually a minimum of 20 good banded metaphases (400 – 500 bands per genome, ISCN 1995) is necessary for diagnostic purposes. The analysed metaphases should originate from different clones of at least 2 different culture flasks.

If the analysed tissue shows mosaicism of two or more cell lines, the number of investigated metaphases must be increased to at least 50 to determine the exact amount of each cell line. If the amount of analysable metaphases is insufficient, an interphase diagnosis by FISH (Chapter 18) is advisable.

Some chromosome analyses performed on solid tissue cultures do not show clear results. The amount of these cases depends on the origin of the analysed tissue, on its vitality, on the medium used for cultivation and on the time of in vitro culture.

In our investigations of more than 2500 chromosome analyses of solid tissues the amount of these cases was 17%. Most of them showed diagnostic irrelevant single pathologic cells (90%), caused by numerical as well as structural aberrations. The second group of questionable pathologic results is comprised by cases showing two or more metaphases with the same aberration seen in only one of several cell clones or one of two or more culture vessels. For diagnostic purposes it is of utmost importance to clarify whether these aberrant cells are in vitro artefacts (pseudomosaicism) or are actually present in the investigated tissue (true mosaicism).

However, the classification of pseudomosaicism or real mosaicism cannot be definitely proven in every case. Here the analysis of additional somatic tissues may help.

To a certain extent, polyploid mitoses (mostly tetraploid) are present in each long-term cell culture. Their frequency ranges from 3 – 5% in our own cultures. Under exceptional circumstances (e.g. slowly growing cultures) this percentage can increase up to 100%.

However, one has to consider that polyploidy as a constitutional chromosome aberration also exists. Therefore, distinguishing between mosaicism versus pseudomosaicism and polyploidy in vitro versus in vivo will remain a problem and needs the awareness of the investigator. It is noteworthy that interphase diagnostics may contribute in solving crucial cases.

Futhermore, the interpretation of a karyotype after tissue long-term culture remains difficult in cases a of discrepancy between phenotype and karyotype.

The most detailed findings on chromosome mosaicism with unequal distribution in different tissues have been published on trisomy 8 mosaicism (Kautza et al., 1991) and in Pallister Killian syndrome (Priest et al., 1992). The pathologic cells were confined to single germ layers or tissue systems. Also, in cases of gonosomal mosaicism, comparable problems are fairly common (see Examples, Case 1).

In these cases it is essential to analyse more than one cell system. This can also be done by interphase diagnostics in non-dividing tissues.

Generally, in cases of mosaicism, conclusions drawn from the investigation of one organ system must be regarded with caution as evidence for the chromosomal constitution of another organ system. Even within one organ the distribution of two cell lines can be inhomogenous.

Troubleshooting

Despite the best care in long-term cell cultures the risk of contamination is ever present. The biopsy itself can be infected or contamination may occur during cultivation. Infection with bacteria or fungi presents the serious risk of cross-contaminating non infected cultures. Cultures with high grade contamination should be discarded. Here success of any treatment is very low and subsequent cell growth is usually abnormal. Our experience shows that in the case of low grade contamination it is possible to salvage cells by washing with medium containing additional penicillin or fungizones (or a mixture of both if the type of contamination can not be identified). We add 4 ml penicillin and/or 2 ml fungizones to 100 ml cultivation medium. The cultures are washed with this special medium. This means the old medium is removed and the vessel is shaken carefully with fresh medium. Replace medium again with fresh medium. Repeat this proceedure after 6 to 16 hours over several days.

Applications

Examples

The following examples demonstrate our approach in cases of equivocal initial results.

Case 1 Chromosome investigations were carried out on a 13-year-old boy with clinical symptoms of Klinefelter syndrome. In lymphocytes all 40 analysed metaphases revealed an isochromosome Yp.

As there was no convincing phenotype-karyotype correlation, a skin biopsy was taken to initiate a fibroblast culture. 80 mitoses were investigated and, in fact, 3 different cell lines were observed: One with monosomy X, a main cell line identical with that diagnosed in the lymphocyte culture and a third one with 2 aberrant Y chromosomes (karyotype: mos46,X,i(Yp)[87%]/47,X,2i(Yp)[10%]/45,X[3%]).

An increased analysis from the first lymphocyte culture by examination of another 50 metaphases showed no additional cell lines.

The findings in the long-term cell culture were of significant prognostic value, as the mixture of cell lines with a male and a monosomy X karyotype (45,X) leads to a highly increased risk of a gonadoblastoma development in the patient and therefore makes regular controls essential.

Case 2 Because of a combination of malformations and facial dysmorphisms in a newborn female karyotype analysis from peripheral blood was required.

The child died before the diagnosis was finished and therefore a biopsy of the achilles tendon was additionally taken postmortem and sent for chromosome analysis in case the peripheral lymphocytes could not be stimulated for cell division. This might occur when samples are taken shortly before the death of the patient. Both cell cultures were successful.

After lymphocyte culture for 72 hours 25 metaphases were analysed. All revealed a normal female karyotype (46,XX,QFQ).

The long-term culture of fibroblasts, which took 3 weeks, revealed a supernumerary marker chromosome in 49 of 55 metaphases. After QFQ-banding and fluorescence in situ hybridization (FISH) it was identified as an isochromosome 12p. This additional i(12p) is the characteristic chromosomal aberration of the Pallister-Killian syndrome. It is characterized by a tissue specific mosaicism. The isochromosome can rarely be identified in lymphocytes, as a selection against the pathologic cell line already takes place prenatally during embryonic and fetal development. Thus in most cases only the investigation of fibroblasts leads to the correct identification of this syndrome.

References

Hayflick L, Moorhead PS: The serial cultivation of human diploid cell strains. Exp Cell Res 25: 585-621, 1961

Kautza M, Schwanitz G, Hosenfeld D, Grote W, Hunze-Fuhrmann D, Brandt I, Schleiermacher E, Gellissen K, Bopp E, Zerres K: Psychomotor development of three children with mosaic-trisomy 8 and literature review. Acta Med. Auxol. 23: 215-226, 1991

Priest JH, Rust JM, Fernhoff PM: Tissue specificity and stability of mosaicism in Pallister-Killian +i(12p) syndrome: relevance for prenatal diadnosis. Am. J. Med. Genet. 42: 820-824, 1992

Chapter 8

Cells from Urine Sample

HANNELORE KÖRNER, HENRIKE DIA AND CHRISTIANE BOMMER

Introduction

Cells found in the sediment of voided urine of newborns, children and adults can be cultured *in vitro*. The samples contain many exfoliative cells with different viabilities. The cells originate from the epithelial layer of the kidney pelvis, ureter, bladder, urethra, vagina, and prostate (urothel). The proliferating cells have an epithelial-like appearance. As demonstrated by specific antigens, most of the cells cultured from normal urine are bladder derived (Koskull et al. 1984).

A variety of papers on culture of urine sediment cells have already been published since 1974. A general view of the topic, historical developments and aspects about the technique and its application can be found in the following communications: Felix et al. 1980; Herz et al 1979, 1985, 1993; Hoehn et al. 1974, 1975; Koskull et al. 1984; Leschot et al. 1988; Shroki-Tabidzadeh et al. 1982; Tsai et al. 1995.

Urine cell cultures can be used for various diagnostic purposes. In clinical cytogenetics the examination of urine derived cells is a further method of chromosomal diagnosis in addition to the well-established chromosomal analysis from blood lymphocyte and skin fibroblast cultures. It combines the opportunity of chromosomal analysis with an easily repeatable and non-invasive procedure of cell sampling. Above all, the urinary cell culture is a promising tool for cytogenetic studies dealing with various aspects of mosaicism.

Correspondence to Hannelore Körner, Klinikum Charité, Institut für Medizinische Genetik der Medizinischen Fakultät der Humboldt-Universität, Luisenstr. 13 a, Berlin, 10098, Germany (*phone* +49-30-2802-3987; *fax* +49-30-2802-1286; *e-mail* hannelore.koerner@charite.de), Henrike Dia, Klinikum Charité, Institut für Medizinische Genetik der Medizinischen Fakultät der Humboldt-Universität, Luisenstr. 13 a, Berlin, 10098, Germany, Christiane Bommer, Klinikum Charité, Institut für Medizinische Genetik der Medizinischen Fakultät der Humboldt-Universität, Luisenstr. 13 a, Berlin, 10098, Germany

Urine cells should be cultured in cases of suspected tissue specific chromosome mosaicism.
Mosaicism can cause serious problems in cytogenetic diagnosis. The limitation of an aberrant cell line to specific tissues or organs can lead to the misdiagnosis "phenotypic abnormalities not attributable to a chromosomal aberration". The examination of an additional tissue - mostly a skin sample - to the regularly used lymphocytes is therefore recommended.

The examination of urinary sediment cell cultures offers a further chance to detect tissue specific chromosomal aberrations and can therefore be a valuable extension to the examination of blood and skin cells. Especially in newborns and small children urine cell cultures can be an alternative to skin biopsy because of a better acceptance by the patients and their parents. Whether or not a skin biopsy should also be carried out, depends on the result of urine cell analysis.

Cytogenetic follow-up studies in liveborn babies with true mosaicism in amniotic fluid cells should include a culture from urine sediment.
As amniotic fluid mainly consists of fetal urine, a high percentage of vital cells in the amniotic fluid derives from the urinary tract. Epithelial layers of the fetal urinary tract provide a substantial proportion of the cells in amniotic fluid (Hoehn et al. 1974, 1975). Two distinct epithelial cell types which are frequently seen in cultures from amniotic fluid originate from the fetal bladder urothelium as may be shown by comparison of cell morphology and growth patterns, as well as by the postive reaction of amniotic fluid cells to urothelium specific antibodies (Koskull et al. 1984). Kidney pelvis, ureter, urethra, prostate, and vagina also exfoliate cells that are found in urine and amniotic fluid.

For this reason it is possible, that an aberrant cell line found in prenatal diagnosis from amniotic fluid cells will appear in urine derived cells as well. A relevant example is the presence of mosaicism for trisomy 20 (T20) diagnosed prenatally on amniotic fluid cells. The incidence of true mosaicism is about 1 per 2000 (Djalali, et al. 1985). T20 cells are usually not detected after birth by lymphocyte or skin fibroblast cultures, which are the standard methods used for postnatal chromosome analysis. In single cases T20 cells could be detected in cultured urine cells (Miny et al. 1989).

A similar situation was reported for mosaic trisomy 12 detected in amniotic fluid cells. The aberrant cell type could be found in the placenta and cultured urine sediment cells, but not in lymphocytes and in skin fibroblasts of the phenotypical normal girl (Leschot et al. 1988). Furthermore, urine cell cultures can be used for research on problems dealing with normal and atypical epithelial cell growth or may even be of prognostic relevance in cancer.

Thus, studies have been published on the ageing process of normal epithelial cells, the effects of carcinogens on bladder epithelium (Felix et al.1980), malignant cell transformation (Herz et al. 1979), the diagnosis, monitoring and drug-sensitivity testing of bladder tumors and prostatic carcinoma cells as well as the control of cytostatic therapy (Herz et al. 1993).

Materials

Urine sediment cells can be cultured following a modified technique applied to amniotic fluid cells. As there is only a small number of viable cells to obtain from the urine sediment we recommend to use multiwell plates with only one ml culture volume to improve cell growth by better cell to cell contacts. Careful urine sampling under aseptic conditions, washing steps immediately after sampling as well as the modified culture method, ensure a relatively low rate of contamination (not more than 5%) and good culture results.

According to our experience in the newborn, even small urine samples (2 to 3 ml) are sufficient for successful cell cultivation. Cell growth can be observed from the fourth day onwards and the cells are suitable for transfer into cell culture flasks after about ten days of cultivation. Approximately three days later chromosome preparation can be done.

For urine sampling

- sterile destilled H_2O
- sterile pads to clean the childrens genitals
- sterile self adhesive urine bags (Braun Melsungen AG, Melsungen, Germany)
- sterile tubes (10 ml or 50 ml) and pipettes (5 or 10 ml)
- sterile gloves
- Refobacin 10 (Merck, Darmstadt, Germany)
- cell culture medium: amniomax 100 (basal medium and supplement: Gibco BRL, Life Technologies, USA)

For tissue culture

- multiwell tissue culture plates (four well multidishes or 24 well plates: Nunc, Denmark and Falcon New Jersey, USA)
- cell culture flasks, 25 cm^2 (Costar, USA)
- cell culture medium: amniomax 100

- Ca^{2+}- and Mg^{2+} – free phosphate buffered saline (Biochrom, Berlin, Germany)
- Trypsin/EDTA solution in PBS (Trypsin 0.05%; EDTA 0.02%: Boehringer Mannheim, Germany)

For chromosome preparation

- pasteur pipettes (Falcon)
- Colcemid in PBS (10µg/ml: Boehringer Mannheim, Germany)
- Ca^{2+}- and Mg^{2+} – free phosphate buffered saline (PBS)
- Trypsin /EDTA solution in PBS (Trypsin 0.05%; EDTA 0.02%: Boehringer Mannheim, Germany)
- KCl 0.45% in distilled H_2O
- Prefixative:
  - 3 ml methanol
  - 5 ml glacial acetic acid
  - ad 100 ml distilled H_2O
- Fixative:
  - 3 parts methanol
  - 1 part glacial acetic acid

Procedure

Urine sampling

1. Clean the children's genitals with sterile distilled water and sterile pads.
2. Fix the self-adhesive urine bag carefully wearing sterile gloves.
3. Transfer the urine sample into sterile tubes immediately after urinating and dilute the urine with cell culture medium 1:1 or 1:2.
4. Add about 100 µl refobacin per ml urine.
5. Centrifuge at 150 g for 10 min.
6. Remove the supernatant and wash the cells twice with 5 ml cell culture medium.

Tissue culture

1. Add complete cell culture medium to the washed urine sediment cells – about the same amount of medium as the original amount of urine for small urine samples, about half or fewer of these amount for larger samples.

2. Resuspend the cells carefully in the medium and inoculate into multiwell plates, 1 ml per well.
3. Culture the cells at 37°C in a humidified atmosphere with 5% CO_2.
4. Leave the cultures undisturbed for at least 48 hours. Then control for cell attachment and outgrowth every second day.
5. Change medium three times a week.
6. When cell density reaches confluence subculture into cell culture flasks.

- Remove the medium and rinse with PBS.
- Add 100 µl trypsin /EDTA solution per well and incubate for about 5 to 7 min.
- Control the detachment of the cells under an inverted microscope.
- When cells are detached add 1 ml medium per well and suspend the cells.
- Transfer the cell suspension into cell culture flasks – in general 2 wells into one flask – and add a further 2ml of fresh medium per flask.
- Change medium next day.

Chromosome preparation

Chromosome preparation can be done according to the method used for amniotic fluid cells. The highest mitotic activity has been observed about 3 days (2 – 4 days) after the passage into cell culture flasks (Figure 2). It is recommended to change the medium 24 hours prior to harvest.

1. Add 100µl colchicine per flask for 3 hours and incubate at 37°C.
2. Transfer the medium from the flask into a centrifuge tube.

Fig. 1. Urine cells after culture initiation.

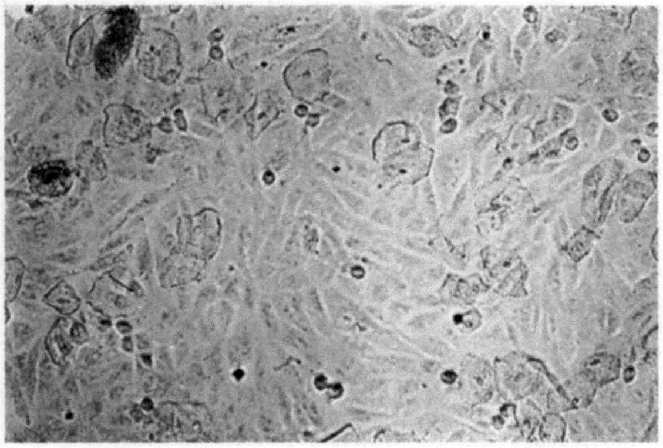

Fig. 2. Cultured urine cells 10 days after culture initiation.

3. Rinse the attached cells in the flask with PBS and add the PBS to the medium in the tube.
4. Add 400 µl trypsin/ EDTA solution per flask and incubate for about 5 to 7 min.
5. Control the detachment of the cells under an inverted microscope.
6. When most of the cells are detached fill the contents of the centrifuge tube into the flask to inhibit the trypsin.
7. Suspend the cells carefully with a pipette and transfer them into the centrifuge tube.
8. Centrifuge at 500 g for 3 minutes.
9. Remove the supernatant, resuspend the cells carefully and, mixing thoroughly, add 8 ml of the KCl-solution – first 2 ml dropwise with a pasteur pipette or a 1ml-pipette next 6 ml slowly with a 5 – or 10 ml pipette.
10. Incubate at 37°C for 20 min.
11. Centrifuge at 500 g for 3 min.
12. Remove the supernatant, resuspend the cells and, mixing thoroughly, add 5 ml prefixative, the first 2 ml dropwise.
13. Centrifuge as above.
14. Remove the supernatant, resuspend the cells and add 5 ml methanol, first 2 ml dropwise.
15. Centrifuge.

16. Remove the supernatant, resuspend the cells and add 5 ml fixative, first 2 ml dropwise.
17. Repeat the fixative washing 3 to 4 times.
18. After the last centrifugation and removing of the supernatant give some drops of fresh fixative to the cells.
19. Resuspend the cells carefully and drop the suspension with a pasteur pipette onto wet, ice cooled slides.

The cell suspension may be stored in the last fixative at -20°C before making slides.

Results

The chromosome quality is demonstrated in Figure 3. There are no differences in the quality whether in the number of countable metaphases nor in the banding resolution compared to chromosomes from amniotic fluid cells.

Troubleshooting

Main problems in chromosome preparation from urine derived cells arise from the failure of the cell culture because of:

- contamination of the urine with fungi or bacteria

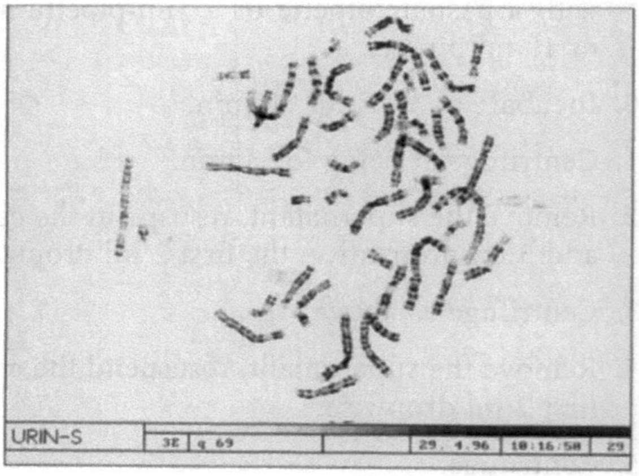

Fig. 3. Metaphase from cultured urine cells (GTG-banding).

- cell damage by toxic urine components
- low number of viable cells in the urine sediment.

Tips

- Before urine sampling clean children's genitals carefully with distilled water only! Disinfectants are neither necessary nor acceptable in newborns and small children!
- Prevent the contamination of the urine bag! Use sterile gloves! Do not touch the bag from inside!
- Do not take urine samples from children with diaperm dermatitis for chromosome analysis! Postpone the test and treat dermatitis first!
- Keep the time between urine collection and cultivation as short as possible to avoid cell damage. Immediate dilution of the urine with medium or, if this is not available, with an isotonic solution is very important, especially when the set up of the cell culture is not possible immediately after sampling. If possible, wash the urine sediment cells after sampling in such situations.
- Offer adults and older children beverages (500 – 1000 ml, 1 hour before sampling) to increase sample volume and to reduce the risk of exposure to toxic urine components.
- It has been reported, that higher cell counts in the urine sediment are achieved after physical exercise of the patient.
- For the parallel cultivation of several urine samples the use of 24 well tissue culture plates is recommended to increase the efficiency of the method. Contamination of (a) single well(s) has no effect on the other samples on the plate! Only remove the medium in the affected well(s), rinse twice with 70% ethanol and refill with 1% $CuSO_4$. Continue the cultivation of the remaining samples.
- Fluorescence *In Situ* Hybridization (FISH) on interphase nuclei in uncultured and cultured urine derived cells can be a valuable extension for the detection of aneuploidies.

References

Djalali M, Steinbach P, Schwinger E, Schwanitz G, Tettenborn U, Wolf M (1985) On significance of true trisomy 20 mosaicism in amniotic fluid. Hum Genet 69:321 - 326

Felix J, Sun TT, Littlefield JW (1980) Human epithelial cells cultured from urine: growth properties and keratin staining. In Vitro 16:866 - 874

Herz F, Schermer A, Koss LG (1979) Short-term culture of epithelial cells from urine of adults. Proc Soc Exp Biol Med 161:153 - 157

Herz F, Gazivoda P, Papenhausen PR, Katsuyama J, Koss LG (1985) Normal human urothelial cells in culture. Subculture procedure, flow cytometric and chromosomal analyses. Lab Invest 53:571 - 574

Herz F, Deitch D, Adler SA, Brijall D (1993) Short term culture of exfoliated cells from urine of patients with bladder tumors. Urol Res 21:23 - 26

Hoehn H, Bryant EM, Karp LE, Martin GM (1974) Cultivated cells from diagnostic amniocentesis in second trimester pregnancies. I. Clonal morphology and growth potential. Mut Res 8:746 - 754

Hoehn H, Bryant EM, Fantel AG, Martin GM (1975) Cultivared cells from diagnostic amniocentesis in second trimester pregnancies. III. The fetal urine as a potential source of clonable cells Hum Genet 29:285 - 290

Hsu LYF, Kaffe S, Perlis TE (1991) A revisit of trisomy 20 mosaicism in prenatal diagnosis - an overview of 103 cases. Prenat Diagn 11:7 - 15

Koskull H von, Aula P, Trejdosiewicz LK, Virtanen I (1984) Identification of cells from fetal bladder epithelium in human amniotic fluid. Hum Genet 65:262 - 267

Leschot NJ, Wilmsen-Linders EJM, Geijn HP Van, Samson JF, Smit LME (1988) Karyotyping urine sediment cells confirms trisomy 12 mosaicism detected at amniocentesis. Clin Genet 34:135 - 139

Miny P, Karabacak Z, Hammer P, Schulte-Valentin M, Holzgreve W (1989) Chromosome analyses from urinary sediment: Postnatal confirmation of a prenatally diagnosed trisomy 20 mosaicism. New Engl J Med 320:809.

Shokri-Tabibzadeh S, Herz H, Koss LG (1982) Fine structure of cultured epithelial cells derived from voided urine of normal adults. Virchow's Arch 39:41 - 48

Tsai YC, Simoneau AR, Spruck III CH, Nichols PW, Steven K, Buckley JD, Jones PA (1995) Mosaicism in human epithelium: Macroscopic monoclonal patches cover the urothelium. J Urol 153:1697 - 1700

Chapter 9

Classical and Molecular Cytogenetics of Tumor Cells

BRIGITTE SCHLEGELBERGER, SIMONE METZKE, SVETLANA HARDER, REINA ZÜHLKE-JENISCH, YANMING ZHANG AND REINER SIEBERT

Introduction

Cytogenetic findings are becoming increasingly important for the management of patients with malignant diseases, especially for those with hematologic neoplasias. The detection of aquired somatic mutations may help to establish the diagnosis of a neoplastic disorder and to rule out reactive changes due to toxic injury, vitamin deficiency or infections. Before, however, a chromosome aberration found in tumor cells can be taken as tumor-associated change it should be ruled out by chromosome analysis on PHA-stimulated blood lymphocytes that this chromosome aberration does not represent a constitutional abnormality. It is now clear that certain so-called primary chromosome abnormalities of tumor cells are associated with distinct clinico-histological disease entities. During tumor evolution additional chromosome aberrations appear and may determine the clinical course of the disease. Even these so-called secondary chromosome aberrations are non-randomly distributed throughout the genome. Therefore, cytogenetic studies are essential to make a specified diagnosis, to classify malignant disorders, to characterize the degree of neoplastic progression, to predict the prognosis, to test for remission, and to establish when relapse occurs. Thus, cytogenetic data can be of great help to select the appropriate treatment strategy.

Correspondence to Brigitte Schlegelberger, Universität Kiel, Institut für Humangenetik, Schwanenweg 24, Kiel, 24105, Germany (*phone* 0431-597-1781; *fax* 0431-597-1880; *e-mail* schlegelberger@medgen.uni-kiel.de) Simone Metzke, Universität Kiel, Institut für Humangenetik, Schwanenweg 24, Kiel, 24105, Germany, Svetlana Harder, Universität Kiel, Institut für Humangenetik, Schwanenweg 24, Kiel, 24105, Germany), Reina Zühlke-Jenisch, Universität Kiel, Institut für Humangenetik, Schwanenweg 24, Kiel, 24105, Germany, Yanming Zhang, Section of Hematology/Oncology, University of Chicago Medical Center, 5841, S. Maryland Ave, MC 2115, Chicago, IL 60637-1470, USA, Reiner Siebert, Universität Kiel, Institut für Humangenetik, Schwanenweg 24, Kiel, 24105, Germany

In addition to its role for daily patient care, cytogenetics has become a powerful tool for scientific purposes, e.g. to identify new homogenous tumor entities, to support the development of biologically relevant classification systems, to follow the differentiation of tumor cells and to find genes that play a critical role in tumorigenesis.

Chromosome analysis is the standard technique to obtain the karyotype of the tumor cells. During recent years this method was supplemented by molecular cytogenetics, especially fluorescence *in situ*hybridization (FISH). While chromosome analysis is able to give an overview over all chromosome aberrations of a tumor cell, FISH can detect certain chromosome aberrations that are specifically looked for with increased sensitivity. Since it can be applied to interphase cells, FISH can overcome the major technical limitation of chromosome analysis, ie the need for spontaneously proliferating tumor cells. Therefore it is best to combine both techniques in order to benefit from their individual advantages.

Subprotocol 1
Chromosome Analysis

Materials

Reagents

- 12-Tetradecanoylphorbol-13-acetate (TPA) (Sigma, P 8139)
- Amphothericin B (50mg) (Bristol-Myers-Squibb, 21004)
- Bromdesoxiuridine (BrdU) (Sigma, B 5002)
- Chromomycin A_3 (Sigma, C 2659)
- Citric acid (Merck, 1.00247)
- Colcemide (1ml Injektionlösung, 10µg/ml) (Boehringer, 295892)
- Collagenase (Serva, 17449 or Sigma, C6885)
- Concanavalin A (ConA) (Sigma, C 0412)
- Dimethylsulfoxide (DMSO) (Sigma, D 5879)
- Disodium hydrogen phosphate (Na_2HPO_4) (Merck, 1.06566)
- EDTA (Pharma Hameln, Infusionen 52298)
- Ethanol absolut (Merck, 1.00983)

- Ethidium bromide (Sigma, E 8751)
- Fetal calf serum (Biochrom, S 0115)
- Fluordesoxyuridine (FdU) (Sigma, F 0503)
- Giemsa (Merck, 9204)
- Glacial acetic acid (100%) (Merck, 1.00063)
- Glutamine (Biochrom, K 0202)
- Glycerol (Merck, 4095)
- GM-CSF (Granulocyte-macrophage colony stimulating factor) (Boehringer, 1087 762 or Biochrom, W 9050)
- Ham's F10 medium (nutrient mixture) (Gibco, 81200-065)
- Hank's solution (HBSS) (Gibco, 041-04170 M)
- Heparine, use preservative free lithium or sodium heparine, for example Liquemin (1000IU/ml) (Roche)
- Histopaque (Sigma, 1077-1)
- L-glutamine (200mM, 100x) (Seromed, K 0202)
- Lipopolysaccharide (LPS) (Sigma, L 8274)
- Lipopolysaccharide from Escherichia coli (LPS) (Sigma, L 2880)
- Magnesium chloride ($MgCl_2$) (Merck, Darmstadt, 8.14733)
- Methanol (Merck, 1.06009)
- Methotrexate solution (MTX)
- Methyl green (Sigma, M 8884)
- Penicillin and Streptomycin (Gibco, 15145-014)
- Phorbol myristat acetate (TPA) (Sigma, P 8139)
- Phorbol-12,13-dibutyrate (P) (Sigma, P 1269)
- Phosphate buffer (pH 6,88) (Merck, 1.07294)
- Phosphate buffer (pH 7,2-7,4) (Seromed, L 182-01)
- Phytohaemagglutinin (PHA) (Wellcome, HA 21)
- Pokeweed mitogen (PWM) (Sigma, L 9379)

- Potassium chloride (KCl) (Merck, 1.04938)
- Potassium di-hydrogen phosphate (KH_2PO_4) (Merck, 1.05108)
- RPMI-1640 medium (Seromed, F 1235)
- Thymidine (Sigma, T 3763)
- Trypan blue (Sigma, T8154)
- Trypsin/EDTA solution 0,05%/0,02% (Seromed, L2143)
- Trypsine 1:250 (Seromed, L 2123)
- Uridine (Sigma, U 3750)

Buffers and solutions

- Carnoy's fixative: methanol:glacial acetic acid = 3:1
- Chromomycin A3 solution: (0.5 mg/ml buffer pH 6.8, stored at 4°C for a few weeks in the dark). The buffer consists of 0.07 M Na_2HPO_4 (9.94 g/l), 0.07 M KH_2PO_4 (9.5 g/l) and $5x10^{-4}$ M $MgCl_2$ (0.102 g/l)

Note: Care: Wear gloves when preparing the solution since Chromomycin A_3 is mutagenic and toxic!

- Colcemide solution: To 1mg (1ml) colcemide add 200ml aqua dest
- Complete culture medium: To 80 ml RPMI medium add 20 ml fetal calf serum, 1ml penicillin/streptomycin stock solution and 1ml L-glutamine
- ConA stock solution: To 100 mg ConA add 14 ml aqua dest (Final concentration: 70µg/ml)
- Giemsa solution: To 80 ml phosphate buffer pH 6.8 add 6ml Giemsa
- GM-CSF stock solution: To 100 µg GM-CSF add 2ml RPMI (Final concentration: 0.5µg/ml) or to 10.000 U add 10 ml aqua dest (Final concentration: 1mg/ml)
- Hypotone solution: 0.075 M KCl
- LPS stock solution: To 100 mg LPS add 10ml aqua dest (Final concentration: 100µg/ml)
- McIlvain's buffer: To 61.45 ml 0.1 M citric acid add 38.55 ml 0.2 M Na_2HPO_4
- Methyl green stock solution: (1.925 g/100 ml McIlvain's buffer pH 4.0, stored at 4°C in the dark for not longer then 1 year)
- Methyl green working solution: (stable for 2-3 days)

To 33 ml 0.1 M citric acid and 66 ml 0.2 M Na_2HPO_4 add 2 ml methyl green stock solution

- PBS: 0,72 g $Na_2HPO_4 2H_2O$, 0,15 g KH_2PO_4, 0,1 g KCl, 4,0 g NaCl to 1 l aqua dest
- PWM stock solution: To 10 mg PWM add 13 ml aqua dest (Final concentration: 7.7 µg/ml)
- TPA stock solution: To 5 mg TPA add 5 ml 100% ethanol and 4 ml aqua dest (Final concentration: 5.5 µg/ml)
- Trypsin/EDTA solution: To 0,05 ml trypsin and 0,02 ml EDTA add 100ml PBS

Procedure

Tumor sample collection

For a successful cytogenetic analysis it is necessary to obtain a sample that contains a sufficient number of tumor cells. Make sure that the sample submitted for cytogenetic analysis is indeed infiltrated by tumor cells. To set up a suspension culture of hematologic neoplasms under routine conditions at least 1-1.5x10^6cells are required. But: the more the better. The sample must be taken under sterile conditions and be placed in a sterile container. Send it to the tumor cytogenetic laboratory immediately. A transport time of up to 24h is acceptable especially for heparinized bone marrow and blood from myeloid disorders.

In addition to the tumor sample, peripheral blood should always be sent to set up a PHA-stimulated culture in order to establish the constitutional karyotype. The technique is described in Part II, Chapter 5.

Leukaemias

Bone marrow is the sample of choice. Collect 2-3ml into a sterile 20ml tube containing 0.3 ml heparine stock solution containing 300 IU and 10ml RPMI medium. Gently mix. In case the bone marrow is hypercellular, 1-2 ml bone marrow is enough, in case the bone marrow is hypocellular, 4-5 ml bone marrow is needed.

Peripheral blood can be investigated if it contains at least 10% blasts or immature cells. Collect 10 -20ml peripheral blood into a sterile 20ml tube containing 0.3 ml heparine stock solution containing 300 IU.

Note: To prevent clotting heparinize the syringe.

Lymphomas
Lymph node biopsy is the sample of choice. Place the tissue, at least $1 cm^3$ in size, in a sterile container with culture medium. If no culture medium is available, RPMI medium with penicillin and streptomycin or physiological solution can be used. The sample must not dry up.

Bone marrow or peripheral blood is only adequate for chromosome analysis if there is a massive infiltration, which is usually present only in advanced disease stage. Other tissue, e.g. pleural effusion, ascites or spleen, can be used if the tissue is infiltrated.

Solid tumors
Tumor biopsy is the sample of choice. Place the tissue, at least $1 cm^3$ in size, in a sterile container with transport medium or physiological solution. The sample must not dry up. Make sure that the biopsy contains a sufficient tumor infiltrate and not only surrounding normal tissue.

Note: If it is not clear whether the tissue has been kept sterile, add higher concentrations of antibiotics and 2.5 mg/ml Amphothericin B to the transport medium.

Note: Attention: Avoid freezing and fixation of the tissue submitted for chromosome analysis. Putting the tissue onto dry ice or into formaline is a frequent mistake! Samples dealt with in this way are no longer useful for chromosome analysis.

Preparation of tissue

The preparation of tissue for chromosome analysis must be performed under sterile conditions.

Note: Attention: All tumor specimens are potentially infectious. Handle under full aseptic conditions.

Peripheral blood

1. Let the tube stand at room temperature until you can collect the buffy coat.
2. Determine the cell concentration.

Bone marrow

1. Wash the bone marrow once in RPMI medium by centrifuging at 200 g for 10 min.
2. Discard the supernatant and resuspend the pellet in RPMI medium.
3. Determine the cell concentration.

Note: If clotted bone marrow arrives, it can be rescued according to Metzke (1995).

1. Transfer the coagulum into 10 ml unsupplemented RPMI 1640 medium.
2. Dissolve it by adding 20 mg trypsin. Digestion of the clot can be aided by gently waving the samples and, if necessary, by incubation at 37°C, until it is completely resolved.
3. Wash the cell suspension twice with RPMI medium.
4. Determine the cell concentration.

If EDTA has been used instead of heparine, the cells can be washed three times in 10 ml RPMI medium without serum. After that, before setting up the culture, the cells are placed into a tube containing lithium heparine.

Separation of mononuclear cells

In some instances, e.g. the preparation of cytospin slides from bone marrow, it may be necessary to separate the mononuclear cells. For routine culturing, cell separation is not necessary, and in our opinion even affects adversely the mitotic rate.

1. Gently overlay 5ml Histopaque® with bone marrow or blood, which had been diluted 1:1 in PBS.
2. Centrifuge at 300g for 20 min without break.
3. Collect the ring of mononucleated cells above the Histopaque® solution with a sterile pipette into another tube and wash twice in RPMI medium.
4. Determine the cell concentration.

Note: PBS, bone marrow and blood must have room temperature.

Preparation of lymph node biopsies

1. Transfer the tissue to a petri dish containing 1-2 ml RPMI and prepare a single cell suspension by cutting or mincing the tissue into small pieces. If the lymph node tissue is soft, the cells burst into the medium giving it a milky appearance. If the lymph node tissue is of harder consistency, put the small pieces on metal gauze which is placed over a petri dish or a glass

container. Release the cells by pressing the pieces of tissue against the gauze using the plough of a syringe. Wash the tissue thoroughly with medium.

2. Collect the obtained cell suspension from the sample with the retained transport medium into a centrifuge tube and centrifuge at 200g for 10 min.

3. Discard the supernatant and resuspend the cells in 10 ml fresh RPMI medium.

4. Determine the cell concentration.

Solid tumors

Only few soft tissues may release single cell suspension by simple mechanical disaggregation as described for the preparation of lymph nodes. For most solid tumors of harder consistency the cells have to be isolated by enzymatic disaggregation. Additionally, fractionation of fibroblasts and epithelial cells by repeated sedimentation may be used. It is recommended to inform about specific techniques successfully used for the tumor to be studied, since the cell isolation technique has to be adjusted to each individual tumor (Czepulkowski et al. 1992, Trent et al. 1986).

Enzymatic disaggregation of solid tumors

1. Transfer the tissue to a petri dish containing 1-2 ml RPMI and cut it into small pieces. Remove all necrotic and fatty tissue. Separate normal tissue, which may be stored for control studies.

2. Disaggregate fragments enzymatically by incubation with 200-400 IU collagenase/ml Hank's solution for 1-4 h. The concentration may be increased up to 1300-1500 IU collagenase/ml and the incubation time may be increased up to 15-24 h. Both the concentration of collagenase and the incubation time have to be adjusted to each tumor type.

Note: To enhance the breakdown of extracellular material, incubation can be performed at 37°C or you may add 0.05% pronase or 0.01% hyaluronidase. To avoid high viscosity due to large quantities of released DNA, DNase I can be added to the collagenase solution at concentrations of about 100μg/ml. To avoid cell aggregation, heparine (2 IU/ml) can be added. Remove fat on top of the supernatant of lipogenic tumors carefully with a pipette, because it interferes with the subsequent attachment of the cells in the flasks.

3. Collect the obtained cell suspension without cell debris and centrifuge at 200g for 10 min.

4. Discard the supernatant and resuspend the cells in 10 ml fresh RPMI medium.

5. Wash twice in culture medium

6. Determine the cell concentration.

Determination of the cell count

Since cells need an optimal environment to grow and divide *in vitro*, it is essential that the cell density in cultures is optimized. The optimum cell density for bone marrow and blood cultures is 1.5-2x10^6 cells/ml. Cell count is most accurately determined by means of a counting chamber. Of course, a Coulter counter can also be used, although cells can not be critically visualized and additional information about the morphology of the cells or the prevailing cell type is lost. If there is no possibility to count cells, cell density can be assessed based on guesswork. If guesswork is relied upon, the suspected disease and the appearance of the sample should be considered. In general, hypercellularity is a feature of newly diagnosed CML and, less extremely, of ALL, but is less common in AML and MDS. Hypocellularity may be expected soon after chemotherapy or bone marrow transplantation.

1. Ascertain the total volume of your sample.

2. To 50 µl of the sample add 500 µl 2% glacial acetic acid. Mix well.

3. Place one drop of this cell suspension to each side of the counting chamber.

4. Count the cells within four different four corner squares and divide the total number of cells by four to obtain the average number of cells per four corner square.

5. Multiply with 1.1x10^5. This gives you the number of cells per millilitre of your sample.

Note: If you suspect a considerable portion of the cells to be defect, you may determine the percentage of viable cells by dilution of the well-mixed sample with 0.4% trypan blue solution 1:1 on an uncovered counting chamber. The viable cells remain clear and unstained whereas the nonviable cells take up the blue stain.

6. Adjust the final concentration to 1.5-2x10^6 viable cells/ ml in complete culture medium.

Set up of cultures and chromosome preparation

Leukaemias and lymphomas

Bone marrow, lymph node and - in case ›10% blasts are present - peripheral blood cells should be preferentially used to set up unstimulated short term cultures. In our opinion, an unstimulated 24h culture is the most important culture for all suspected hematologic malignancies. We have good experiencies with the addition of conditioned medium to 24h and 72h cultures, and therefore we set up these cultures in each case, given there are enough cells. If there are not enough cells, you can set up 5ml instead of 10ml cultures, but the 10ml cultures usually give better results. Growth factors and mitogens can be used to increase the mitotic activity of the leukaemic cells, but keep in mind that usually the growth of normal cells is also increased. If possible it is helpful to set up a series of cultures with different culture times or with different growth factors and mitogens, because the response of tumor cells to different culture conditions varies significantly.

Table 1. Preference of cultures for setting up of samples in routine tumor cytogenetics:

diagnosis	preference of cultures				
	direct preparation	24 h without mitogens	24h with conditioned-medium	48 h without mitogens	2 – 5 days with mitogens
ALL	2.	1.	3.	4.	-
AML, s-MDS, RAEB-T	4.	1.*	2.	3.	-
CML, MPS	-	1.	3.	2.	
high grade NHL	3.	4.	1.	2.	-
low grade B-NHL	-	3.	1.	-	2. LPS+PWM 4. LPS+TPA
low grade T-NHL	-	3.	1.	-	2. PHA 4. Con A 5. TPA

* in cases with expected M2 or M3 48 h culture is most informative
Samples from patients under or closely after chemotherapy, cryoconserved or clotted samples after trypsination and low proliferating cells as it is expected in CLL and early stages of MDS need one or two days longer cultivation times, respectively.
rating from 1 to 5, **1.**: highest preference

Table 1 summarizes the cultures we set up for routine analysis of samples according to the suspected malignancy.

Direct preparation

Direct preparation is beneficial for acute lymphoblastic leukaemias, especially in childhood, and for malignant effusions. However, the chromosome preparations are often of poor quality.

1. Place 20x10^6 cells in 10 ml culture medium in a centrifuge tube.
2. Add 0.3 ml colcemid solution.
3. Incubate at 37°C in a humified 5% CO_2 atmosphere for 10 min.
4. Centrifuge at 200g for 10min.
5. Remove the supernatant, resuspend the cell pellet and add 10 ml of pre-warmed hypotonic solution.

Note: If the culture contains a high amount of erythrocytes, divide it into two 10 ml cultures and perform chromosome preparation separately.

6. Incubate at 37°C for 20 min
7. At the end of the hypotone treatment add some drops of freshly prepared ice-cold Carnoy's fixative.
8. Centrifuge at 200 g for 10 min.
9. Remove the supernatant, resuspend the cell pellet and add, drop by drop, 10ml freshly prepared ice-cold Carnoy's fixative.
10. Repeat steps 8 and 9 several times until the cell pellet appears white and the supernatant is clear. Usually it is necessary to repeat one or two times for lymph node cultures and four or five times for bone marrow and peripheral blood cultures. Store the cell suspension at -20°C for at least 24 h prior to slide preparation.

Note: The cell suspension can be stored for several years. It can be used for fluorescence *in situ* hybridization studies and for RNA extraction.

11. Centrifuge at 200 g for 10 min.
12. Discard the supernatant and dilute the pellet with freshly prepared Carnoy's fixative to a light milky appearance. Usually you need 0.5 - 1 ml Carnoy's fixative.
13. Hold a wet slide horizontally with forceps and drop 2-4 drops of the cell suspension with a pasteur pipette onto the slide. Hold the pipette about

10cm above the slide, but you may increase the distance to get a better spreading of the metaphases. Make sure that the cell suspension is equally distributed over the slide. This is especially important if you want to localize the metaphases automatically.

14. Lay the slides horizontally and air dry.

Note: It is essential to clean the slides carefully. This can be achieved by cleaning with chromic acid, ether, absolute alcohol containing a few drops of HCl or water. Store the cleaned slides at 4°C in distilled water.

Note: It is often difficult to obtain well-spread metaphases of good quality from tumor cells. If the metaphases are not well-spread, you can increase the incubation time of the hypotonic shock. To prevent cell clumping and ensure proper fixation especially in blood samples, gentle dropwise fixation is necessary. When you prepare the slides, take the ambient temperature and humidity into account. If the prevailing conditions are cold and wet, warmed or flamed slides may be helpful. If it is hot and dry, cold and wet slides and breathing on the slides may facilitate spreading. For spreading of hyperploid metaphases, as it is suspected in ALL and solid tumors, hold the freshly prepared, wet slide over a flame for only a few seconds; this technique may produce an unequal morphology of the chromosomes hampering G-bands (Williams et al. 1984). Fixation and spreading are also facilitated by placing the tubes with the fixed cells in the freezer for 24-72 h before slide-making.

Unstimulated short-term cultures

1. Place 10-15x10^6 cells in 10ml culture medium in a 25ml tissue culture flask and loosen the cap. You can also use centrifuge tubes; put them into the incubator in a 30° angle. It appears that a large surface area promotes the cell growth.
2. Incubate at 37°C in a humified 5% CO_2 atmosphere for 24h.
3. If there are enough cells, you can incubate other cultures for 48h and 72h.
4. Add 0.3 ml colcemid solution. Mix gently.
5. Incubate at 37°C in a humified 5% CO_2 atmosphere for 30 min.
6. Continue with cell harvesting as described under direct preparation, step 4-12.

Addition of growth factors and mitogens

Different conditioned media may be added to short term cultures in a concentration of 10%. For myeloid leukaemias, GM-CSF in a final concentration of 10% can be used. For chronic lymphocytic leukaemias and other low grade lymphomas, we always set up cultures with different mitogens. B-lymphocytes can be stimulated by Epstein Barr Virus, B-cell growth factor or by polyclonal B cell activators like LPS, PWM and TPA (Gahrton et al. 1979, Ross et al. 1982, Autio et al. 1979), T-lymphocytes can be stimulated by PHA, ConA and TPA (Zech et al. 1986, Pirc-Danoewinata et al. 1995). If these mitogens are used, the culture time is 3 to 6 days. We adjusted the stock solutions of these mitogens so that the appropriate final concentration is achieved if you add 100µl to a 10ml culture. For the stimulation of solid tumors a large number of additives like insulin, epidermal growth factor, hydrocortisone or dihydrotestosterone have been described.

Note: You can produce your own conditioned medium by pooling cell-free supernatant of at least four PHA-stimulated blood cultures of healthy donors.

Synchronization techniques

Long chromosomes of good morphology that enable the identification of subtle structural chromosome aberrations as they are common in malignant cells may be achieved by the application of synchronization techniques (Yunis et al 1982, Hagemeijer et al. 1979, Garipidou and Secker-Walker 1991). There are a number of protocols for high resolution banding, which can be integrated into the laboratory time-tables with little additional effort (Table 2). Our experiences with synchronization techniques, however, led us to leave these techniques for routine use.

Chromosome analyses on solid tumors

Chromosome analyses on solid tumors are often problematic because of the difficulties to obtain sufficient metaphases and because of the complexity of the karyotypes. Another problem is how to remove or reduce the number of stromal fibroblasts in the sample and how to counteract their growth in culture. Therefore, chromosome analyses on solid tumors are rarely performed for clinical purposes. Keep in mind that the optimal culture conditions are different from one tumor type to the other and even from one tumor to the other (Czepulkowski et al. 1992, Trent et al. 1986).

1. Place 10-20x10^6 cells in 10ml Ham's F10 medium with 10-20% inactivated FCS and penicillin/streptomycin into a 25ml culture flask or onto coverslips, if you want to set up an *in situ* culture.

Note: If you have enough cells, set up cultures with different culture media, e.g. Ham's F10 and RPMI, or with different antibiotics, e.g. gentamycin.

2. Incubate at 37° C in a humified 5% CO_2 atmosphere. Alternatively: Place little pieces of tumor tissue onto the bottom of a culture flask, onto a feederlayer or onto a coverslip, if you want to set up an *in situ* culture. Incubate at 37° C in a humified 5% CO_2 atmosphere overnight. The next morning, carefully overflow with medium.

Table 2. High resolution techniques.

	incorporation of		Synchronisation of cell division with		
laboratory procedure	thymidine	ethidium bromide	FdU	MTX + thymidine	FdU+BRdU
Inoculate the appropriate amount of cells in a sterile container with complete low serum culture medium to a final volume of 10 ml and incubate at 37 °C.					
Prior to harvesting (recommended in the late afternoon), add	4 µl thymidine I	0.2 ml ethidium bromide + 0.2 ml colcemid	0.2 ml FdU+ 0.2 ml uridine	0.1 ml MTX	0.2 ml BRdU/FdU/ Uridine cocktail
Incubate at 37 °C for 14 – 17 h.					
Transfer the cell suspension to a centrifuge tube and spin at 200 g for 10 min.	↓	↓	↓		
Discard the supernatant and resuspend the pellet in 10 ml prewarmed low serum culture medium.	↓	↓	↓		
Add	↓	↓	0.2 ml thymidine II	0.2 ml thymidine II	0.4 ml thymidine II
Incubate at 37 °C for	↓	↓	5-6.5 h	5-6.5 h	5-6.5 h
Add 0.2 ml colcemid and incubate at 37 °C for 15 min.		↓			
Follow the harvesting protocol in Chapter 7.		↓			

⟶ skip the step and continue with the following

3. Check cell growth every two days under an inverted microscope. Change the medium after approximately one week or if you observe that the color of the medium has turned to blue.

4. If the tumor cells cover most of the bottom of the culture flask, you have to divide them into subcultures. Remove the culture medium, wash once with Hank's solution and add 5ml trypsin solution or 5ml trypsin/EDTA solution. Observe under the inverted microscope when most cells are detached. Wash the cells in culture medium and set up two or three cultures as described in step 1.

Note: Alternatively to trypsin, EDTA or collagenase may be used.

5. If the tumor cell growth is sufficient for harvesting, add 0.2ml colcemide and incubate for 2h. Longer incubation time up to 17h may increase the number, but may also decrease the quality of metaphases.

Note: Usually you can not obtain metaphases from the primary culture, but you have to wait until the tumor cells show a significant mitotic rate which usually occurs only after several passages. Mitotic cells have a round appearance.

6. Remove the medium, wash once with Hank's medium or PBS and add 5ml trypsin solution. Incubate at 37°C for 10-15 min. Check under the inverted microscope that most of the cells are detached.

7. Add 5 ml culture medium containing 10-20% FCS and transfer the cell suspension to a centrifuge tube. Rinse the culture flask with medium and transfer it to the centrifuge tube, too. Remaining cells can be cultivated further.

8. Follow the protocol described for direct preparation, steps 4-12. For *in situ* preparations follow the protocol described in Part III, Chapter 12.

Long term storage of cells

Save all remaining cells! They may be required for reanalysis if the patient suffers from a relapse and they may be very valuable for molecular genetic investigations, e.g. for cloning of translocation breakpoints.

1. Mix 1ml of the cell suspension containing 10^5-10^6 cells and 1ml inactivated sterile filtrated FCS with 10% DMSO in an Eppendorf tube.

2. Freeze the sample immediately in liquid nitrogen or at -80° C for up to one month.

3. Thaw the sample rapidly at 37° C in a waterbath. As soon as the cell suspension is fluid, transfer it to a centrifuge tube, centrifuge it at 200g and wash twice with Hank's medium or PBS.

Chromosome staining and banding

Giemsa staining

To make it easier to score the quality of the slides or to localize the metaphases automatically, e.g. by the metafer systems (Metasystems), it is helpful to stain the chromosomes homogenously.

1. Stain slides in a Coplin jar with 5% Giemsa solution for 10-15min.
2. Rinse in tap water and air dry.
3. For destaining immerse the slides briefly in 50%, 70%, 90% and absolute ethanol.

Fluorescence R-banding

Fluorescence R-banding with Chromomycin A_3 / Methyl green modified according to Sahar and Latt (1978).
Of course every banding method adequate for lymphocytes or amniotic cells can also be used for tumor cytogenetics. In tumor cytogenetics, where the metaphases often are of poor quality, it is extremely important to be familiar with the specific banding pattern of a certain banding method. We prefer fluorescence R-banding, because even fuzzy chromosomes in badly spread metaphase can be banded reproducibly. This is in our opinion a great advantage over G-banding where usually only part of the metaphases on a slide are sufficiently banded. Moreover, many break events in tumor cells, e.g. in the Philadelphia translocation, occur in light G-bands and therefore can be better recognized in R-banded metaphases.

1. Place 3-4 drops of chromomycin A_3 solution onto the slide, coverslip and incubate for 30-120 min at room temperature in a closed metal box.
2. Rinse with tap water.
3. Incubate the slide in methyl green working solution for 4 min at room temperature in a Coplin jar wrapped in aluminium foil.
4. Rinse with pure glycerol.
5. Mount with the remaining glycerol and press out surplus glycerol between paper towels.
6. View with a fluorescence microscope within 3 days.

Note: Attention: Avoid contact between glycerol and immersion oil since the view will become less sharp. Should glycerol and immersion oil have mixed, soak off the coverslip, rinse with glycerol and repeat steps 5 and 6.

Results

Evaluation

To make a tumor cytogenetic diagnosis it is essential to analyze at least 20 metaphases thoroughly. Of course it is not sufficient to only detect a recurrent chromosome aberration. Keep in mind that karyotypic evolution may occur and that secondary chromosome aberrations that appear in addition to primary chromosome aberrations may be of prognostic significance. Try to analyze even bad metaphases; they may contain a characteristic chromosome abnormality which may not be present in high-quality metaphases. Sometimes it can be possible to verify the clonality of an aberration by identifying the same aberration in a bad metaphase.

Nevertheless, make any effort to get metaphases of good quality. Many characteristic chromosome aberrations are very fine and easy to be overlooked (Figure 1). Never be sure to have detected all chromosome aberrations: a cell with trisomy 8 or trisomy 22 can have an inversion 16 that is easily overlooked. The responsible interpretation of tumor cytogenetic findings is only possible if you have several years of training in a tumor cytogenetic lab to get familiar with typical chromosome aberrations and to be able to interpret the consequences of tumor cytogenetic findings.

In your report, you must give the detailed cytogenetic findings and discuss associations between certain chromosome abnormalities and morphologic changes of the tumor as well as the typical clinical course of patients with the observed chromosome aberrations. It is of great help to keep close contact with the clinician and the pathologist - they can give you valuable information. Cytogenetic findings can be the major diagnostic criterion a therapeutic decision like bone marrow transplantation relies upon. The report should indicate:

- the cultures set up
- the number of metaphases analyzed, specified for metaphases from unstimulated and stimulated cultures as well as for normal and aberrant metaphases
- the banding level (bands per haploid chromosome set)
- a description of the karyotype according to ISCN
- a detailed interpretation of the meaning the cytogenetic findings have in the present clinical situation.

The report should be sent as soon as possible. In certain situations, ie if acute myeloid leukaemia M3 is suspected, the tumor cytogenetic study should be finished within 24-72h. Usually it is sufficient if the report is received before the end of induction therapy, i.e. within two to four weeks.

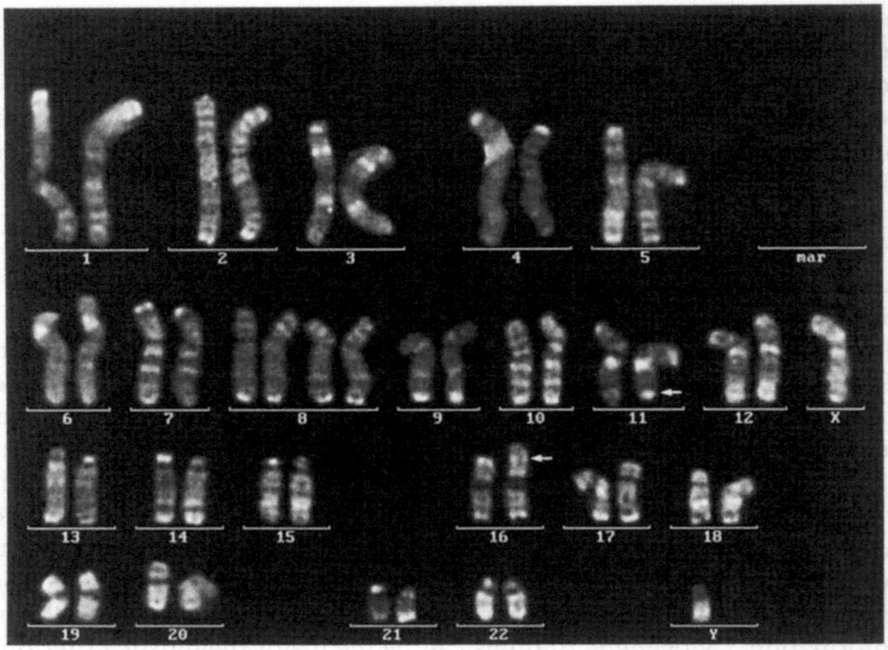

Fig. 1. Karyotype of a patient with secondary acute myeloid leukaemia, FAB type M4: 48, XY,+8,+8, t(11;16)(q23;p13). Fluorescence R-banding.

Subprotocol 2
Fluorescence in Situ Hybridization

Making up a reliable cytogenetic diagnosis is, in a considerable number of tumor samples, hampered by a low yield and poor quality of metaphases. Additionally, due to the destruction of cell membranes in the preparative process neither a correlation of genetic alterations to morphologic nor to immunophenotypic features of single tumor cells is possible by conventional cytogenetics. At last, the determination of the percentage of aberrant cells in a primary tumor specimen by chromosome analysis is impossible due to growth differences of some cell populations *in vitro*. The application of molecular cytogenetics in the tumor cytogenetic lab not only helps to clarify dubious karyotypes but also overcomes some of the major limitations of conventional cytogenetics. Therefore, molecular cytogenetic techniques like FISH (Fluorescence *in situ* hybridization) should at least in part be available in a tumor cytogenetic lab. It is beyond the scope of this chapter to give a detailed review of molecular cytogenetic methods (for review see

Lichter and Word 1990, Pinkel et al. 1986, Weber Matthiesen et al. 1996). Protocols for generating probes, preparing tissue samples, hybridizing and detecting probes and evaluating FISH results are provided in Chapter 18 of this manual. Nevertheless, some universal protocols fitting the specific properties of FISH on tumor specimens will be described. Additionally, the FICTION (*F*luorescence *I*mmunophenotyping and *I*nterphase *C*ytogenetics as a *T*ool for *I*nvestigation *o*f *N*eoplasms) technique, which combines immunophenotyping and FISH and which was developed particularly for the diagnosis of genetic alterations in neoplastic cells expressing certain antigens, will be detailed.

Materials

Reagents

- AMCA-conjugated avidin (Jackson 003-150-083)
- Biotin-labelled goat anti-avidin antibody (Vector BA-0300)
- class-matched mouse antibodies IgG1, IgG2a, IgG2b, IgM (Immunotech 0571, 0572, 1266, 1268)
- Human Cot-1 DNA (Gibco BRL/Life Technologies 15279-011)
- Cy3-conjugated avidin (Jackson 003-160-083)
- Cy3-conjugated goat anti-mouse antibody (Jackson 115-165-068)
- Cy3-conjugated rabbit anti-goat antibody (Jackson 305-165-045)
- Cy3-conjugated donkey anti-rabbit antibody (Jackson 711-165-152)
- Cy3-conjugated rabbit anti-mouse antibody (Jackson 315-165-003)
- DABCO (1,4-Diazabicyclo-(2,2,2)-octane) (Sigma D2522)
- DAPI (4,6-Diamidino-2-phenylindole) (Sigma D9542)
- Deionized formamide (Gibco BRL/Life Technologies 15515-018)
- Dextran sulphate 50% (Oncor S4030)
- Digoxigenin-conjugated sheep anti-mouse antibody (Boehringer 1214624)
- FITC-conjugated donkey anti-mouse antibody (Jackson 715-095-151)
- FITC-conjugated sheep anti-digoxigenin antibody (Boehringer 1207741)
- Mouse anti-FITC antibody (monoclonal) (Dakopatts M878)

- Mouse anti-digoxigenin antibody (monoclonal) (Boehringer 1333062)
- Normal mouse serum (Jackson 015-000-001)
- Pepsine (3900U/mg solid) (Sigma P-6887)
- Rubber cement: Fixogum (Marabu 2901 10 000)
- Sonicated salmon sperm DNA (10 mg/ml) (Sigma D9156)

Buffers and solutions

- **Antifade solution :** To 90ml glycerol add 10ml PN-buffer containing 23 mg/ml DABCO.
- **DAPI stock solution:** 0.2mg/ml DAPI in A. bidest
- **Digestion solution:** 5mg pepsine, 1ml 1N HCl, 99ml A. bidest
- **Paraformaldehyde solution (1%):** Solve 1g paraformaldehyde powder (harmful!) in 60ml A. bidest.; add 5 drops 10N NaOH. Heat to 70-100°C until the solution is clear under a bench. Cool down and add 10ml 100mM $MgCl_2$. Adjust pH to 7-7.5 and fill up to 100ml with distilled water. Filter the solution. Store at 4°C.
- **PN buffer:** 0.1 M NaH_2PO_4, 0.1 M Na_2HPO_4, pH 8.0
- **PNM buffer:** Add 5% dry milk powder to PN-buffer; heat to 50°C overnight while stirring; add 0.03% NaN_3 (poisonous); PNM buffer is stable at 4°C for at least 6 months.
- **20xSSC:** 3M NaCl, 0.3M sodium citrate; pH 7.0
- **Mastermix for centromeric probes:** Mix 6 ml deionized formamide, 2 ml 50% dextran sulphate, 50µl sonicated salmon sperm DNA (10 mg/ml) and 0.5 ml 20xSSC. Vortex thoroughly, adjust to pH 7.5 and make up to a final volume of 9ml with A.bidest.
- **Mastermix for single copy probes:** Mix 5 ml deionized formamide, 2 ml 50% dextran sulfate and 1 ml 20xSSC. Vortex thoroughly, adjust to pH 7.5, fill up to 9 ml with A.bidest.
- **Mastermix for single copy probes plus Cot-1 DNA:** Add 250µl Ethanol 100% to 100µl Cot-1 DNA (1µg/1µl) in a 1.5ml Eppendorf tube and spin 30min at 14.000rpm in a lab centrifuge. Remove supernatant, add 500µl 70% Ethanol and vortex. Spin again 15min at 14.000rpm. Remove supernatant, dry pellet in a speed vac or allow to air dry. Add 9µl prewarmed (37°C) mastermix for single copy probes and shake for at least 30min.

All mastermixes can be aliquoted and stored at -20°C.

Probes and probe preparation for FISH

For a series of questions addressed in a tumor cytogenetic lab by FISH or FICTION, probes directly labelled with a fluorescent dye or indirectly labelled with biotin or digoxigenin are commercially available (Oncor, Vysis, etc.). For the detection of numerical aberrations, satellite probes specific for centromeric regions of distinct chromosomes should be applied. Chromosome painting probes can be used for the identification of marker chromosomes and structural chromosome aberrations in metaphase spreads but are generally not recommendable for any use in interphase nuclei. For some recurrent translocations, e.g. for the t(9;22), the t(15;17), or for some deletions, e.g. of the RB gene or the p53 gene, probes are also commercially available. For other structural aberrations probe sets have been published which can be obtained from the scientific community. The latter can be prepared according to routine protocols and labelled by random primed labelling or nick translation using commercially available kits according to the manufacturers' instructions (Life Technologies, Eggenstein, FRG; Boehringer, Mannheim, FRG).

In this paragraph a series of protocols developed for easy and universal application of commercial and non-commercial probes are described. For other commercial probes follow the recommendations of the supplier.

- **Hybridization mixture for indirectly labelled centromeric probes (Oncor):** Add 1µl of biotin labelled probe (Oncor Cat. No. P5000-BIO -P5090-BIO) or 1µl of digoxigenin labelled probe (Oncor Cat. No. P5000-DG – P5090-DG) or 0.5µl of biotin labelled probe (Oncor Cat. No. P5000-BIO - P5090-BIO) and 0.5µl of digoxigenin labelled probe (Oncor Cat. No. P5000-DG – P5090-DG) as supplied by the manufacturer to 9 µl of "mastermix for centromeric probes". Non-commercial DNA probes should be adjusted to 10 ng/µl in TE-buffer (1 mM EDTA, 10 mM Tris-Hcl, pH8) before being added to the mastermix.
- **Hybridization mixture for indirectly labelled commercial single copy probes (Oncor):** Indirectly labelled single copy probes (Oncor), e.g. probes for translocations or microdeletions, can be applied without prior modifications. For the combination of single copy probes with centromeric probes a 10:1 mixture, for the combination of two single copy probes a 1:1 mixture is recommended. If hybridization with the directly applied single copy probe or a mixture of two probes produces a high background noise a 1:2 or 1:4 (up to 1:10) dilution of the probe/probe mixture in "mastermix for single copy probes plus Cot-1 DNA" may enhance the hybridization quality.

- **Hybridization mixture for indirectly labelled non-commercial single copy probes with or without combination with indirectly labelled centromeric probes (Oncor):** For non-commercial single-copy probes the mixture of 1µl labelled single-copy probe (40-100ng/µl) and 9µl "mastermix for single copy probes plus Cot-1 DNA" is recommended. If the probe produces a high background noise or results in a low signal intensity 1-5µl labelled single-copy probe can be coprecipitated with 100µl Cot-1 DNA using conventional Ethanol precipitation (as described for "mastermix for single copy probes plus Cot-1 DNA") and resuspended in 10µl "mastermix for single copy probes". For combining a single-copy probe with a centromeric probe (Oncor) add 1µl of the latter to the respective probe mixture.
- **Hybridization mixture for indirectly labelled painting probes (Oncor, AGS):**Painting probes can be applied directly or can be handled as described for commercial single-copy probes. In general, predigestion of the slides is recommended.
- **Hybridization mixture for directly labelled probes (Vysis):** Directly labelled probes should be premixed in the supplied buffers according to the manufacturers' instructions. Alternatively or for the combination with indirectly labelled probes dilution in the mastermixes described for indirectly labelled centromeric or single copy probes is possible.

Procedure

Slide preparation for FISH

Preparing high quality slides is one of the most critical steps for successfully performing FISH. In general, cytogenetic preparations as described above, conventional unstained peripheral blood or bone marrow smears, touch preparations and cell suspensions from fresh and frozen tissue are applicable for FISH. The use of formaline-fixed paraffin-embedded tissue, tissue sections and stained blood or bone marrow smears is not recommended for routine use and should be restricted to labs experienced with the technical considerations of these techniques. In order to facilitate the correlation of the hybridization signals to individual nuclei, cells should not be packed too densely. The number of cells per cytospin slide should not exceed 3.000 cells (diameter of the area containing cells = 0.5 cm). In imprints and smears cells often overlap, but there are usually areas in the periphery where the cells lie separated from each other. Only such areas should be evaluated. The erythrocytes should lie flat on the slide and not be standing on end. Sometimes, this is only possible at the end of the smears. Prior to hybridization the quality of slides should be checked by phase contrast microscopy. Areas qualifying for hybridization are characterized by a high density of clearly separated greyish nuclei lacking any surrounding cytoplasm and should be marked on the back of the slide with diamond pencil. In general, all specimens can be hybridized without any pretreatment. Nevertheless, in order to enhance the hybridization efficiency predigestion of slides according to the following or similar protocols is recommended for most tumor specimens.

Pepsine-digestion of FISH slides

1. Incubate slide in a Coplin jar for 5min at 37°C with freshly prepared (not prewarmed!) digestion solution.
2. Wash slide once in A. bidest.
3. Fix slide for 10min in paraformaldehyde solution (1%).
4. Wash slide once in A. bidest.
5. Dehydrate in 70%, 85% and 100% Ethanol for each 1min.
6. Air dry slide.

For different samples, optimal pepsine concentration and incubation duration may vary slightly.

Slide preparation for cytogenetic samples, imprints and blood/bone marrow smears: Slides from cytogenetic samples are prepared as described above (Subprotocol 1). Blood and bone marrow smears as well as imprints are prepared as for routine morphologic evaluation.

Cytospin preparations

Cytospin preparations of cytogenetic specimen: The hybridization efficiency for interphase FISH from cytogenetic specimen can be enhanced by preparing cytospin preparations. This procedure is especially recommended for samples containing few cells or displaying no signals or a high background noise after conventional preparation.

1. Determine the cell number.
2. Add 1xPBS ad 10ml to 1-3 ml cell containing Carnoy's fixative in a 15ml tube.
3. Spin 10min at 1800rpm in a lab centrifuge.
4. Discard supernatant.

5. Resuspend pellet in 1xPBS to a cell count of $1x10^4$ cells/ml.
6. Spin 200µl of this solution (2.000 cells/slide) in a cytocentrifuge for 5min at 800rpm.
7. Air dry slide.
8. Proceed with Pepsine-digestion of FISH slides.

Cytospin preparations of cell suspensions from solid tumors: This protocol is intended for FISH analyses of frozen specimens from lymph nodes and solid tumors, in which neither cytogenetic preparations nor freshly prepared imprints are available.

1. In a cooled petri dish cut a piece of tissue (10-40mg) from a tumor block stored in liquid nitrogen or at -80°C. Refreeze remaining tumor material immediately!
2. Per 10mg tumor tissue add 60-120µl 1xPBS (tissue-dependent).
3. Unfreeze tissue in 1xPBS for 15-30min at RT.
4. Disaggregate tissue mechanically by cutting and scraping with a scalpel.
5. Collect cell containing 1xPBS solution from the dish and transfer it to a 1.5ml Eppendorf tube.
6. Disaggregate cells in solution by pipetting several times through a 10-100µl pipette.
7. Count the number of cells in solution.
8. Transfer 3000 cells/slide into a new Eppendorf tube.
9. Add 200µl digestion solution. (Do not prewarm!)
10. Incubate at 37°C for 10min.
11. Immediately transfer tube on to ice; cytocentrifuge the solution for 5min at 800rpm.
12. Air dry slides for 10min.
13. Fix slides for 10min in paraformaldehyde solution (1%).
14. Wash slides once in A. bidest.
15. Dehydrate in 70%, 85% and 100% Ethanol for 1min each.
16. Air dry slides.

Note: The remaining cell solution can be stored for several months. Add 2 – 3 Vol. 100% Ethanol and freeze at 20°C. For preparing slides from this stored cell solution transfer required amount of cells into an Eppendorf tube, pellet cells by short spinning in a centrifuge, discard supernatant, add 40µl 1xPBS per $1x10^5$ cells and incubate on ice for at least 15min. Proceed with step 9.

Hybridization and post-hybridization washes

Hybridization

1. Apply 1.5µl of the respective hybridization solution on the cell containing area of the slide.
2. Cover with a round 10mm-coverslip.
3. Seal with rubber cement.
4. Place the slides at the bottom of a metal box and a wet paper towel, close the box and denature for 7 min in a waterbath at 75°C.
5. Immediately transfer the hot metal box into a 37°C incubator. According to the probe, hybridize for 1h to 3 days.

Post-hybridization washes

1. Prewarm three Coplin jars containing 0.1xSSC at 60°C in a waterbath.
2. Remove rubber cement with a needle from the slides.
3. Shake slides in the first Coplin jar at 60°C in 0.1xSSC until all coverslips are removed.
4. Wash slides by shaking in each of the three 0.1xSSC containing Coplin jars for 5min at 60°C.
5. Equilibrate slides in PN buffer for 2 min at RT.

Counterstaining

If directly labelled probes have been applied, the cells can now be counterstained blue by incubating for 5-10min in DAPI-Solution (e.g. 6µl DAPI stock solution in 60ml 2xSSC) followed by washing for 5-10min in 2xSSC and mounting in antifade solution. Alternatively, if green (e.g. FITC) and/or blue (e.g. AMCA) fluorochromes have been applied, the slides can be mounted with antifade solution containing 500ng/ml propidium iodide (red).

Simultaneous detection of biotin- and digoxigenin-labelled probes with Cy3 (red) and FITC (green)

The following protocol allows the simultaneous detection of biotinylated probes with the red fluorochrome Cy3 and of digoxigenin labelled probes with the green fluorochrome FITC (Figure 2 A-C). In analogy, a series of alternative cascades allowing the detection of indirect labelled probes by other colors or fluorescent dyes can be designed. The respective solutions can be incubated under coverslips followed by three washings for 2min each in PN buffer in a Coplin jar. Nevertheless, the use of Shandon holders not only facilitates manual detection but also provides the prerequisite of automatisation by pipette roboters (e.g. Tecan SLT).

Simultaneous detection

1. Incubate slides with a mixture of Cy3-conjugated-avidin (1:400 in PNM buffer) and mouse anti-digoxigenin antibody (1:100 in PNM buffer) at RT for 30 min.
2. Wash once with PN-buffer.
3. Incubate with a mixture of biotinylated goat anti-avidin antibody (1:200 in PNM buffer) and digoxigenin-conjugated sheep-anti mouse antibody at RT for 30 min. (1:100 in PNM buffer).
4. Wash once with PN-buffer.
5. Incubate slides with a mixture of Cy3-conjugated avidin (1:400 in PNM buffer) and FITC-conjugated sheep anti-digoxigenin antibody (1:100 in PNM buffer) at RT for 30 min.
6. Wash once in PN-buffer and for 5-10min in 2xSSC.
7. For counterstaining blue incubate the slide for 5-10min in DAPI-Solution (6µl DAPI stock solution in 60ml 2xSSC) (According to preferred staining intensity, DAPI-concentration and/or incubation time can be adjusted).
   For counterstaining red – if the hybridization signals are visualized by AMCA and FITC-mount the slides with antifade solution containing 500ng/ml propidium iodide (red).
8. Wash for 5-10min in 2xSSC.
9. Mount the slide with antifade solution.

Slide preparation for FICTION

The quality of slides is even more important for successfully performing the FICTION technique than for FISH. For FICTION, cytospin slides, smears, imprints and cryostat sections may be used. Freshly prepared slides should be air-dried overnight at room temperature. Thereafter, slides can be processed immediately or stored at -80°C. Slides preserved in this way can be used for years; when single slides are taken out of the freezer, it is very important to make sure that the remaining slides in the freezer do not thaw. This is possible by taking the slides out of the container inside the freezer. If slides are stored at -20°C, the quality of the immunophenotyping may be diminished after a few months. However, the quality of the hybridization is not affected. Before starting the FICTION procedure slides should be examined for the morphological quality of the cells by phase contrast microscopy. If many slides are available, only those with good cell morphology should be processed.

FICTION protocol for combined immunophenotyping and two-color FISH (modified according to Weber-Matthiesen et al. 1992)

Immuno-phenotyping

1. Fix slides in fresh acetone for 10min at room temperature and air-dry for 10 min.

Note: Cryo-preserved slides should be air-dried prior to the fixation for at least one hour after being taken out of the freezer.

2. Apply 100µl of monoclonal mouse antibody (against the antigen that is to be stained) diluted in PNM buffer to the area of the slide containing the cells or the section. Incubate for 30 min at RT.
3. Wash once in PN buffer.
4. Incubate with Cy3-conjugated goat anti-mouse antibody (1:200 in PNM buffer) for 30 min.
5. Wash once in PN buffer.
6. Incubate with Cy3-conjugated rabbit anti-goat antibody (1:200 in PNM buffer) for 30 min.
7. Wash once in PN buffer.
8. Incubate with Cy3-conjugated donkey anti-rabbit antibody (1:100 in PNM buffer) for 30 min.
9. Wash once in PN buffer.

This four-step immunostaining results in very strong fluorescent staining. If strongly expressed antigens, such as CD3, have to be stained, it is possible to omit steps 8 and 9 or even steps 6 to 9.

Monitoring of immunophenotyping

At this point the success of the immunostaining can be monitored. Mount the slides with PN buffer (not with glycerol!) and examine them under the fluorescence microscope.

Note: The fluorescence intensity is relatively low if the slides are mounted in PN buffer. However, mounting in glycerol prior to *in situ* hybridization may cause hybridization artifacts. Monitoring of immunophenotyping is always recommended if the processed slides have been stored for a long period or if the cells show poor morphology prior to the staining procedure.

Hybridization

10. Place the slides in 1% paraformaldehyde for 1 min at 4°C.
11. Wash in A. bidest for 2 min at RT.
12. Put the slides into freshly prepared, ice-cold Carnoy's fixative for 10 min.
13. Wash in A. bidest for 2 min at RT.
14. Dehydrate the slides in a series of 70%, 85% and 100% ethanol at RT, each for 2 min. Air-dry the slides for 10 min at RT.
15. Hybridize with indirectly labelled probes (as described above).

Detection

Simultaneous detection of biotin- and digoxigenin-labelled probes by AMCA (blue) and FITC (green):

16. Incubate slides with a mixture of AMCA-conjugated avidin and FITC-conjugated monoclonal mouse anti-digoxigenin antibody at RT for 30 min. [both 1:200 in PNM buffer].
17. Wash once in PN buffer.
18. Incubate with a mixture of biotinylated goat anti-avidin antibody and mouse anti-FITC antibody at RT for 30 min. [both 1:200 in PNM buffer].
19. Wash once in PN buffer.
20. Incubate with a mixture of AMCA-conjugated avidin and FITC-conjugated donkey anti-mouse antibody at RT for 30 min. [both 1:200 in PNM buffer].
21. Wash once in PN buffer.

Note: The fluorescence intensity can be amplified by multiple repetitions of steps 18 to 20.

22. Mount the slides with antifade solution

FICTION protocol for combined double-immunophenotyping plus single-color FISH

Immuno-phenotyping

Immunophenotyping against antigen A by Cy3:

1. Incubate with monoclonal mouse antibody against antigen A (e.g. CD5) for 30 min at RT.
2. Wash once in PN buffer.
3. Incubate with Cy3-conjugated rabbit anti-mouse antibody (1:200 in PNM buffer) for 30 min at RT.
4. Wash once in PN buffer.
5. Incubate with Cy3-conjugated donkey anti-rabbit antibody (1:200 in PNM buffer) for 30 min at RT.
6. Wash once in PN buffer.
7. Incubate with 20% normal mouse serum diluted in PNM buffer for 15 min at RT.
   By incubating with normal mouse serum containing high concentrations of mouse immunoglobulins, the free binding-sites of the Cy3-conjugated rabbit anti-mouse antibody (step 2) are blocked. This prevents the monoclonal mouse antibody against antigen B, which is applied in step 5, from being caught by free binding sites of the rabbit anti-mouse antibody.

Note: Attention: Do not wash the slides after this step

Immunophenotyping against antigen B by AMCA:

8. Incubate with biotinylated monoclonal antibody against antigen B (e.g. CD19) in PNM buffer containing 20% normal mouse serum for 30 min at RT.
9. Wash in PN buffer.

Note: The detection system is very similar to that used for the detection of biotinylated DNA probes.

10. Incubate with AMCA-conjugated avidin (1:200 in PNM buffer) for 30 min at RT.

11. Wash in PN buffer

12. Incubate with biotinylated goat anti-avidin antibody (1:200 in PNM buffer) for 30 min at RT.

13. Wash in PN buffer

Note: Multiple amplifications of the AMCA staining are possible by repeating steps 10 to 13.

Hybridization

14. Fix in 1% paraformaldehyde for 1 min at 4°C.

15. Wash in A. bidest for 2 min at RT.

16. Put the slides into fresh ice-cold Carnoy's fixative for 10 min.

17. Wash in A. bidest for 2 min at RT.

18. Dehydrate the slides in a series of 70%, 85% and 100% ethanol at RT for 2 min each.

19. Air-dry the slides for 10 min at RT.

20. Apply hybridization mixture containing digoxigenin labelled probe to the slide.

Note: Attention: In this protocol a digoxigenin labelled probe is applied. Do not use biotin-labelled probes, because they bind unspecifically to the antibody cascade of antigen B.

Detection

Detection of the digoxigenin labelled probe by FITC (green):

21. Incubate with a FITC-conjugated monoclonal mouse anti-digoxigenin antibody at RT for 30 min (1:200 in PN buffer).

22. Wash in PN buffer.

23. Incubate with FITC-conjugated donkey anti-mouse antibody at RT for 30 min (1:200 in PNM buffer).

24. Wash in PN buffer.

25. Incubate with monoclonal mouse anti-FITC antibody for 30 min at RT (1:200 in PNM buffer).

26. Wash in PN buffer.

27. Incubate with FITC-conjugated donkey anti-mouse antibody for 30 min at RT (1:200 in PNM buffer).

28. Wash in PN buffer.

Note: If required, the fluorescence intensity can be amplified by repeating steps 25 to 28.

29. Mount the slides with antifade solution.

Storage of stained slides

Mounted slides can be stored for several weeks at 4°C and for several months or up to years at -20°C. Always keep slides in the dark.

Other FICTION protocols

The FICTION technique can generally be performed with any kind of FISH procedure (Figure 2 D). If immunophenotyping and, most important, paraformaldehyde fixation after immunophenotyping is done according to the FICTION protocol, then it is possible to proceed with individual FISH protocols as established in other laboratories (Siebert and Weber-Matthiesen et al. 1997). This way, all types of probes for detecting numerical or structural aberrations can be employed. The only restriction is that proteolytic treatment must be avoided because it would considerably impair the quality of immunophenotyping.

Results

Evaluation of FISH and FICTION-Analyses

Technical prerequisites: In general, the different fluorescent colors of the slides can be evaluated separately under each fluorescence microscope equipped with specific filter sets for AMCA, FITC and Cy3. Alternatively, if triple-dye fluorescence filter sets are used all three fluorescent colors can be evaluated simultaneously. Appropriate filter sets are available from microscope-manufacturers and from filter-producers. Documentation is possible by conventional photography using a 400ASA film and exposure times up to 20s. Digital imaging using commercially available software like the ISIS system supplied by MetaSystems Hard & Software GmbH is an easy alternative.

Evaluation criteria and cut-off levels: For inter-observer-comparability the evaluation criteria have to be defined exactly. This is particularly necessary for translocation probes, in which a colocalisation of two signals indicates a translocation. As to the organization of the nucleus in interphase cells a spatial separation of the two colocalised signals is possible. Thus, it has to be defined up to which distance between the signals (calculated in relative signal size, ie once or twice the signal diameter) a real colocalisation is assumed (Figure 2 C).

In each laboratory, diagnostic thresholds have to be defined for each probe. This is done by analysing at least five control cases with a minimum of 200 interphase nuclei each. For each slide preparation process these controls have to be performed independently as there are significant differences in the false positive and negative rates according to the protocols used. The cut-off levels are usually defined as mean percentage of false positives in these controls plus three standard deviations. Positive controls on tumor specimens with known karyotype have to be performed to verify the specificity of each assay.

Evaluation of tumor specimen: For the evaluation of tumor specimen the same rules have to be applied as for the controls. At least 100 interphase nuclei and/or 15 metaphases have to be evaluated by the observer. Only good quality slides and clearly analysable cells can be counted. In cases with low percentages of aberrant cells (in the range of the thresholds) at least 200 nuclei have to been counted. Each result should be confirmed independently by a second observer.

Evaluation of FICTION-analyses: For controlling the specificity of the immunophenotyping process in FICTION, class-matched mouse antibodies, non-reactive with human tissue, should be applied on a control slide. A major problem of evaluating FICTION-slides is autofluorescence, which must not be confused with non-specific antibody binding. Typically, autofluorescence can be observed with all different filter sets appearing dull red (Cy3 filter), yellow (FITC filter) or white (AMCA filter), whereas true Cy3 fluorescence is not visible if the AMCA filter set is used.

For the evaluation of FICTION slides the same rules have to be observed as for conventional FISH. Controls have to be performed for each immunofluorescence as this may interfere with probe binding and signal evaluation.

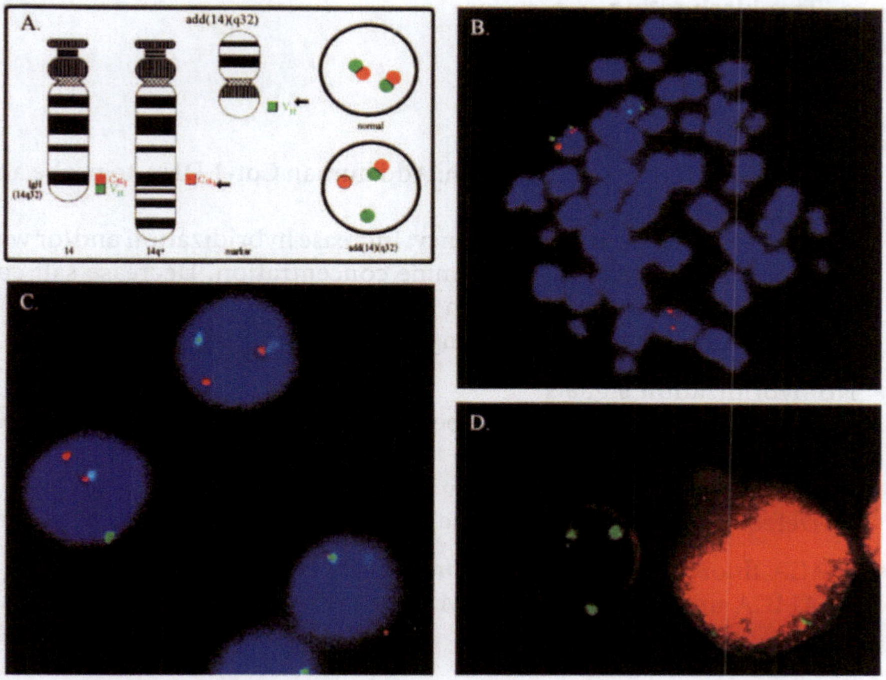

Fig. 2. FISH-assay for the detection of chromosomal translocations involving the IgH-locus in 14q32 in B-cell lymphomas (A-C) and FICTION study on breast carcinoma cell line MCF-7 (D). **A** Ideogram representing the expected localization of signals by two-color FISH in 14q+-negative (left) and -positive (right) metaphase chromosomes and interphase cells. Red color: Cosmid-probe Cos-Cα1 hybridizing to the constant region of the IgH-Locus centromeric of the typical breakpoint region in 14q32. Green color: Pooled V_H-cosmid-probes hybridizing to the variable region of the IgH-locus telomeric of the breakpoint region in 14q32. Interphase cells lacking a chromosomal aberration affecting the IgH-locus show two red-green hybrid signals. Cells carrying a translocation involving the typical breakpoint region of the IgH-locus display one red-green hybrid signal and each one isolated red and green signal indicative for the translocation. **B** Metaphase cell of a B-cell lymphoma with a translocation affecting the IgH-locus: The normal chromosome 14 without aberration of the IgH-locus is indicated by a colocalisation of each one red and one green signal in chromosome regions 14q32. The 14q+-marker only displays a real signal for the probe hybridizing to the constant region of the IgH-locus. The variable region indicated by a green signal is translocated to a small marker chromosome. **C** Interphase nuclei of a t(14;18)-positive B-cell lymphoma investigated with the 14q+-assay. The cells contain one red-green hybrid signal derived from a normal chromosome 14 and each one isolated red and green signal indicative for a translocation affecting the IgH-locus. **D.** FICTION study on the breast carcinoma cell line MCF-7. Combined immunophenotyping with monoclonal mouse anti-human estrogen receptor antibody visualized by Cy3 (red) and FISH with the YAC 19I11F containing the ESR gene visualized by FITC (green). One positive cell showed strong nuclear staining (red) while the adjacent cell was negative. Both cells showed three hybridization signals (green), indicating both ESR positive and negative cells to contain three copies of the ESR gene.

Troubleshooting

In situ hybridization

- Too much background:
  - Unspecific cross hybridization: Add human Cot-1 DNA to probe and/or coprecipitate.
  - Too low hybridization stringency: Increase hybridization and/or wash temperature. Increase formamide concentration. Decrease salt concentration. Prepare new wash solutions.
  - Signal amplification too strong: Omit some amplification steps.
- No hybridization signals
  - Defect DNA probe: Test probe on freshly prepared slides. If unsuccessful, try a new probe.
  - Paraformaldehyde solution too old (pH!): Replace.
  - Defect detection reagents: Use another detection cascade as control.
- AMCA fluorescence intensity consistently too low
  - AMCA unstable: Buy new AMCA reagents. Store AMCA reagents in 50% glycerol at -20°C.

Immunophenotyping

- No staining/high background staining
  - Inappropriate slide preservation: Use freshly prepared slides. Preserve slides correctly.
  - Defect monoclonal primary antibody: Test antibody on different freshly prepared slides.
  - Defect secondary antibodies: If unsatisfactory immunostaining occurs on different freshly prepared slides all antibodies have to be tested separately. Exchange all antibodies by new ones (only one per test).
- Immunostaining is sufficient prior to FISH but less intensive after FISH.
  - Paraformaldehyde solution is too old. Replace.

References

Autio K, Turunen O, Pentilla P, Eramaa E, de la Chapelle A, and Schroeder J, 1979, Cancer Genet Cytogenet, 1, 147

Czepulkowski BH, Bhatt B, and Rooney DE in : Rooney DE and Czepulkowski BH: Human Cytogenetics 1992, 11 ff.

Gahrton G, Zech L, Robert, KH, and Bird AG, 1979, New Engl J Med 301, 438

Garipidou V, and Secker-Walker LM, 1991, Cancer Genet Cytogenet 52, 107-111

Hagemeijer A, Smith EME, and D Bootsma, 1979, Cytogenet Cell Genet 23, 208-212

Lichter P, and Ward DC, 1990, Nature 345, 93-95

Metzke S, 1995, Leukaemia 9, 1413-1414

Pinkel D, Straume T, and Gray JW, 1986, Proc Natl Acad Sci USA 83, 2934-2938

Pirc-Danoewinata H, Onderka E, Porenta G, Kundi M, Nowotny H, Schlögl E, Heinz R, Kreiner G, and Marosi C, 1995, Cancer Genet Cytogenet, 80, 129-134

Ross FM, and Robert KH, 1982, Cancer Genet Cytogenet, 25, 109

Sahar E, and Latt SA, 1978, Proc Natl Acad Sci USA, 75, 5650-5654

Siebert R, and Weber-Matthiesen K, 1997, Histochem Cell Biol, 108, 391-402

Trent J, Crickard K, Gibas Z, Goodacre A, Pathak S, and Sandberg AA, 1986, Cancer Genet Cytogenet, 157ff

Yunis JJ, 1982, Cancer Genet Cytogenet 7, 43-50

Zech L, Godal T, Hammarström L, Mellstedt H, Smith CIE, Tötterman T, and Went M, 1986, Cancer Genet Cytogenet 21, 67-77

Weber-Matthiesen K, Winkemann M, Müller-Hermelink A, Schlegelberger B, and Grote W, 1992, J Histochem Cytochem, 40, 171-175

Weber-Matthiesen K In: Clark M (ed.) *In situ* hybridization. Chapman & Hall, Weinheim, 1996 pp 67-90

Williams DL, Look AT, Melvin SL, Roberson PK, Dahl G, Flake T, and Stass S, 1984, Cell, 36, 101-109

Chapter 10

Cytogenetics of Meiotic Cells

REINER JOHANNISSON

Introduction

It is evident that the cytogenetic approach has been enormously fruitful in contributing to our understanding of the meiotic process. Cytogenetics of meiotic cells represents a field that permits the light and electron microscopic visualisation of chromosomes during different stages of the reductional process. Chromosome studies have made very major contributions to our knowledge of chromosomally derived infertility and they begin to play a significant role in our efforts to detect and evaluate the consequences of exposure to environmental mutagens of male and female germ cells (Allen et al. 1987, Backer et al. 1991, Johannisson et al. 1994, 1996; Johannisson and Ocker 1997).

This chapter contains an overview of light and electron microscopic methods that are particular useful to the infertility researcher. Though female germ cells may be prepared with modified methods used for male germ cells there is, in my opinion, no use for the laboratory diagnostic. Thus, oocyte spreading techniques are not included in this comprehensive description of meiotic preparation.

Principle and applications The aim of meiotic studies is to evaluate the structure and behaviour of chromosomes during the meiotic process and to analyse different processes as segregation of chromosomes in meiosis I and II, chiasmata formation, pairing, and to determine chromosomal aberrations and mutations, respectively. These techniques should virtually never be used without prior investigation of mitotic chromosomes.

Reiner Johannisson, Institut für Pathologie, Ratzeburger Allee 160, Lübeck, 23538, Germany (*phone* +49-451-500-2722; *fax* +49-451-500-3328; *e-mail* reiner.johannisson@t-online.de)

The principle of visualisation of meiotic chromosomes is to disrupt the germ cell nuclei and to spread the chromosome sets. Two methods have made it possible to visualise and to analyse full meiotic chromosome sets: air-drying and surface spreading. The air-drying techniques gives information about chromosomes of prophase I, diakinesis-metaphase-I, metaphase-II, and also spermatogonia. The surface microspreading which visualise the synaptonemal complexes (SC) has been established for analysis of synapsis and desynapsis of prophase chromosomes. As pointed out by Moses (1980) chromosomal axes and the SC may be taken as paradigms of the chromosomes themselves. Three-dimensional reconstructions of SCs from electron micrographs of serial sections have the advantage of retaining spatial relationship of chromosomes in cells, but it is a slow procedere with practical limits on sample size. For methodology see Holm and Rasmussen (1977).

For diagnostic purposes in male infertility and subfertility meiotic cytogenetic methods have been widely used. Analysis of air dried meiotic chromosome sets gave information on altered chiasmata frequency which is discussed in relation to impaired fertility (eg Hultén et al. 1985, Goldman et al. 1992). In many studies chromosomally derived male infertility was analysed with SC spreadings showing the specific behaviour of translocation chromosomes (eg Chandley et al. 1986, Gabriel-Robez et al. 1988, Saadallah and Hultén 1985, Johannisson et al. 1983, 1993, Forejt et al. 1996). Specifically, the flowering intracytoplasmic sperm injection (ICSI) method, developed during the last few years (van Steirteghem 1994), sheds a new light on the application of the SC-microspreading techniques. The group of infertile males asking for this type of assisted reproduction technique (ART) includes 4.3% with abnormal karyotype showing a wide range of chromosomal anomalies (Mennicke et al. 1997). If sperm cannot be taken by micro-epididymal sperm aspiration (MESA), the testicular sperm extraction (TESE) procedure gives biopsies which may be used partly for analysis of chromosomal behaviour by SC spreading.

Preparation of meiotic cells may also be obtained from ejaculates as described in a few papers, however, there are many limitations. Nevertheless, procedures and protocols are given to test and to adjust it to the own laboratory.

Methods

Meiotic chromosome preparations can be made according to several rather simple basic methods in which only some steps are essential. However, like cookery, meiotic cytogenetics have many variations on each of these techniques. Thus, it is mainly my own experience which is expressed in this overview. It is often valuable to develop one's own modification on a laboratory animal before starting on the human being, from whom it is almost impossible to get biopsies twice. In my experience the techniques are useful in males if they work in the mouse, but minor modifications are often necessary. One distinct advantage of the chromosome preparations described in this chapter is that it is possible to run the technique in even a small laboratory.

Materials

Equipment

- Air drying technique: Light microscope, equipment of a normal histology laboratory.
- Surface spreading technique: Stereo microscope, light microscope, electron microscope, equipment of an electron microscopic laboratory.

Obtaining the material

Testicular biopsies are taken by open incision under local or complete anaesthesia from one or, preferably, from both testes. Tubes containing about 3 ml of eg Ham's F-10 medium should be taken to the operation theatre before the surgery begins so that the biopsies can be placed into the media immediately after removal; the choice of medium is a matter of experience and personal preference. One problem is that often only very small pieces of testicular material are available for histological analysis and meiotic studies. Though primary spermatocytes are normally numerous, in many cases a severe reduction of prophase germ cells impedes the meiotic technique in respect to an efficient analysis.

Subprotocol 1
Air-Drying Method

In the following a method is described, which is a slight modification of the air-drying technique described by Evans et al. (1964) and Ford and Evans (1969) which gives very high quality, permanent chromosome preparations.

Materials

Solutions

- Hypotonic solution: Dissolve 1g of sodium citrate (Merck, Darmstadt, Germany, order no. 6447) in 100 ml distilled water.

- Fixative: Mix 3 parts methanol (Merck, Darmstadt, Germany, order no. 6009) to 1 one part glacial acetic acid (Merck, Darmstadt, Germany, order no. 63). Add 2 drops of chloroform (Merck, Darmstadt, Germany, order no. 2445) per 100 ml of fixative.

Procedure

Preparation of the cell suspension

1. Transfer the biopsy in the tube to the laboratory as quickly as possible.
2. In the laboratory, transfer the biopsy to about 3 ml of the hypotonic solution contained in a Petri-dish and chop it gently. Hold the mass of tubules with fine curved forceps and thoroughly squeeze out their contents repeatedly with the aid of a round needle.
3. Transfer the cell suspension and the remains of the tubules into a test-tube and agitate the solution with a pipette to flush out the remaining cells from the tubules.
4. Leave the solution for 5 to 10 minutes to allow the tubules fragments to settle.
5. Finally transfer the supernatant fluid into a 15 ml centrifuge tube. The hypotonic treatment should last only 15 to 17 minutes. The complete procedure so far should not have exceeded 30 minutes. When adapting the methods with the mouse-model to your own laboratory, you may use 2.2% sodium citrate.

Fixation

1. Centrifuge the cell suspension obtained with 40g for 10 minutes. This generally leaves the majority of sperm in suspension and sediments the larger cells, including the spermatocytes.
2. Discard most of the supernatant fluid and add about 1 ml of fixative. Remove the supernatant fixative and add 3 ml fresh fixative. Resuspend the cells by flicking the tube gently with thumb or forefinger to ensure thorough mixing of the solutions.
3. Leave the cells in the fixative for 10 minutes at room temperature.
4. Centrifuge the cells at 500 rpm for 10 minutes and discard the supernatant leaving only a small volume in the tubule.
5. Resuspend the cells in the remainder by flicking the tube with thumb or forefinger and add about 1 ml fresh fixative. Again flick the tube to ensure thorough mixing of the solution.

Air-drying

1. Clean slides in absolute ethanol.
2. Store for 30 minutes in a freezer at -10 °C.
3. Moisturise the slide by breathing.
4. Take up an aliquot of the cell suspension into a small pipette. Allow 1 or 2 droplets to fall from about 10 cm on each slide. The following step is the most critical point during the whole procedure. If the slide is thoroughly clean and the fixation has been satisfactory, the droplet will expand evenly, reach a maximum and then begin to retract. When Newton's rings appear blow gently onto the slide. This will rapidly evaporate the fixative, leaving the cells "air-dried" on the slide.

Note: If the chromosome sets show under the microscope a condensed appearance, blow later. If the chromosomes appear fuzzy, blow a little bit earlier.

5. The slide is allowed to dry in air at room temperature.
6. Store slides in a freezer or refrigerator (for months).

Chromosome staining techniques

A number of conventional methods used for staining of somatic chromosomes can be used for meiotic chromosomes without modification. Modern chromosome banding methods were developed initially for application in clinical cytogenetics (see Chapter 2) and G-Banding is still the primary method applied to a chromosome sample in a clinical cytogenetics laboratory. C-banding is valuable for drawing attention to the heterochromatin. C-banding stains centromeric or constitutive heterochromatin, hence the name, and allows a good analysis of chromosome orientation. G- and C-banding may be performed according to Gallimore and Richardson (1973) and Sumner (1972), respectively. Microscopy of banded chromosomes requires a high quality microscope with the highest solution.

Subprotocol 2
Surface Spreading Method Using Light Microscopy

For spreading and staining of pachytene chromosomes, techniques were developed based on the work of Counce and Meyer (1979) in Drosophila. Adaptations for human material have been worked out in different groups (eg Johannisson et. al. 1983, Hultén et al. 1985, Chandley 1991, Gabriel-Robez 1986). The SC spreading represents an elegant tool to analyse meiotic chromosomal behaviour (Figures 1 and 2). Studies of SC spreads can provide much information from meiotic chromosomes unobtainable from air drying techniques, including a better identification of chromosome rearrangements. The basic principle of pachytene spreading is the visualising of the synaptonemal complex, a proteinecous structure. For a detailled description of this meiotic structure see eg the review of von Wettstein et al. (1984).

Again, it is helpful to work out the technique carefully on a laboratory animal before starting on human material because of the unique availability of the patient's material. The mouse represents an experimental system appropriate to practise the micro spreadings, since the technique is the same as that used for human material. Normally, biopsies are undertaken for histological analysis only. Sometimes a follow up analysis of the proband's somatic karyotype reveals a chromosomal aberration suspected as being the cause of impaired spermatogenesis. Only in very rare cases the patients agree to a second testicular biopsy for meiotic chromosome preparations (Johannisson et al. 1993). Thus, I highly recommend the analysis of the somatic karyotype before the operative procedure.

Though by our experience light microscopic analyses of pachytene spreads are practicable and effective (see eg Johannisson et al. 1994), the exact knowledge of the ultrastructural configuration is a precondition since

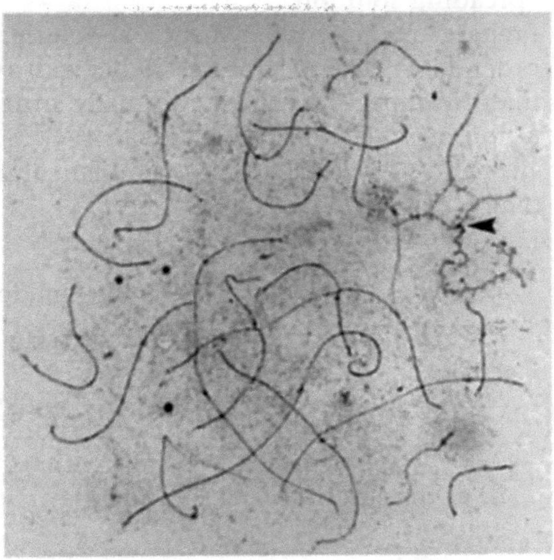

Fig. 1. Light microscopy. Silver staining. 46,XY,t(14;21)(q13;p13) karyotype. Complete set of chromosomes. Arrow head indicates association of the quadrivalent with the X-chromosome.

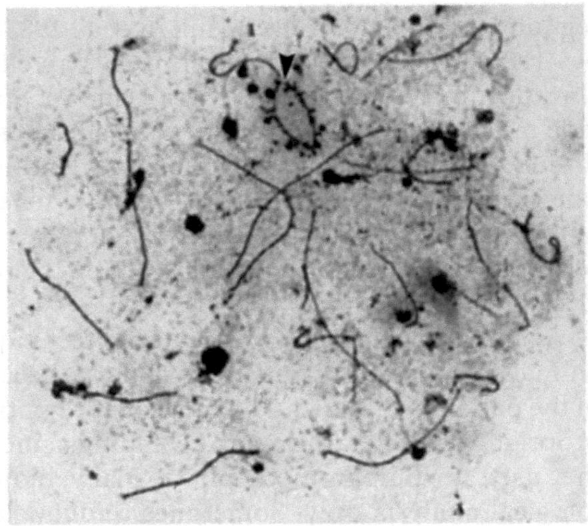

Fig. 2. Light microscopy. Silver staining. 46,XY,-fra(X)(q) karyotype. Complete set of chromosomes. Sex chromosomal bivalent shows secondary association of Xq and Yq (arrow head).

the evaluation of complex chromosomal configurations is limited in light microscopy (Johannisson and Ocker, 1997). For both methods, the testicular material is obtained and prepared in the same way. Furthermore, most of the solutions, the material and the procedures are essentially the same.

Materials

Solutions

- Spreading solution
  Prepare an aqueous 0.2 M sucrose-solution (Sigma, St. Louis, MO, USA, order no. S-9378) in aqua dest. Filter with a 0.22μm "Millex GS" Millipore filter (Millipore, Eschborn, Germany, order no. SLGSD 025 BS). The solution may be stored for weeks at 4 °C. Adjust the solution to pH 10.0, 10.5, and 11.0 with 0.01 borate buffer solution (Merck, Darmstadt, Germany, pH 9,22).
- Fixative
  - Dissolve 3.4 g sucrose in about 90 ml double distilled water, add 4.0g paraformaldehyde (Serva, Heidelberg, Germany, order no. 31628) and make up to 100ml with water.
  - Heat slowly to 60-80 °C while stirring, and add about 6 drops 1.0 N NaOH. Continue to stir until solution is clear.
  - Allow to cool, adjust to pH 8.5 with 0.01 borate buffer solution, and filter the solution. The solution may be kept a few days under refrigeration.

- Detergent solution
  Kodak Photoflo is a widely used solution in pachytene spreading, however, to my own experience it may easily precipitate. Best results were obtained with a Joy detergent solution (dish washing liquid: Procter & Gamble, Cincinnati, Ohio, USA, order no. 0840332-1) - which was introduced by Miller and Beatty (1969) for spreading techniques. The Joy, as a surface wetting agent, promotes even-drying.
  Stir 0.4 ml Joy in double distilled water and adjust to pH = 8.5 with borate buffer. Use only fresh solutions.
- Silver nitrate solution
  Prepare with silver nitrate ($AgNO_3$ - Merck, Darmstadt, Germany, order no. 1512) a 33% aqueous solution. Filter the solution using a 0.22µm "Millex GS" Millipore filter. The solution should be stored in a brown bottle in the refrigerator.

Note: You should always wear gloves when silver staining.

- Colloidal developer
  - Dissolve 2 g gelatine powder (Merck, Darmstadt, Germany, order no. 4078) in 100 ml aqua dest, warm up a little bit and add 1 ml formic acid pure (Merck, Darmstadt, Germany, order no. 264). The solution may be stored in the refrigerator for about two weeks.
  - Immediately before use heat to about 40 °C.

Equipment

- Petri dish
  A small plastic petri dish is used for droplet spreading. The type Nunclon ®Delta (Nunc, Roskilde, Denmark, order no. 1-50288) is well suited due to its hydrophily which forms a well rounded droplet. Use the top of the upper part as its hydrophily is more effective than the lower one.
- Plastic coated slides:
  - Dissolve 0.4g of small pieces of FALCON "Optilux"-petri dishes (Becton Dickinson, Heidelberg, Germany, order no. 3003) overnight in 100 ml chloroform while gently stirring. The next day filter three times with paper filters. The solution may be used for months.
  - Clean a slide with 96% ethanol and afterwards clean scrupulously with lens paper. Dip the clean slide vertically into the solution and lift it out carefully. The slower the lifting the thinner the plastic film. Leave the slide in a vertical position to dry. The use of a special apparatus (Johannisson et al. 1994) modified according to a technique of Kölbel (1976) allows easy and reproductive handling.
  - Store the plastic coated slides in a dustfree container.

- Embryoglass:
  An embryoglass should be carefully cleaned in a detergent and ethanol and stored in distilled water to avoid dust and specifically a fatty surface. Prior to adding the cells suspension, sweep the surface clean with lens paper.
- Pipette:
  For applying the testicular suspension on the spreading solution hypophase, prepare a micro pipette from a 50μl blood pipette (Brand, Wertheim, Germany, order no. 708733) by drawing carefully over a gas-jet. The diameter of the outlet should be about 0.2 mm. Smooth the tip of the pipette on a small grindstone.
- Syringe:
  Use the pipette in combination with a AGLA-micrometer all-glass syringe (Wellcome Reagents Ltd., England). The advantage of this syringe is eg the possibility to adjust exactly the size of the droplet of the testicular suspension.

Procedure

Sample transportation and/or storage

In contrast to the air-drying procedure, the testicular biopsies obtained for SC-spreadings can be used as fresh or frozen samples.

- Fresh material may be used after transferring directly to the cytogenetic laboratory, or it may be sent by mail or air freight (see Goldman et al. 1992) in cell culture medium placed in small glass tubes to the laboratory, if the operation theatre is located in a different place from the laboratory. The SCs remain stable for a relatively long period of time. The tissue can thus be successfully transported in cell culture medium for a period of 48 hours.
- The material can easily be frozen directly in the laboratory in liquid nitrogen and stored at -80 °C for a long time, even for months. Sudman (1989) and our group (Metzler 1995) have shown that cryopreservation of meiotic germ cells does not influence the frequency or type of SC abnormalities. Cryopreservation may be carried out with or without freezing media. Cryopreservation of patient's testicular material without freezing media is routinely used in our laboratory. It represents a very simple and rapid method. In our own experience cryopreserved ma-

terial may be transported from one laboratory to another on dry ice and used for spreadings, even after renewed storage in a freezer or liquid nitrogen.

Cryopreservation without freezing medium

1. Put pieces of the testicular material obtained in cell culture medium with long, fine forceps into a small container (about 20 ml) with liquid nitrogen for about one minute.
2. Precool small pieces of aluminium foil (about 2 cm x 2 cm) in liquid nitrogen. Wrap the frozen testicular pieces carefully in the aluminium foil using long, solid forceps.
3. Place the wrapped material again for a minimum of one minute in liquid nitrogen and put it in small plastic tubes. Transfer the tube to an ultra-low freezer for storage at -80 °C or to a liquid nitrogen cell storage freezer for storage in the liquid phase (-190 °C).
4. The material may be used for spreading even after months.

Cryopreservation with freezing medium

1. Mince the testicular material in a few drops of freezing medium (1 : 9 either glycerol : Ham's F-10 cell culture medium or DMSO : Ham's F-10)
2. Dilute with additional freezing solution (5ml/1g testicular material)
3. Place the solution in cryogenic tubes and freeze in liquid nitrogen or in a freezer at - 80 °C.
4. The material may be used for SC analysis after some days or even after months.

Cryopreservation for TESE-ICSI: the specific protocol for freezing biopsies for the TESE-ICSI procedure does not impede a successful SC-spreading procedure (Johannisson et al. unpubl.).

Thawing cryogenically preserved cells

Remove a sample of testicular tissue from the freezer or liquid nitrogen cell storage freezer and place it in an embryo glass (Hecht, Sondheim/Rhön, Germany, order no. 2021) at room temperature. Note: the testicular material tends to stick during the spreading procedure. This problem can be easily overcome by using different pH's of the spreading solution (s. below).

Preparation of the testicular cell suspension

As mentioned above the material may be used fresh, frozen and thawed, or transported by mail etc. Any material, fresh or frozen, should be placed in cell culture medium (Ham's F-10) at room temperature. In my experience the type of medium has no significant effect on the spreading procedure.

1. The material should be sliced off with scissors and chopped up with two fine forceps. Large pieces of the tubules are removed and the remaining suspensions of cells taken up in a 1 ml syringe without needle.
2. Make up the volume in the syringe to 0.5 to 1.0 ml with cell culture medium and allow the syringe to stand upright for one minute. Thus, the large pieces of debris settle into the tip. Then, the debris is discarded and the remaining suspension expressed into a glass centrifuge tube (10 ml).
3. At room temperature, centrifuge the suspension for about 5 min with about 70g.
4. The supernatant is carefully drawn off, except for a small amount twice the volume of the pellet, in which the pellet is resuspended. The density of the suspension is important for a good quality of spreadings. If the suspension is too concentrated, cell-aggregates tend to sink when layered on the top of the hypophase; if the concentration is too low the amount of spreaded germ cell is not sufficient for analysis. The concentration can easily be checked by flicking the tube gently with the forefinger. If the suspension splashes like pure cell culture medium the concentration is too low, if it move very slowly the concentration is too high. The colour is like an asparagus cream soup.
5. Keep the solution on ice until processing for spreading.

Spreading procedures

For the spreading procedure we routinely use a solid black embryoglass obtaining the spreading solution or a Petri dish on which small droplets of spreading solution are placed. Both variations do have their own advantages. A specific advantage of the embryoglass is that the spreading process on the hypophase can easily be observed and controlled. The droplet method offers the advantage of an easy handling system. For both procedures work with a stereo microscope (WILD M5) with glass fibre reflected light in a ring-like arrangement around the objective (Leica). An even simpler method is spreading directly on a plastic coated slide.

Slide method

Use for this technique slides which are coated with Optilux (see Subprotocol 2, Materials). The spreading procedure takes place on the slides.

1. Place the slides horizontally and transfer one drop of the filtered spreading solution with a Pasteur pipette to the centre of the slide.

2. To this drop, add one drop of the testicular cell suspension with the AGLA syringe. Mix the two drops carefully together with a fine, rounded glass stick. Care must be taken not to disrupt the film while mixing the solutions.
   a) High concentration of germ cells: leave the mixture to spread for about 10 minutes. Add 5 drops of fixative to the slide. Spread the fixative and the mixture over the slide using a glass stick. Leave the slides lying on the bench for about 1h.
   b) Low concentration of germ cells: Leave the mixture to spread for about 60 minutes. Add 5 drops of fixative to the slide. Leave the mixture for about 10 minutes on the slide.

3. Wash the slides by placing them in a Coplin jar containing 0.4% solution of Joy for a minimum of 30 sec. Transfer to a second and a third Coplin jar. After five slides have been washed, discard the Joy as fixative will accumulate in it.

4. Remove the slides from the Coplin jar. Leave upright to air-dry at room temperature.

Droplet-spreading

1. Place four droplets of spreading solution with a 50µl BRAND pipette on the Petri dish.

2. Apply a small droplet of testicular suspension laterally to the spreading solution using the AGLA micrometer syringe.

3. After the spreading process carefully touch the droplet with an Optilux coated slide.

4. Place the slides in a Coplin jar with fixative for about ten minutes.

5. Transfer the slides into a Coplin jar containing Joy solution. Follow the procedure as described for the slide-spreading.

Black embryoglass spreading

1. Fill the embryoglass with the spreading solution and adjust the liquid level so that it is slightly concave. First use a spreading solution with a pH = 10. The selection of the pH value depends on the specific testicular probe. A low pH value gives spreadings showing nuclei with non-sufficient dispersion of chromosomes. A high pH value affects the nuclei by "overspreading" showing incomplete SCs sets or nearly "empty" grids.

2. A small drop of the testicular cell suspension is applied with the AGLA-micrometer syringe. The best size is 1.0 to 1.5 mm in diameter. The hypophase can be reached under different angles. The clean, mirror-like surface allows an easy application of the syringe to the surface. Immediately after contact to the solution the cell suspension spreads like a flash on the surface. The cells are distributed evenly over the hypophase, and hypotonic shock ruptures the plasma membrane, dispersing the cytoplasm and affecting the nuclei. Wait for about 1 minute for the spreading to settle.
3. Touch an Optilux coated slide carefully on the convex surface of the spreading solution and remove it slowly.
4. Renew the spreading solution or clean with lens paper while surfing over the surface and add some fresh solution.
5. Transfer the slides to a Coplin jar with fixative and follow the procedure as described above.

Staining

Slides may be stained with ammoniacal silver (Goodpasture and Bloom, 1975) or AgNO3/gelatine developer (Howell et al., 1980). The latter is by far the simpler and more reliable method. Though silver staining has been applied for many years it nevertheless gives different results every time. This rapid method is described as follows. One advantage of this staining procedure is that it lasts not more than one hour.

Silver staining

1. Flood the spreading area on slides with two drops of gelatine solution and add four drops of silver nitrate solution. Thoroughly mix both solutions. Then top with a coverslip.
2. Transfer slides to a 50 °C hot plate for 2-3 minutes.
3. Rinse the slides three times with distilled water. Dry the slide.
4. Repeat steps 1-3. When heating, again control intensity of staining with white paper under the slide. Stop the staining process when spreadings have developed a golden brown colour. Rinse the slides well in distilled water. Examine under microscope and repeat staining if necessary.

Note: Avoid metallic forceps, use plastic forceps.

Troubleshooting

Preparation of the testicular cell suspension

If only a little piece of material is available prepare the germ cells by a time-consuming, however, effective method.

1. The mass of tubules is gently stripped out into a drop of medium on a wax plate or a small glass plate.
2. Small pieces of tubules may be isolated from the mass of tubules. Then, the cells are squeezed out carefully with fine curved forceps. The content from some tubule segments are sucked with a small syringe. For spreading procedure see below. This technique allows a very fine presentation of SCs.

Staining

Good staining is preferably judged under a 63 or 100 oil objective. To do so under lower magnification requires experience. If the staining is correct, the preparation is made permanent in eg Eukitt with a coverslip top after drying for a couple of hours or days.

Note: Though staining protocols have been proved since many years the staining intensity should be individually controlled for each slide.

Subprotocol 3
Surface Spreading Using Electron Microscopy

Choice of techniques: SC-Visualisation with silver-nitrate vs. phosphotungstic acid. The visualising of the SCs for electron microscopy may be undertaken by two different staining methods, namely silver-staining and phosphotungstic acid (PTA) staining, which implies variations of the pachytene microspreading procedure. Silver-staining is a simple and rapid method for visualising the SCs, however only lateral elements become visible (eg Chandley et al. 1986). One distinct advantage of PTA-staining is to identify centromeres and central elements (see Schmid et al. 1987 and Figures 3 to 9).

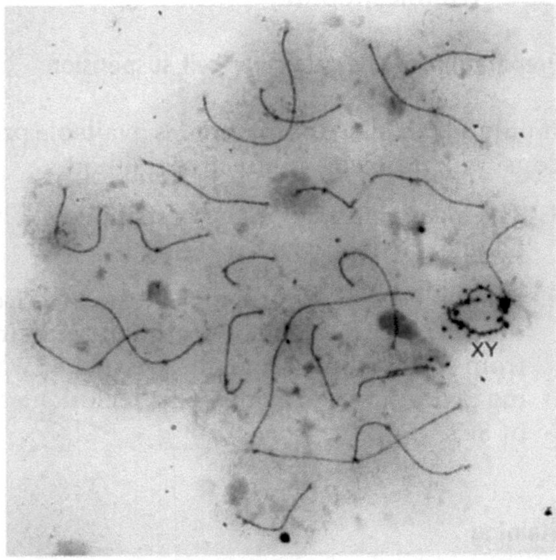

Fig. 3. SC spreading of a human primary spermatocyte. Electron micrograph. PTA-staining. Complete set of 22 autosomal bivalents and one pair of sex chromosomes.

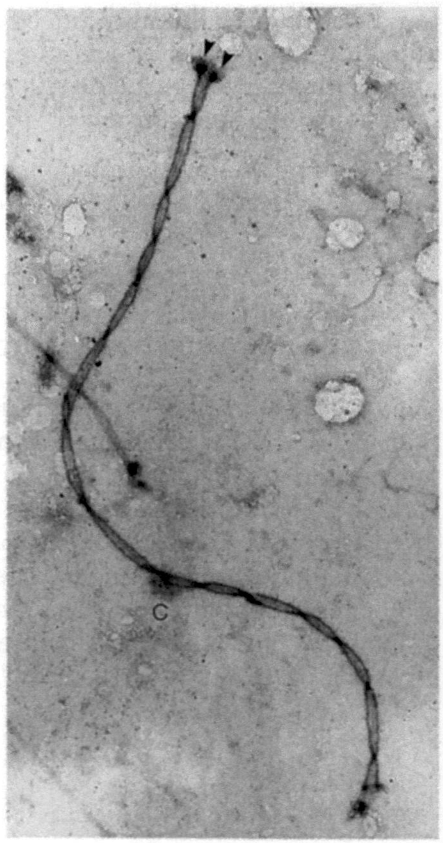

Fig. 4. Electron micrograph of an autosomal bivalent showing two lateral elements, one central element, the centromere (C), and telomeric knobs (arrow heads). PTA-staining.

Fig. 5. Electron micrograph. XY-pair with characteristic associated granules. Pachytene stage II. PTA-staining.

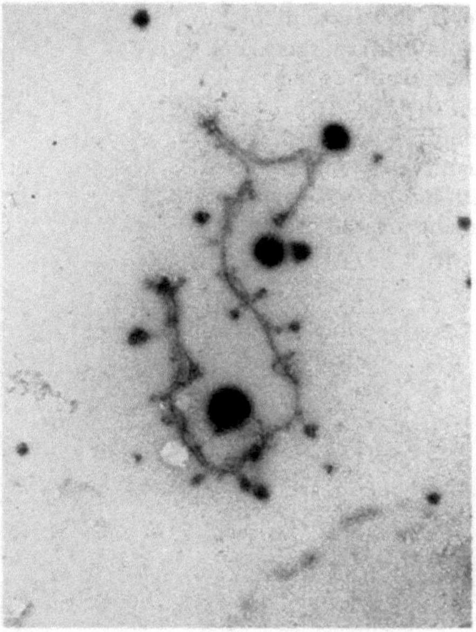

Fig. 6. Electron micrograph. Detail of a first meiotic prophase by EM surface spreading. Chromosome no. 9 with prominent 9qh+ (arrow head). PTA-staining.

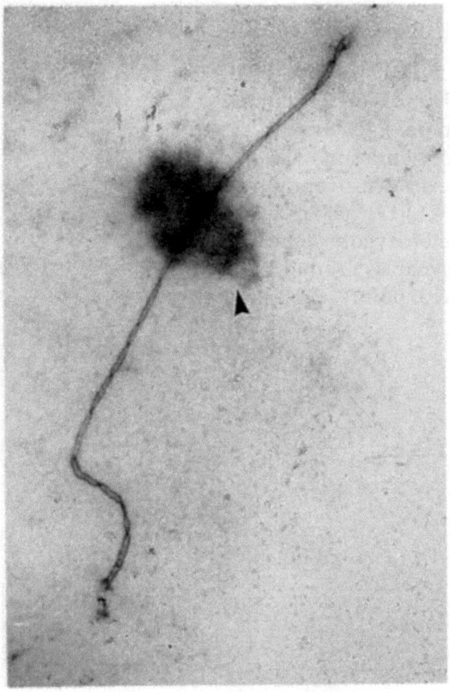

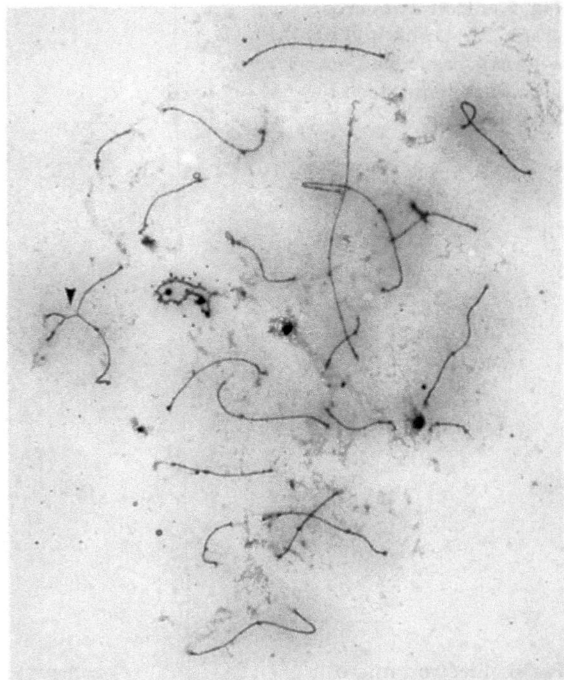

Fig. 7. Electron micrograph. Primary spermatocyte spreading from a t(9;15)(p22;q15) heterozygote showing a quadrivalent (arrow head) with nonhomologous pairing around breakpoints. PTA-staining

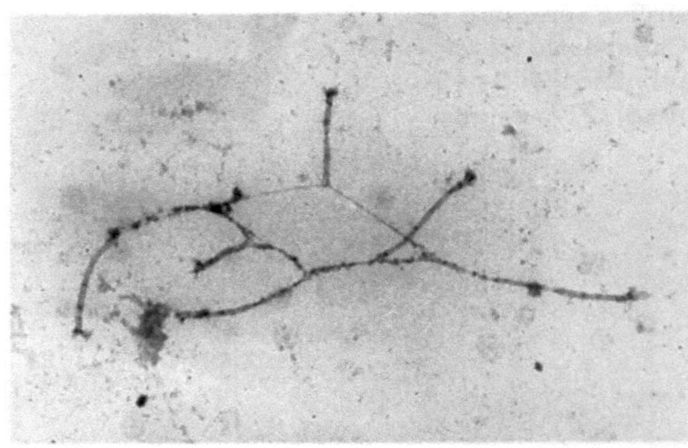

Fig. 8. Electron micrograph. Pachytene configuration of a hexavalent from a t(9;12;13)(q22;q22;q32) heterozygote showing asynapsis around the breakpoints. PTA-staining.

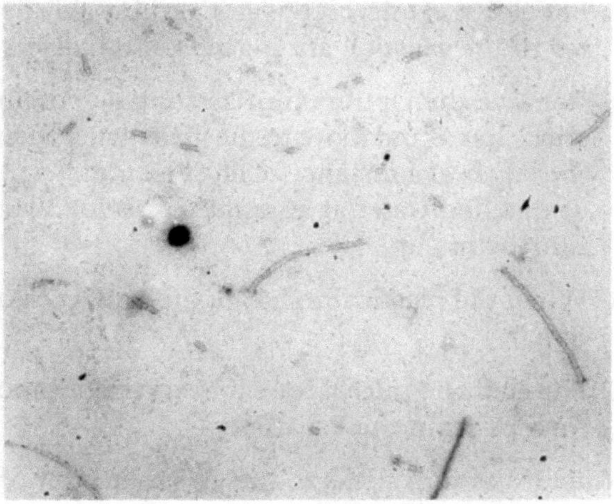

Fig. 9. Electron micrograph. Fragmentation of autosomal SCs. Normal karyotype. PTA-staining.

Materials

Equipment

- Electron microscopic grids: Use hexagonal 100-mesh copper grids on one side coated with palladium (AGAR, type hexagonal 100 mesh Cu/Pd; Plannet, Wetzlar, order no. G 2410PD)

Solutions

- For PTA-staining: Prepare a 2% aqueous PTA solution (Merck, Darmstadt, Germany, order no. 583) with double distilled water immediately before use. Filter through a 0.22µm "Millex GS" Millipore filter". Immediately before staining mix 95% ethanol and PTA-solution 3:1.
- Spreading solution, fixative, and detergent solution are essentially the same as described for light microscopy (see Subprotocol 2).

Procedure

Silver-staining

For this method, exactly the same procedures as described for light microscopy can be used, including the silver staining. Do not make permanent slides. To check the sample area for quality of spreading, light microscopy of unmounted slides can be made with a Zeiss 16x water immersion planneofluar for overviews or for details with a Zeiss 63x water immersion planneofluar, with a correction collar adjusted for optimum resolution.

1. For electron microscopy grids, shiny side down (this surface adheres better to the plastic), are placed on the dried sample area.
2. Remove the Optilux film as soon as possible from the slide as the film sticks more and more to the glass when ageing. Cut the plastic-film along the edges at a distance of about 2 mm. Dip the slide carefully into a glass of distilled water at an angle of about 30°. Watch the surface when the film swims up.
3. When the plastic film has floated off, pick up on a piece of Parafilm and dry.
4. For electron microscopic observation remove the grids slowly with fine forceps from the Parafilm.

Note: If the film sticks to the glass slide, mordant the glass, use a 0.1% hydrofluoric acid (Merck, Darmstadt, Germany, order no. 329) instead of plain water according to Kölbel (1976).

PTA-staining

Preparing the supporting film

Grids must be coated with plastic film to pick up the spread germ cell from the spreading solution. Different chemicals are available for producing support plastic films. Pioloform (Plannet, Wetzlar, order no. R1275) and Formvar (Plannet, Wetzlar, Germany, order no. R 1202) are widely used for films in electron microscopy. For SC spreadings, the grids should be carbon-coated. Carbon-coating is a necessary procedure to stabilise the plastic film with a thin fine-granular carbon film before glow discharging. It is necessary to glow discharge the grids to make them hydrophilic: this process is always a problematic step which decides on the success of the spreading event. Though the procedure gives good results, these problems and time-consuming steps make the procedure inconvenient for routine spreading. The advantage of Butvar-98 (Agar Scientific Ltd, Stansted, Essex, UK., distributed by Plannet, Wetzlar, Germany, order no. R1276) is its excellent hydrophily - without carbon-coating and glow discharging - , however staining has to be performed with aqueous PTA-solution (10%) as ethanolic solutions disrupt the film. It may be used immediately after preparation. The disadvantage of this type of staining is a "rough" visualisation of the SCs. I used Butvar-98 only for problematic cases.

Glow discharging and carbon-coating may be overcome by using Optilux which was introduced to the electron microscopy by Felluga and Martinucci (1976) and for meiotic preparation by Moses (1981). The advantage of Optilux is its easy handling and reliability in respect to its hydrophily. Its disadvantage is a sometimes poorer quality compared with the carbon-coating/glow-discharging techniques. Grids coated with this plastic film provide satisfactory substrate without additional treatment. However, choice of plastic is important. The polystyrene of Falcon brand Optilux petri dishes is the only suitable source that is encountered. Ordinary plastic dishes are not suitable.

Coating of grids with plastic film

1. Prepare slides coated with Optilux exactly as described for light microscopy in Subprotocol 2.
2. Cut the Optilux-film along the edges at a distance of about 2 mm with eg a scalpel. Dip carefully into a glass of water at an angle of about 30°. Watch the surface illuminated with a lamp when the film swims up. The interference colour should be silvery or gray. Place the grids with fine forceps with their surface downwards on the floating plastic film. Use hexagonal 100-mesh copper grids on one side coated with palladium (AGAR, type hexagonal 100 mesh Cu/Pd; Plannet, Wetzlar, order no. G 2410PD). This surface adheres better to the plastic. Take a small piece of filter paper or Parafilm®(American Can Company) and cover it up. Take it out and let them dry in a Petri dish.
3. Use the grids after a few days. The older the plastic film the better the stability.

Droplet spreading

1. Spread the testicular material on a small droplet exactly as described for light microscopy.
2. Place 3 to 4 grids plastic-coated side down on the surface of the droplet.

Black embryoglass spreading

1. Fill the embryoglass completely with the spreading solution and adjust the liquid level so that it is slightly *convex*.
2. Follow exactly the procedure described for light microscopy.
3. Touch approximately 10 clean grids at one time to different regions of the surface of the spread, preferably in the centre of the spreading event.

(When the grids are transferred to the fixative, renew the spreading solution or clean with lens paper while surfing over the surface.)

Fixation

Transfer the grids from the droplets or the black embryoglass to the fixative in lucid embryoglasses. Float the grids on the surface of the fixative for 10 min. Move the grids gently two or three times with a fine needle on the surface of the solution.

Washing The fixation leaves the nucleoprotein sufficiently non-rigid for the nuclei to be flattened by surface tension upon drying.

1. Transfer the grids from the fixative with eyelets to a lucid embryoglass containing the Joy solution. Wash the grids by placing them on the surface of the solution for 1 minute. Remove the grids from the surface, transfer and gently rinse again in a second embryoglass with a fresh solution. Repeat this step using a third embryoglass.
2. After 10 grids have been washed, discard the Joy solutions as fixative will accumulate in them.

Drying Transfer the grids with scrupulously cleaned, curved fine forceps to a Petri dish and store on filter paper or Parafilm until staining. Doing so, place with one hand a small piece of filter paper between the grips of forceps soaking up the surplus liquid while opening the grips to lay down the grids. Staining may be followed after some minutes or later, even weeks.

Staining with phosphotungstic acid

1. Place the grids with curved forceps with the surface downwards on the PTA-solution in lucid embryoglasses.
2. Stain for 2-3 minutes. Transfer with eyelets to the next embryoglass.
3. Wash 3 times in 95% ethanol by transferring to new embryoglasses.
4. Transport the grids with a platinum or silver eyelet used for microbiological inoculations slightly larger in diameter than the 3 mm grids.
5. Collect the grids on a layer of filter paper or Parafilm in a Petri dish which provides a convenient way to store the grids.

Troubleshooting

Care must be taken during the whole procedure to minimise the transfer of contaminants to the grids. All problems with contamination may be easily overcome with extreme cleanliness of all glass wares, forceps, and solutions. It cannot be overemphasised that cleanliness is very important.

Microscopy Micrographs from a transmission electron microscopy are suitable for initial magnification from 500 to 2000 x at 60 KV. Frequently, spreadings without staining that were observed immediately after preparations under electron micrograph, revealed sufficient contrast for micrographs.

Subprotocol 4
Ejaculate

As meiotic studies of germinal cells obtained from testicular biopsies involve a surgical procedure, Sperling and Kaden (1971) draw attention to the observation that samples of semen in ejaculates from normal males contain not only mature spermatozoa, but also a varying proportion of immature germ cells. The authors found that in smear preparations of the ejaculate of patients with a normal sperm count nearly 3% of the cells consisted of spermatogonia, spermatocytes, and spermatids, the maximum percentage being 5%. As the sperm count decreases the percentage of immature germ cells increases up to a maximum of about 40% (Vasterling 1960, cited in Sperling and Kaden, 1971). Based on the findings and observations of Sperling and Kaden (1971) modified procedures are described for the study of meiotic cells in the ejaculate (Templado et al. 1986; Vidal et al. 1986).

Materials

Obtaining the material

Collect the semen samples in large plastic jars and leave the semen sample at room temperature for 1h. After liquefaction the semen sample may be divided in two parts for both the air drying method and the SC-procedure.

Solutions

Air-drying:

- Hypotonic solution: Dissolve 0.075g potassium chloride (Merck, Darmstadt, Germany, order no. 4933) in 100 ml distilled water.
- Fixative: Mix 3 parts methanol to 1 part acetic acid.
- Colcemid solution: Dissolve 1mcg Colcemid (Demecolcine; Sigma/Aldrich Chemie, Deisenhofen, Germany, order no. D 6279) in 1 ml distilled water.

Surface spreading:

- Isotonic solution: Dissolve 0.9g of sodium chloride (Merck, Darmstadt, Germany, order no. 6404) in 100 ml distilled water.

Procedure

Air-drying

1. Add 0.25 ml of the Colcemid solution to the semen sample.
2. Incubate at 37 °C for 30 min and add 1:1 (vol:vol) of hypotonic solution. Incubate at 37 °C for 30 min again.

3. Place the sample in a conic centrifuge and allow to sediment for 30 min at room temperature.
4. Remove the supernatant to a centrifuge tube and centrifuge at 800 prm for about 10 min. Resuspend the pellet.
5. Fix the solution for 30 min at 4 °C. Wash the material several times in fixative until clean.
6. Follow the procedure of the Subprotocol 1.

Surface spreading

1. Add 10-15 ml of the isotonic solution to the semen sample.
2. Leave the semen sample for 18-20 hrs at 37 °C.
3. Centrifuge at 600g for 10 min.
4. Resuspend the pellet in the isotonic solution. Repeat this step 4-5 times.
5. At the last wash resuspend the pellet only in a small volume of the isotonic solution.
6. Proceed following the protocols for light and electron microscopy in Subprotocols 2 and 3.

Results

The air-dried technique was successfully applied in about 46% of all cases according to Templado et al. (1986). The frequency of SC preparations is described from Vidal et al. (1986) as useful for diagnostic tool in about 23% of the cases. The procedure to obtain meiotic information from ejaculates is reported by the authors as acomplementary method to the meiotic studies in testicular biopsies.

References

Allen JW, DeWeese GK, Gibson JB, Poormann PA, Moses MJ (1987) Synaptonemal complex damage as a measure of chemical mutagen effects on mammalian germ cells. Mutation Res 190:19-24

Backer LC, Sontag MR, Allen JW (1991) Stage-specific damage to synaptonemal complexes and metaphase chromosomes induced by X rays in male mouse germ cells. Radiat Res 125:187-196

Chandley AC, Speed RM, McBeath S, Hargreave TB (1986) A human 9;20 reciprocal translocation associated with male infertility analyzed at prophase and metaphase I of meiosis. Cytogenet Cell Genet 41:145-153

Counce SJ, Meyer GF (1973) Differentiation of the synaptonemal complex and the kinetochore in Locusta spermatocytes studied by whole mount electron microscopy. Chromosoma 44: 231-253

Evans EP, Breckon G, Ford CE (1964) An air drying method for meiotic preparations from mammalian testes. Cytogenetics 3:289-294

Felluga B, Martinucci GB (1976) A simple method for karyotyping by transmission electron microscopy. J Submicr Cytol 8:347-352

Forejt J (1996) Hybrid sterility in the mouse. TIG 12:412-417

Gabriel-Robez, O., C. Ratomponirina, B. Dutrillaux, F. Carre-Pigeon, and Y. Rumpler (1986) Meiotic association between the XY chromosomes and the autosomal quadrivalent of a reciprocal translocation in two infertile men, 46,XY,t(19;22) and 46,XY,t(17;21). Cytogenet Cell Genet 43:154-160

Goldman ASH, Martin RH, Johannisson R, Gould CP, Davison EV, Emslie JE, Burn J, Hultén MA (1992) Meiotic and sperm chromosome analysis in a male carrier of an inverted insertion (3;10)(q13.2;p14p13). J Med Genet 29:460-464

Holm PB, Rasmussen SW (1977) Human meiosis I. The human pachytene karyotype analyzed by three dimensional reconstruction of the synaptonemal complex. Carlsberg Res Commun 42:283-323

Howell WM, Black DA (1980) Controlled silver staining of nucleolus organizer regions with a protective colloidal developer: a 1 step method. Experientia 36:1014-1015

Hultén M, Saadallah N, Wallace BMN, Cockburn DJ (1985) Meiotic investigation of aneuploidy in the human. In: Dellarco VL, Voytec PE, Hollander A (eds). Aneuploidy. Etiology and mechanisms. Plenum Press, New York London, pp 75-90

Johannisson R, Gropp A, Winking H, Coerdt W, Rehder H, Schwinger E (1983) Down's syndrome in the male. Reproductive pathology and meiotic studies. Hum Genet 63:132-138

Johannisson R, Löhrs U, Schwinger E, Wolff HH (1987) Two different XY-associations and impairment of fertility in men. Cytogenet Cell Genet 45:222-230

Johannisson R, Löhrs U, Passarge E (1988) Pachytene analysis in males heterozygous for a familial translocation (9;12;13)(q22;q22;q32) ascertained through a child with partial trisomy 9. Cytogenet Cell Genet 47:160-166

Johannisson R, Schwinger E, Wolff HH, vom Ende V, Löhrs U (1993) The effect of 13;14 Robertsonian translocations on germ-cell differentiation in infertile males. Cytogenet Cell Genet 63:151-155

Johannisson R, Mörmel R, Brandenburg B (1994) Synaptonemal complex damage in fetal mouse oocytes induced by ionising irradiation. Mutation Res 311:319-328

Johannisson R, Ocker H, Mörmel R (1996) Effekte von Noxen auf die Chromosomenpaarung während der Meiose. Der synaptonemale Komplex als In-vivo-Keimzellassay. Fertilität 12:152-164

Johannisson R, Ocker H (1997) Effects of cyclophosphamide on pachytene chromosomes in female mice. Mutat Res 374:185-192

Kölbel HK (1976) Kohleträgerfilme für die hochauflösende Elektronenmikroskopie - Verbesserung von Eigenschaften und Herstellungstechnik. Mikroskopie 32:1-16

Mennicke K, Diercks P, Schlieker H, Bals-Pratsch M, Al-Hasani S, Diedrich K, Schwinger E (1997) Molecular cytogenetic diagnostics in sperm. Int J Androl 20, Suppl 3: 11-19

Metzler C (1995) Einfluß von Hyperthermie auf die Meiose und die Spermatohistogenese von Rattus norvegicus. Eine experimentelle Studie zur 1. Reifeteilung durch Darstellung der Synaptonemalen Komplexe und zur morphologischen Differenzierung der Keimzellen. Med Diss Schr Lübeck

Miller OL, Beatty BR (1969) Visualization of nucleolar genes. Science 164:955-957
Moses MJ (1979) The synaptonemal complex as an indicator of chromosomal damage. Genetics 92 (Suppl):73-82
Moses MJ (1980) New cytogenetic studies on mammalian meiosis. In: M. Serio, Martini L (eds) Animal models in Human Reproduction. Raven Press, New York, pp. 169-190
Saadallah N, Hultén M (1985) A complex three breakpoint translocation involving chromosomes 2,4, and 9 identified by meiotic investigations of a male ascertained for subfertility. Hum Genet 71:312-320
Schmid H, Johannisson R, Haaf T, Neitzel H (1987) The chromosomes of Micromys minutus (Rodentia, Murinae). II Pairing pattern of X and Y chromosomes in meiotic prophase. Cytogenet Cell Genet 45:121-131
Speed RM, Chandley AC (1990) Prophase of meiosis in human spermatocytes analysed by EM microspreading in infertile men and their controls and comparisons with human oocytes. Hum Genet 84:547-554
Sperling K, Kaden R (1971) Meiotic studies of the ejaculated seminal fluid of humans with normal sperm count and oligospermia. Nature 232:481
Sudman PD (1989) Cryogenic preservation of mammalian testicular material for synaptonemal complex analysis. Cytogenet Cell Genet 52:88-89
Sumner AT (1972) A simple technique for demonstrating centromeric heterochromatin. Exp Cell Res 75:304-306
Templado C, Vidal F, Navarro J, Egozcue J (1986) Improved technique for the study of meiosis in ejaculate: results of the first consecutive cases. Hum Genet 72:275-277
Van Steirteghem AC (1994) IVF and micromanipulation techniques for male-factor infertility. Curr Opin Obst Gynecol 6:173-177
Vidal F, Navarro J, Templado C, Egozcue J (1986) Study of synaptonemal complexes in human semen: results in the first consecutive cases. Human reproduction 1:121-123
von Wettstein D, Rasmussen SW, Holm PB (1984) The synaptonemal complex in genetic segregation. Annu Rev Genet 18:331-413

Part III

Prenatal Diagnosis

Chapter 11

Prenatal Diagnosis - An Introduction

ROLF-DIETER WEGNER

Introduction

Nowadays, a broad spectrum of prenatal tests are available for pregnant women to assure the well-being of the unborn child. The various methods fall into two groups: the non-invasive and the invasive procedures. Here only the latter, namely amniocentesis, chorionic villi sampling and fetal blood sampling, should be considered. Tissue sampling is performed under sonographic control by specialized gynecologists who provide the specimen for genetic analysis. Importantly, all invasive techniques carry a significant procedure-related risk for an abortion.

Generally, it is agreed upon the necessity of a pre-test genetic counselling of the parents or the pregnant woman. The aim of this talk is to achieve a well-informed choice of the counselee(s) by considering the indication(s) as well as the risks and the limitations of each of the prenatal procedures. In particular, the pregnant woman should know precisely what she has to expect when opting for an invasive method. Preceding the protocols describ-

Table 1. Common methods in prenatal diagnosis. Time: week of pregnancy; risk: percentage of procedural abortions; result: weeks until the report is issued, AC: amniocentesis, CVS: chorionic villi biopsy, FBS: fetal blood sampling

Method	AC	CVS	FBS
Time (week)	15. - 17.	11. - 12.	> 18.
Risk (%)	0,5 - 1	1 - 2	1 - 3
Result (weeks)	~ 2	~ 2	0,5

Rolf-Dieter Wegner, Charité Campus Virchow-Klinikum, Institut für Humangenetik, Augustenburger Platz 1, Berlin, 13353, Germany (*phone* +49-30-450-66123; *fax* +49-30-450-66904; *e-mail* rolf-dieter.wegner@charite.de)

ing cytogenetic analysis of amniocytes and chorionic villi it seems to be appropriate to give some basic information about such tests (Table 1).

Amniocentesis (AC)

AC is performed mostly between the 15. - 17. week of pregnancy. Early AC in the 13. - 14. week has been applied successfully by a number of groups (for review see Wilson, 1995), however, very early procedures between 11+0 - 12+6 weeks should be considered experimental (Wilson, 1995). The procedure related risk to suffer an abortion is stated to be between 0.5 - 1 % (Tabor et al., 1986; Bauer-Hansmann and Golbus, 1993). On average, tissue culture needs 10 - 14 days before starting the chromosome analysis. Cytogenetically, problems may arise by level-II-mosaicism, ie multiple cells with an identical aberration in a single culture flask or in a single colony, and by level-III-mosaicism, ie multiple cells with an identical aberration in multiple culture flasks or multiple colonies. Level-II and Level-III mosaicism occur in a frequency of about 0,7 and 0.25 %, respectively (Hsu and Perlis, 1984). That report shows, that level-II-mosaicism has been a culture artefact in almost all cases. In contrast, level-III-mosaicism indicates the presence of an aberrant cell line which might, however, be confined to the extraembryonic/extrafetal tissue or be present in the unborn. While all such cases need an extensive post-test genetic counselling of the parents a further testing, eg a repeat AC or a fetal blood sampling, is taken up by the parents most exclusively in cases of level-III-mosaicism.

Chorionic villi sampling (CVS)

CVS is performed mostly in the 11. to 12. week of pregnancy. It seems to be justified to give a procedure-related risk for an abortion in the order of 1 - 2 % although there are marked differences between various studies (for review see Wegner, 1993). Thus, in two randomized studies (Lippman et al., 1992; Smidt-Jensen et al., 1992) the abortion risk of CVS as compared to AC was increased not more than 1 % while a third report stated an additional risk by CVS of 2.9 % (MRC Working Party, 1991). It became obvious that the skill of the gynecologist seems to play a substantial role in the abortion rate and that there exists a marked learning curve (Kuliev et al., 1993).

Two culture systems are the gold standard for cytogenetic analysis of chorionic villi: short term culture (STC) and long term culture (LTC). Results of STC can be reported in 4 to 48 hrs while results of the complementing LTC need between 6 days and 2 weeks. The advantage of an early report in CVS has to be weighed against a higher number of equivocal cytogenetic findings. Thus, 1,5 % of cases amongst more than 92,000 CVS collected by the European Collaborative Research on Mosaicism in CVS (EUCROMIC) showed confined placental mosaicism (CPM), ie the pathological cell line(s) was (were) not present in the embryo proper (Hahnemann and Vejerslev, 1997). For explanation of these findings see Chapter 13. The finding of CPM leads to anxiety in the parents and intensive genetic counselling is needed. Additional tests might be required or requested by the parents. An impact of CPM on the placental function is still a matter of controversial discussion but a doubling of the number of cases with low birth weight is seen in the group with CPM as compared to controls (DeLozier-Blanchet et al., 1997). Anyway, a more frequent sonographic control of fetal development might be helpful in an early recognition of fetal distress and in decisions concerning the birth management. A publication considering the predictive value of CVS as compared to the other PD techniques reported that CVS is less reliable in very-high- and very-low -risk pregnancies (Kennerknecht et al., 1993).

Fetal blood sampling (FBS)

FBS by cordocentesis is performed mainly after the 18th week of pregnancy. Reports referring to unbiased study groups, eg pregnancies with normal sonographic findings, show a procedural abortion rate in the range of 1 - 3 % (Daffos, 1991). For high risk groups or in early pregnancy (12. - 18. week) the abortion rate may reach 5 % or even more (Nicolini and Rodeck, 1992). A numerical result of the chromosome analysis can be reported within two days, a final result applying high resolution banding is possible within three to four days. In principle, analysis of fetal blood cells follows the protocol of normal lymphocyte cultures. Therefore, a separate chapter dealing with fetal blood cell analysis is omitted and the reader is referred to Chapter 5.

In summary, any particular invasive technique brings with it its own constellation of advantages and disadvantages. Parents seeking PD have to be taught carefully about the procedures so that they can thoroughly weigh up all facts and relate them to their own individual situation. It must be considered that amongst other factors the cultural background, the religious bindings, and the degree of anxiety will influence the final decision.

References

Bauer-Hansmann D, Golbus MS (1993) Prenatal diagnosis. In: Stevenson RE, Hall JG, Goddman RM (eds) Human malformations and related anomalies Vol I, Oxford University Press, New York, Oxford.

Canadian collaborative CVS-amniocentesis clinical trial group (1989) Multicentre randomised clinical trial of chorionic villi sampling and amniocentesis. Lancet I: 1-6.

Daffos F (1991) Fetal blood sampling. In Harrison MR, Golbus MS, Filly RA (eds), The unborn patient. Saunders, Philadelphia.

DeLozier-Blanchet CD, Pellegrini B, Hahnemann JM, Pampallona S, Vejerslev LO (1997) The impact of CPM on fetal growth and development. Data from EUCROMIC (Abstract). 1st post EUCROMIC satellite meeting on prenatal diagnosis. Genova.

Hahnemann JM, Vejerslev LO (1997) European collaborative research on mosaicism in CVS (EUCROMIC): fetal and extrafetal cell lineages in 192 gestations with CVS mosaicism involving single autosomal trisomy. Am J Med Genet 70: 179-187.

Hsu LYF, Perlis TE (1984) United States survey on chromosome mosaicism and pseudomosaicism in prenatal diagnosis. Pren Diagn 4: 97-130.

Kennerknecht I, Barbi G, Wolf M, Djalali M, Grab D, Terinde R, Vogel W (1993) Cytogenetic diagnosis after chorionic villus sampling are less reliable in very-high- and very-low-risk pregnancies. Pren Diagn 13: 929-944.

Kulier AM, Modell B, Jackson L, Simpson JL, Brambati B, Rhoads G, Froster U, Verlinsky Y, Smidt-Jensen S, Holzgreve W et al. (1993) Risk evaluation of CVS. Prenat Diagn 13: 197-209.

Lippman A, Tomkins DJ, Shime J, Hamerton JL, and Canadian Collaborative CVS-Amniocentesis clinical trial group (1992) Canadian multicentre randomized clinical trial of chorionic villus sampling and amniocentesis. Final report. Pren Diagn 12: 385-476.

MRC Working Party on the evaluation of chorionic villus sampling (1991) Medical Research Council European Trial of chorionic villus sampling. Lancet 337: 1491-1499.

Nicolini U, Rodeck CH (1992) Fetal blood and tissue sampling. In Brock, Rodeck, Ferguson-Smith (eds) Prenatal diagnosis and screening, Livingstone, Edinburgh.

Smidt-Jensen S, Permin M, Philip J, Lundsteen C, Zachary JM, Fowler SE, Gruning K (1992) Randomized comparison of amniocentesis and transabdominal and transcervical chorionic villus sampling. Lancet 340: 1238-1244.

Tabor A, Madsen M, Obel EB, Philip J, Bang J, Noergard-Peterson B (1996) Randomized controlled trial of genetic amniocentesis in 4606 low-risk women. Lancet I: 1287-1293.

Wegner RD (1993) Chorionic villi analysis. In Obe G (ed) Advances in mutagenesis research 4. Springer Verlag Berlin, Heidelberg, New York.

Wilson RD (1995) Early amniocentesis: a clinical review. Pren Diagn 15: 1259-1273

Chapter 12

Amniotic Fluid Cell Analysis

INGO KENNERKNECHT, MAHMOUD DJALALI, GOTTHOLD BARBI,
WALTER JUST AND WALTHER VOGEL

Introduction

Amniocentesis (AC) is by far the most often used invasive prenatal diagnostic procedure. Amniotic fluid cells were first used for prenatal diagnosis by Fuchs and Philipp (1963) for sex chromatin determination in a pregnancy at risk for an X-linked disease. After successful establishing of amniotic fluid cell culture techniques by Steele and Breg (1966), 2 years later the first prenatal diagnoses of a chromosome aberration ie trisomy 21 (Valenti et al. 1968), and of a metabolic disorder ie galactosaemia (Nadler 1968) were reported.

AC can be performed safely by many gynaecologists, and amniotic fluid samples can be sent easily to cytogenetic or metabolic diagnostic laboratories. Through the years a variety of methods have been evaluated including, eg early tapping of cells (review Wilson 1995), optimized formulas for growth media (Chang et al. 1982), precoated surfaces of culture flasks (Chang et al. 1991), pipette method (Claussen et al. 1986, 1994) for rapid karyotyping results, and culture synchronization for high-resolution chromosome banding techniques (Qu et al. 1989).

Based on our experience with more than 30.000 chromosome analyses after AC, we describe two simple but reliable basic methods for successful karyotyping, the most often used flask method and the *in situ* technique. These procedures are standard protocols and do not require special skills

Correspondence to Ingo Kennerknecht, Westfälische Wilhelms-Universität Münster, Institut für Humangenetik, Vesaliusweg 12-14, Münster, 48149, Germany (*phone* +49-251-8355412; *fax* +49-251-8356995; *e-mail* kennerk@uni-muenster.de, Mahmoud Djalali, Universität Ulm, Abteilung Medizinische Genetik, Parkstrasse 11, Ulm, 89073, Germany), Gotthold Barbi, Universität Ulm, Abteilung Medizinische Genetik, Parkstrasse 11, Ulm, 89073, Germany, Walter Just, Universität Ulm, Abteilung Medizinische Genetik, Albert-Einstein-Allee 11, Ulm, 89069, Germany, Walther Vogel, Universität Ulm, Abteilung Medizinische Genetik, Albert-Einstein-Allee 11, Ulm, 89069, Germany beyond a trained cytogeneticist. They were originally established for amniotic fluid samples at the 16th week of gestation and later. Without any modification these methods can also be used for early AC, even at the 11th completed week (Djalali et al. 1992, Kennerknecht et al. 1992). Taking into account a mean culture time for amniotic fluid cells of 8 to 10 days plus the time to complete the cytogenetic analysis, there is - in the case of an aberrant finding - still enough time for an optional termination of pregnancy by the less problematic dilatation and evacuation technique. However, before more data are available, procedures earlier than 12+0 weeks should be considered experimental, because of an increased rate of culture failure. And also the impact of the relatively high reduction in amniotic fluid volume on the abortion rate is still not clear in very early amniocenteses. A sample of 20 ml represents only 10% of the total amount of amniotic fluid at 16 weeks, whereas the same amount taken at 10 weeks represents nearly 70% (Elejalde et al. 1990). Filtration of amniotic fluid cells and refunding the amniotic fluid to the amniotic sac might overcome the problem until there are more reliable data. It remains to be shown whether short-term reduction in amniotic fluid volume increases the risk of abortion or may be responsible for embryonic and fetal malformations (eg akinesia syndrom), an increased prematurity rate, and (consecutive) respiratory distress (Sundberg et al. 1991, 1995, 1997, Kennerknecht et al. 1993, Byrne et al. 1995). To evaluate literature data is difficult as there is no unique definition for "early" AC. We therefore recommend to use the definitions given by Evans et al.1994:

- very early AC $\leq$ 11.6 weeks
- early AC 12.0 to 13.6 weeks
- standard AC (midtrimester) 14.0 to 25.6 weeks
- third-trimester AC $\geq$ 26.0 weeks

Materials

Equipment

Flask method:

- Laminar air flow
- Glass culture flasks, square bottle, capacity 180 ml, 48 mm diameter, according to Breed-Demeter, Duran™, Schott Glaswerke, D-55014 Mainz, Germany or polystyrol culture flasks, angled neck, 80 cm², 50 ml, Nunc GmbH D-65203 Wiesbaden, Germany (DK-4000 Roskilde, Denmark)

- Tissue culture tubes TC™, sterile, 14 ml, Greiner Labortechnik, D-72636 Frickenhausen, Germany
- Glass Pasteur pipettes, approximately 150 mm
- Slides (76 x 26 mm), cover glasses (24 x 60 mm)*In situ* technique:
- Flaskette™, Nunc GmbH

Media and solutions

Flask method
- complete medium ready for use in cell culture:
  1000 ml DMEM: Dulbecco's modified Eagle medium with Glutamax ™ (= L-alanyl-L-glutamine), GibcoBRL, Life Technologies GmbH, D-76339 Eggenstein, Germany (Europe: GB-Paisley PA4 9RF, Great Britain; worldwide: Gathersburg, MA 20 884 - 9980, U.S.A.) or Dulbecco's MEM (1x) ™ (includes N-acetyl-L-alanyl-L-glutamine), Biochrom KG, D-12247 Berlin, Germany plus
- 200 ml fetal calf serum, Biochrom KG or ICN-Biomedicals GmbH, D-5340 Meckenheim (ICN-Biomedicals, Costa Mesa, CA 92626, U.S.A.); note: Independent of the suppliers the biological activity of fetal calf sera differs significantly between the lots and should therefore always be tested individually plus
- 10 ml antibiotics: Penicillin 5,000 IU/ml + Streptomycin 5,000 µg/ml™, ICN Biomedicals plus
- 1.2 ml, Amphotericin B 250UG/ml, Fungizone™, GibcoBRL Note: As the antibiotica and the antimykoticum are stored at – 20 °C avoid repeated thawing by freezing aliquots of 10 ml and 1.2 ml, respectively, ready for supplementing1000 ml culture medium

In situ technique
- Chang medium D™, Irvine Scientific, Santa Ana, CA 92702, U.S.A.
- chromosome harvest:
  Colchicine™ 20 mg/l, eurobio GmbH, D-65479 Raunheim, Germany, or Colcemide™ (10 mg/l), Biochrom
- Trypsin stock solution: 1000 ml Aqua dest. + 9.55 g Instamed 9.55 g/l PBS Dulbecco's™ + 20 ml EDTA (Versen) 1% in PBS™, Biochrom
  solution to use: 100 ml stock solution + 10 ml Trypsin sol 2.5% (1:250) ™, GibcoBRL

- chromosome preparation:
  hypotonic solution: Na-citrate / KCl 1/1 v/v (2 g Na-citrate + 2 g KCl (or NaCl)/ 1,000 ml)
- fixative: methanol / glacial acetic acid 3/1 v/v (75 ml methanol (100 %) + 25 ml acetic acid (99%))
- mounting solution: Eukitt™, Kindler GmbH, D-Freiburg, Germany)

Procedure

Cell culture initiation and maintenance

Routinely 15 ml amniotic fluid are aspirated transabdominally by a 20-gauge spinal needle under real-time ultrasound guidance. This is sufficient for routine chromosome studies, but in the case of a suspected metabolic or molecular defect more amniotic fluid and/or cell cultures may be advantageous. Amniotic fluid can be transported - also over long distances - directly in the syringe used to obtain the specimen. Extreme temperatures during transport should be avoided. Our experience with transport and storage of amniotic fluid cell samples shows that long-term transport or storage in a refrigerator (4 °C) even up to 3 – 4 days prior to culture initiation do not considerably affect the culture quality and proliferation time. The whole information identifying patient's name, relevant anamnestic data, gestational age, number of previous pregnancies, miscarriages, previous child with a chromosome aberration should be recorded. Also the basic information concerning amniotic fluid specimens including volume, colour (clear or cloudy), grossly or dark blood-spilled, or any unusual circumstances of tapping, such as prolonged transport to the laboratory should be recorded.

0,5 – 1 ml unspun amniotic fluid is pipetted under sterile conditions into small snap-cap tubes for alpha-feto-protein determination. If the amnion sample arrives in two separate culture tubes and only one is bloody, use unbloody fluid sample for alpha-feto-protein examination.

For routine procedure three amniotic fluid cell cultures are initiated. Each amniotic fluid sample should be set up with batches of complete medium labelled A, B, C. We use whole amniotic fluid for routine culturing. Concentration of the cells by centrifugation is not necessary.

Flask method

On a clean bench with laminar airflow 5 ml native amniotic fluid is directly transferred from the syringe (without needle) into a 80 cm^2 plastic tissue flask complemented by 5 ml of complete medium under sterile conditions. Mix cell suspension by gentle shaking. Add CO_2 (after reduction of the pressure to 1 – 3 bar) through a sterile filter using a 20 ml syringe and take care that the pH-indicator of the culture medium shows a pink colour. More conveniently, a dosis syringe adapted for the use of gas can be used. Flasks are placed in a horizontal position with caps tightly closed and incubated at 37 °C. Alternatively, cultures can be kept in an open system in a 5% CO_2-incubator with humid atmosphere. We prefer the closed system, as the cell cultures are better protected from contamination and because an expensive CO_2 -incubator is not needed.

Though this is more labour intensive, we always use the same set of glass flasks for culture. We experience that the cells attach faster and better to a glass than to a plastic surface. After 5 – 6 days of undisturbed culture the cell growth is monitored under an inverted microscope (magnification 20x – 50x). Primary colonies of epithelial (E) cell type may appear within 3 – 4 days, but this type of cell yields only few metaphases with poor quality after preparation. Growth of other cell types, predominantly fibroblast cell type (F) or amniotic fluid cell type (AF), appear usually after 4 – 6 days (Hoehn et al. 1974, Johnston et al. 1982).

Growth is monitored and noted twice a week. Care has to be taken that the first complete change of medium can only be performed, if at least 3 – 5 colonies are found. Amnion cell cultures are fed twice a week, medium is poured off, the rim of the flask is quickly flamed to break any bubbles, and 5 ml fresh medium is added.

When using the flask method, cytogenetic analyses should rely exclusively on primary cultures with trypsinization only for cell harvest, since this procedure maintains optimal information on number, size and morphology of colonies. In cultures which have been trypsinized this information is lost. Subculturing should only be applied, if a structural aberration is detected and needs rapid and more precise characterization. In this case, at least 2 – 3 new cultures should be started, one can be harvested and the others kept as reserve in the case of further technical or diagnostic problems. In the case of mosaicism, information of real percentage of different cell lines is lost after subculturing.

Subculture

After appearance of sufficient cell growth with multiple areas of cell colonies and a diameter of 0.3 - 0.5 mm, subcultures can be established. Make sure that an adequate amount of amniotic fluid cell type (AF) and fibroblast cell type (F) is available in the primary culture. The epithelial cell types (E) cannot be maintained in serial culture after trypsinization, since they have the lowest growth potential. The cell growth of F cell type in subculture is very fast, and these are the cells mainly used for chromosome analysis. The F cell types have the highest growth potential in serial culture (10 – 50 population doublings) and predominate in older amniotic fluid cultures. The F cell type is not predominantly represented in primary culture but does usually overgrow during subculturing. The F cell type can still be maintained in culture, when the growth potential of other cell types is exhausted and the culture growth is declining.

1. Pour off medium from primary culture flask into a sterile centrifuge tube in order to recover the released mitotic cells.
2. Centrifuge at 800 – 1000 rpm (radius of ca. 10 cm) for 3 – 5 minutes. Discard the supernatant quickly - with one movement - leaving approximately 1 ml in the tube.
3. During step 2, add 3 – 5 ml trypsin to 80 cm^2 culture flask, make sure that the trypsin solution covers the entire bottom of the culture flask, and incubate at 37 °C for several minutes.
4. Observe the trypsinized culture under an inverted microscope. If the cells are not released from the flask surface after several minutes repeatedly knock at the flask. If the cells are detached from the container bottom, pipette the cells into the centrifuge tube (ie from step 2) containing 1 ml medium sufficient to inhibit the activity of trypsin.
5. Spin the centrifuge tube again at 800 – 1000 rpm for 5 – 7 minutes.
6. Quickly - with one movement - discard supernatant and add fresh medium to the centrifuge tube to reach a final volume of 1 – 1.5 ml.
7. Resuspend by careful and slow pipetting
8. Add 5 ml of fresh medium to each 80 cm^2 flask, put additional 1 – 2 drops of mixed cells into a culture flask and check the amount of cells under an inverted microscope. If this is sufficient (ie empirically evaluated), seed one drop more and one drop less of the cell suspension to the next two flasks, respectively.
9. Incubate the cultures at 37 °C.
10. Change the medium on the next day, and the best culture can be harvested two days after subculture.

If a confluent culture has to be split, pour off the medium, add 1 – 2 ml trypsin to the culture flask und wait for 2 – 4 minutes. If not enough cells are released, knock against the flask gently. The cells released can be used for subculture as mentioned above. The original culture flask can be fed after the floating cells have been removed.

Chromosome harvesting and preparation

Cultures to be harvested are fed the afternoon before. Feeding at late afternoon partially synchronize the cells allowing optimal harvest with respect to the cell cycle time (20 – 24 hours) in late morning/early afternoon.

1. Add 0.15 – 0.25 ml colchicin or colcemide to a 80 cm^2 culture flask, mix gently, and incubate the culture for 1 – 1.5 h.
2. The culture flasks are then monitored under an inverted microscope. If there is an adequate amount of rounded-up cells (ie cells in M-phase which undergo changes in their cytoskeleton) remove the medium completely (to allow full trypsin activity in step 4) with a pasteur pipette to centrifuge tube A. Note: If there are not enough mitotic cells visible after colcemide treatment, pour off the medium and feed the culture with fresh medium. The harvesting procedure can be performed the next day. Alternately the cultures are trypsinized under sterile conditions till about half of the cells are detached. Then remove the cell suspension for further procedure to a centrifuge tube and continue from step 7 onward. Rinse and feed the culture flasks with fresh medium and continue with culturing of the remaining cells. This step can be repeated several times until sufficient mitotic cells are obtained. To our experience this needs up to 2 days, respectively.
3. Spin centrifuge tube A at 800 rpm 3 – 5 minutes and pour off the supernatant to a centrifuge tube B.
4. During step 3 add 4 – 6 ml trypsin to each culture flask and make sure that the cell layer is completely covered. Incubate at 37 °C for 5 – 7 min.
5. Watch the culture flask under an inverted microscope. If the cells are detached from the flask surface, the cell suspension should be gently collected by a pasteur pipette and also transferred to centrifuge tube A. Note: If the trypsinized cells are not yet detached from the container surface, knock the flask slightly with the hand, this action can be repeated a few times.
6. Rinse the culture flask with the medium from tube B to inhibit trypsin activity and also add this suspension to tube A.

7. Spin tube A at 800 rpm for 5 min.
8. Remove supernatant carefully without disturbing the cell pellet and leave a remnant of 0.5 ml above the pellet.
9. Resuspend the pellet by gently tapping with a finger.
10. Add 2 – 4 ml (depending on the size of the pellet) hypotonic solution (0.2 % Na-citrate and 0.2 % KCl) drop by drop under gentle agitation. Note: It is crucial that the first 1 ml is added very carefully. The thoroughly mixed cell suspension in the hypotonic solution should be placed in a waterbath for 20 minutes at 37 °C.
11. Centrifuge at 800 rpm for 5 min.
12. Remove the hypotonic solution as completely as possible without disturbing the cell pellet. Note: If a significant amount of floating cells are accidentally removed resuspend them, repeat centrifugation, and decantation. If the cells are still not released from the flask surface repeatedly knock slightly at the flask.
13. Cell pellet is resuspended by gentle tapping, add 3-4 ml freshly prepared and ice cold fixative (methanol : acetic acid, 3:1), the first 0.5 ml very slowly drop by drop along the inner wall of the tube, and shake slightly. Then add additional fixative in portions of 2 – 3 drops to a final amount of 3 – 4 ml and mix by tapping. Leave the tube for 20 – 30 min at room temperature.
14. Centrifuge at 800 rpm for 5 min.
15. Supernatant is removed carefully to a minimum of 0.5 ml. Resuspend the pellet by tapping and add 3 – 4 ml fresh fixative all at once and mix gently. Leave the tube for 10 min.

Note: At this stage the cells can be stored at 4 °C for several days or at – 20 °C for several months.

16. Centrifuge at 800 rpm for 5 min.
17. Supernatant is pipetted off leaving a volume between 0.1 ml and 0.3 ml (depending on the amount of cells) without disturbing the cell pellet.
18. Resuspend the pellet by gently blowing air bubbles at the bottom of the tube. Draw up the cell suspension only filling the narrow tip of the pipette.

19. Three drops are spotted beside each other on a wet, cooled slide from a distance of 20 – 30 cm. The slides have to be precleaned overnight in fixative, and be kept in cooled water (4 °C).
20. Air-dry the slides for 15 min and eventually postfix another 2 – 3 min with a hot hairdryer.
21. Label the slides with identification number and date

In situ technique

We do not use the *in situ* technique (Philip et al. 1974) routinely because cell culturing is more laborious, and the quality of the mitoses is not always optimal for fine structural analyses. In addition it is more expensive than the flask method. However, we use it in case of verification of chromosomal mosaicism detected in CVS (Cheung et al. 1990). Individual amniotic fluid cells can be assigned to the colony they stem from. Among a vast spectrum of methods for *in situ* culturing (eg Tabor et al. 1984) we use the following procedure.

1. Add 1 – 2 ml amniotic fluid and the same amount of Chang medium D™ into a Flaskette™, add CO_2 to adjust the pH as described above (section 4), close the cap tightly, and incubate at 37 °C.
2. After 4 – 5 days of undisturbed incubation cell growth is monitored under an inverted microscope. If there are 3 – 5 cell colonies with a diameter of 0.1 – 0.2 mm with an adequate amount of mitotic cells, add 0.1 ml colcemid to the culture and incubate 1h at 37 °C. Otherwise the cell proliferation should be checked the following days.
3. Remove the medium with a pasteur pipette, add very gently 2 ml of KCl solution (0.56 g/100 ml) and wait for 30 minutes at room temperature. Note: Chromosome spreading only by hypotonic treatment does not have the same quality as dropping the cell suspension on to the slide, as in the flask method.
4. Remove the KCl solution with pasteur pipette, add 3 ml of fresh fixative drop by drop with a pasteur pipette, and wait for 30 minutes.
5. Pour off fixative and add 3 ml of fresh fixative all at once.
6. Pour off fixative and disconnect the bottom from the flaskette™.

Results

The reliability of a normal diagnosis is sometimes restricted by undetected chromosomal mosaicism and maternal cell contamination. According to our standard routine procedure (flask methods) cytogenetic diagnosis is based on at least 20 mitoses from a minimum of six clones. This approach allows exclusion of 41% chromosome mosaicism at the 95% confidence level (Hook 1977, Claussen et al. 1984). It must be emphasized that the confidence level of exclusion of mosaicism in AC depends on the number of colonies studied and not on the number of cells. It should also be stressed that the individual laboratories have to take their own decision on which is the minimal number of cells analyzed for a satisfactory diagnosis depending on the tissue culture method used (Richkind and Risch 1990) and on their own experience.

Chromosomal mosaicism

The observation of a single metaphase or of several metaphases with a deviant karyotype makes interpretation more difficult. Such an observation may reflect a true mosaic (which may or may not represent the fetus), a cultural artefact (pseudomosaicism) or contamination with maternal cells. In such cases additional work-up is necessary. Pseudomosaicism should be differentiated from true mosaicism. **Pseudomosaicism** is suggested when one (or more) cell(s) with an identical aberration has derived from a single clone, as can be observed by the *in situ* culture technique. In most cases pseudomosaicism represents a cultural artefact and some of them might depend on medium type (Krawzun et al. 1989).When using the flask method pseudomosaicism may present a single cell aberration (level I mosaicism according to Worton & Stern 1984), or as the observation of several or more cell(s) (level II mosaicism) with an identical aberration confined to only one culture flask. In these cases investigation of further colonies - in case of the *in situ* technique - or of all 3 primary cultures - in case of the flask method - might better help to detect the very rare instances of low percentage but true mosaicism.

Extension of work-up in cases with **true (or level III) mosaicism** (ie identical chromosome aberration in more than one culture flask) is indicated but should be guided by the biological significance of the individual finding: viable (mosaic) trisomies (eg trisomy 21, 18, 13, 9, 8) are more likely to affect the fetus than "exotic"/non-viable trisomies (eg 19, 17, 16). However, in each case of mosaic trisomy the possibility of uniparental disomy (UPD) after trisomy "rescue" should be taken into account (Ledbetter and Engel 1995). In general, true mosaicism detected in cultured amniotic fluid cells is confirmed in the fetus in approximately 70% of cases (Hsu and Perlis

1984). This figure differs significantly when one differentiates between sex chromosome mosaicism (in ca. 90% confirmed), marker chromosome mosaicism (ca. 80%), or autosomal chromosome mosaicism (ca. 50%). Nearly 40% of the fetuses with confirmed autosomal chromosome mosaicism show an abnormal phenotype. Sex chromosome mosaicism is in most cases (ca. 90%) of no phenotypic relevance as is mosaicism of autosomal structural aneuploidies. In conclusion, **extensive** additional work-up is recommended in those autosomal mosaic aneuploidies which are compatible with livebirths. This should also include level II mosaicism (for detailed guidelines see Hsu et al. 1992). Generally accepted is a two-step approach, an effective compromise regarding sensitivity of cytogenetic finding and workload: Two flasks are harvested initially and only in case of an aberrant cell line in one of the culture vessels is the third flask processed.

Maternal cell contamination

Amniotic fluid cell mosaicism of XX/XY type is usually due to maternal cell contamination. The figures given in the literature vary considerably between 0% and 0.84% (Bui et al. 1984, Hsu and Perlis 1984, Moertel et al. 1992). Although maternal cell contamination usually poses no major problems one should consider some facts. Contaminations will in general only be detected in pregnancies with a XY fetus but only a rough minimum estimate can be made by doubling those numbers obtained from pregnancies with XY fetuses. Provided a proper sampling technique of amniotic fluid cells exists (Nuß et al. 1994) the different published frequencies of maternal contamination are rather a function of adequate detection and interpretation than a biological property of the different collections studied. In our experience prolonged culture time and blood-spilled amniotic fluid samples have an elevated risk of karyotyping maternal cells. Apart from XX-males (incidence 1 : 20.000) wrong sex assignment is considered to be very rare and is rather due to laboratory cross-contamination of cases or to secretarial errors. Only when studying uncultured amniotic fluid cells by FISH technique maternal cell contamination might pose serious problems.

Troubleshooting

The methods described here are reliable. For good results only some simple but crucial points should be regarded.

Cell culture, chromosome harvest

Optimal cell culture conditions (tested fetal calf serum, adjustment of pH) and proper judgment of the time point of cell harvest (degree of cell confluence, number of mitotic cells) are equally important. This can only be learnt by experience.

To allow proper hypotonic and fixative treatment large cell pellets should be split into 2 – 3 tubes. To avoid cell loss during preparation never submerge the tip of the pasteur pipette when adding solutions. In order to gently resuspend a cell pellet submerge only the small tip of the pipette and blow only a few air bubbles. For chromosome spreading aspirate only as much cell suspension as necessary for one slide, ie 3 drops which will not exceed the narrow tip of the pasteur pipette.

Early amniocentesis

Despite a much lower cell content of amniotic fluid samples from early gestational ages cell cultures can also be successfully initiated fom 5 ml amniotic fluid. A lower amount of amniotic fluid can be used but the failure rate of culture increases. Increased rates of fetal loss are observed in early AC also by experienced gynecologists. This observation rather seems to be a function of gestational age than of prenatal diagnostic procedures.

Amnifiltration offers an excellent approach for delineating the intensively discussed potential disharm of (short-term) lack of amniotic fluid. It also allows shortening of culture time by enriching the primary cell number. But as this method is more labor intensive it is only of minor benefit for routine procedures. We therefore consider this method only as experimental.

Interpretation of results

This is the most challenging part on the way from AC to a definite diagnosis. The guidelines discussed above are very helpful. However, the definite diagnosis reported is always an individual consideration. The following example demonstrates that also single cell aberrations (level I mosaicism), which are usually neglected can be of prognostic relevance. Because of a hydrops fetalis and hygroma colli in the 16th week of gestation CVS, AC and PUBS were indicated. CVS short-term culture (n=100) and long-term culture (n=100) as well as lymphocytes (n=8) only represented monosomy X. In contrast, among cultured amniotic fluid cells only 1 among 19 cells from the first flask harvested represented the aberrant cell line. This is an extreme example of mosaicism which shows that all available anamnestic and clinical data should be known by the cytogeneticist and taken into his consideration.

References

Bui TH, Iselius L, Lindsten J (1984) European collaborative study on prenatal diagnosis: Mosaicism, pseudomosaicism and single abnormal cells in amniotic fluid cell cultures. Prenat Diagn 4:145-162

Byrne DL, Penna L, Marks K, Offley-Shore B (1995) First trimester amnifiltration: technical, cytogenetic and pregnancy outcome of 104 consecutive precedures. Brit J Obstet Gyn 102:220-223

Chang HC, Jones OW, Masui H (1982) Human amniotic fluid cells grown in hormone-supplemented medium: Suitability for prenatal diagnosis. Proc Natl Acad Sci USA 79:4795-4799

Chang HC, Jones OW (1991) Enhancement of amniocyte growth on a precoated surface. Prenat Diagn 11:357-370

Cheung SW, Spitznagel E, Featherstone T, Crane JP (1990) Exclusion of chromosomal mosaicism in amniotic fluid cultures: Efficacy of in situ versus flask technique. Prenat Diagn 10:41-57

Claussen U, Schäfer H, Trampisch HJ (1984) Exclusion of chromosomal mosaicism in prenatal diagnosis. Hum Genet 67:23-28

Claussen U, Klein R, Schmidt M (1986) A pipette method for rapid karyotyping in prenatal diagnosis. Prenat Diagn 6:401-408

Claussen U, Ulmer R, Beinder E, Voigt H-J (1994) Six years' experience with rapid karyotyping in prenatal diagnosis: Correlations between phenotype detected by ultrasound and fetal karyotype. Prenat Diagn 14:113-121

Djalali M, Barbi G, Kennerknecht I, Terinde R (1992) Introduction of early amniocentesis to routine prenatal diagnosis. Prenat Diagn 12:661-669

Elejalde BR, de Elejalde MM, Acuna JM, Thelen D, Trujillo C, Karrmann M (1990) Prospective study of amniocentesis performed between weeks 9 and 16 of gestation: Its feasibility, risks, complications, and use in early genetic prenatal diagnosis. Am J Med Genet 36:188-196

Evans MI, Johnson MP, Holzgreve W (1994) Early amniocentesis. What exactly does it mean? J Reprod Med 39:77-78

Fuchs F, Philip J(1963) Mulighed for antenatal andersogelse at fosterets kromosomer. Nord Med 9:69

Hoehn H, Bryant EM, Karp LE, Martin GM (1974) Cultivated cells from diagnostic amniocentesis in second trimester pregnancies. I.Clonal morphology and growth potential. Pediatr Res 8:746-754

Hook EB (1977) Exclusion of chromosomal mosaicism: Tables of 90%, 95%, and 99% confidence limits and comments on use. Am J Hum Genet 29:94-97

Hsu LYF, Perlis TE (1984) United States survey on chromosome mosaicism and pseudomosaicism in prenatal diagnosis. Prenat Diagn 4:97-130

Hsu LYF, Kaffe S, Jenkins EC, Alonso L, Benn PA, David K, Hirschhorn K, Lieber E, Shanske A, Shapiro LR, Schutta E, Warburton D (1992) Proposed guidelines for diagnosis of chromosome mosaicism in amniocytes based on data derived from chromosome mosaicism and pseudomosaicism studies. Prenat Diagn 12:555-573

Johnson JAM, Wilson RD, Winsor EJT, Singer J, Dansereau J, Kalousek DK (1996) The early amniocentesis study: A randomized clinical trial of early amniocentesis versus midtrimester amniocentesis. Fetal Diagn Ther 11:85-93

Kennerknecht I, Baur-Aubele S, Grab D, Terinde R(1992) First trimester amniocentesis between 7th and 13th week. An experimental pilot study for evaluation of earliest possible genetic diagnosis. Prenat Diagn 12:595-601

Kennerknecht I, Krämer S, Grab D, Terinde R (1993) Evaluation of amniotic cell filtration: An experimental approach to early amniocentesis. Prenat Diagn 13:247-255

Krawczun MS, Jenkins EC, Masia A, Kunaporn S, Stark SL, Duncan CJ, Sklower SL, Rudelli RD (1989) Chromosomal abnormalities in amniotic fluid cell cultures: A comparison of apparent pseudomosaicism in Chang and RPMI-1640 media. Clin Genet 35:139-145

Ledbetter DH, Engel E (1995) Uniparental disomy in humans: Development of an imprinting map and its implications for prenatal diagnosis. Hum Mol Genet 4:1757-1764

Moertel CA, Stupca PJ, Dewald GW (1992) Pseudomosaicism, true mosaicism, and maternal cell contamination in amniotic fluid processed with in situ culture and robotic harvesting. Prenat Diagn 12:671-683

Nadler HL (1968) Antenatal detection of hereditary disorders. Pediatr 42:912-918

Nuß S, Brebaum D, Grond-Ginsbach C (1994) Maternal cell contamination in amniotic fluid samples as a consequence of the sampling technique. Hum Genet 93:121-124

Philip J, Bang J Hahnemann N, Mikkelsen M, Niebuhr E, Rebbe H, Weber J (1974)Chromosme analysis of fetuses in risk pregnancies. Acta Obstet Gynec Scand 53:9-14

Qu JY, Dallaire L, Lemieux N. Drouin R, Richer CL (1989) Synchronization of amniotic fluid cells for high resolution cytogenetics. Prenat Diagn 9:49-56

Richkind KE, Risch NJ (1990) Sensitivity of chromosomal mosaicism detection by different tissue culture methods. Prenat Diagn 10:519-527

Steele MW, Breg WR (1966) Chromosome analysis of human amniotic fluid cells. Lancet I:383-385

Sundberg K, Smidt-Jensen S, Philip J (1991) Amniocentesis with increased cell yield obtained by filtration and reinjection of the amniotic fluid. Ultrasound Obstet Gynecol 1:91-94

Sundberg K, Bang J, Brocks V, Flemming RJ, Schmidt-Jensen S, Philip J (1995) Early sonographically guided amniocenteses with filtration technique: Follow up on 249 procedures. J Ultrasound Med 14:585-590

Sundberg K, Bang J, Schmidt-Jensen S, Brocks V, Lundsteen C, Parner J, Keiding N, Philip J (1997) Randomised study of risk of fetal loss related to early amniocentesis versus chorionic villus sampling. Lancet 350 (9079): 697-703

Tabor A, Lind AM, Andersen AM, Philip J (1984) A culture vessel for amniotic fluid cells allowing faster preparation of chromosome slides. Prenat Diagn 4:451-455

Valenti C (1968) Chromosomal study of throphoblastic tissue. Am J Obstet Gynecol 92:211-214

Wilson RD (1995) Early amniocentesis: A clinical review. Prenat Diagn 15:1259-1273

Worton RG, Stern R (1984) A Canadian collaborative study of mosaicism in amniotic fluid cell cultures. Prenat Diagn 4:131-144

Chapter 13

Chorionic Villi Sampling

ROLF-DIETER WEGNER AND HOLGER TOENNIES

Introduction

Since the mid 80s chorionic villi sampling (CVS) has become an important and powerful procedure in prenatal diagnosis. In this chapter, the abbreviation CVS is used in the common, broader sense including also the analysis of the extracted tissue.

In 1983, Simoni et al. showed that chorionic villi sampled in the first trimester of pregnancy could be reliably processed for chromosome analysis. In the following years a large series of studies confirmed these findings. However, it became obvious that cytogenetic discrepancies between the extraembryonic tissue and the fetus proper exist in about 1,5 % of cases (Hahnemann and Vejerslev, 1997). In the majority of cases this is due to an abnormal chromosome constitution restricted to the extraembryonic tissue while the fetus possesses a normal chromosome set, a constellation termed confined placenta mosaicism - CPM (Kalousek, 1983). Most frequently, only one layer of the chorionic villi - the cytotrophoblast (Figure 1) - shows trisomy mosaicism while the other layer - the mesenchymal core - and the fetus exhibit a normal chromosome constitution.

For an understanding of this phenomenon, inherent to chorionic villi, the structure of chorionic villi and the early embryonic development must be known: At the end of the first trimenon a typical villus consists of three layers: syncytiotrophoblast, cytotrophoblast, and mesenchymal core (Figure 1). The syncytiotrophoblast does not possess any mitotic potential and is of no importance for the cytogenetic analysis. The cytotrophoblast is characterized by a high spontaneous rate of cell division and provides al-

Correspondence to Rolf-Dieter Wegner, Charité Campus Virchow-Klinikum, Institut für Humangenetik, Augustenburger Platz 1, Berlin, 13353, Germany (*phone* +49-30-450-66123; *fax* +49-30-450-66904), Holger Toennies, Charité Campus Virchow-Klinikum, Institut für Humangenetik, Augustenburger Platz 1, Berlin, 13353, Germany most exclusively the metaphases seen in direct preparations or short term culture (STC). Cells of the mesenchymal core generally show no spontaneous mitotic activity but they resume the cell cycle when kept in culture either after mincing the tissue or after enzymatic digestion to provide a single cell suspension (Wegner, 1993). Such long term cultures (LTC) need about 1 - 3 weeks of growth to obtain a sufficient number of cells for chromosome preparation.

A schematic drawing of early embryonic development (Figure 2) helps to explain cytogenetic discrepancies between the extraembryonic tissue and the embryo. In particular, the higher number of discrepancies between the cytotrophoblast and the embryo proper as compared to the number found between the mesenchymal core and the embryo will become obvious. Two chromosome constellations of the zygote must be considered: A normal chromosome set and a pathological chromosome set.

- The cytotrophoblast lineage separates earliest from the embryonic lineages. Chromosome aberrations occurring in the cytotrophoblast immediately after its differentiation do not affect the embryonic lineages

Fig. 1. Schematic drawing of a longitudinal section through a chorionic villus showing all three cell layers.

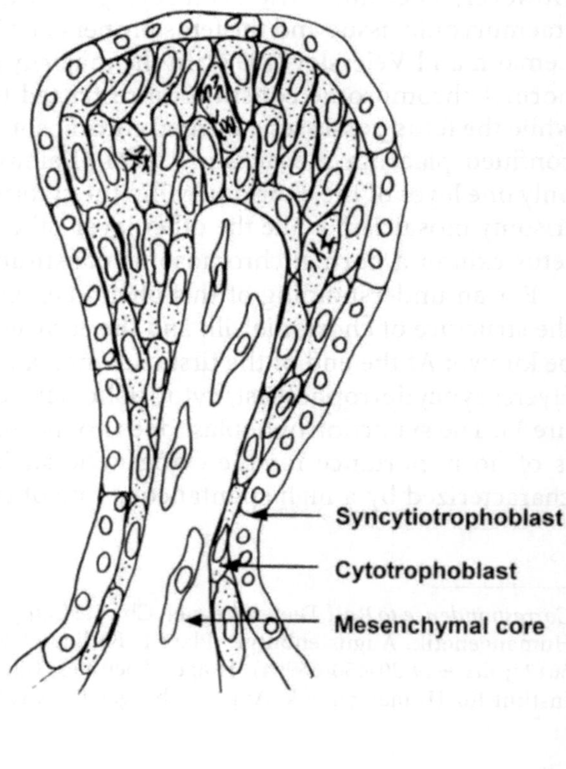

(containing also cells determined to become the mesenchymal core). The probability of suffering a mutation is probably about 20 times higher for the trophoblast as compared to the inner cell mass when the ratio of cells are taken into consideration. So, studies in mice show that the fetus proper is derived from just three progenitor cells - the inner cell mass - of the 64-cell blastocyst while all other cells are dedicated to become the trophoblast (Markert and Peters, 1978).

- More frequently, cytogenetic discrepancies may be due to a phenomenon called trisomy rescue (for review see Ledbetter and Engel. 1995): In cases of a trisomic zygote, mitotic non-disjunction or anaphase lag may lead to loss of the supernumerary chromosome leading to mosaicism. The strength of selection against a pathological cell line is much higher in the developing embryo than in the extraembryonic tissue. Thus, conditions where the inner cell mass contains only the derived diploid cells are favored. The various possible mosaic constellations depend on the time in embryogenesis and on the location of trisomy rescue. Experimental studies with chimeric mouse embryos (James and West, 1994) further support the assumption that natural selection for the embryo with only the normal cell line may be a common mechanism in creating the cytogenetic discrepancies disclosed by CVS.

As mentioned before the limited cytogenetic reliability of CVS-STC can be compensated for by karyotyping cells from the LTC. However, it is obvious that the phenomenon of CPM requires an intensive genetic counselling of the pregnant women. Examples of uncommon cytogenetic discrepancies and practical consequences are given in the section "Trouble shooting".

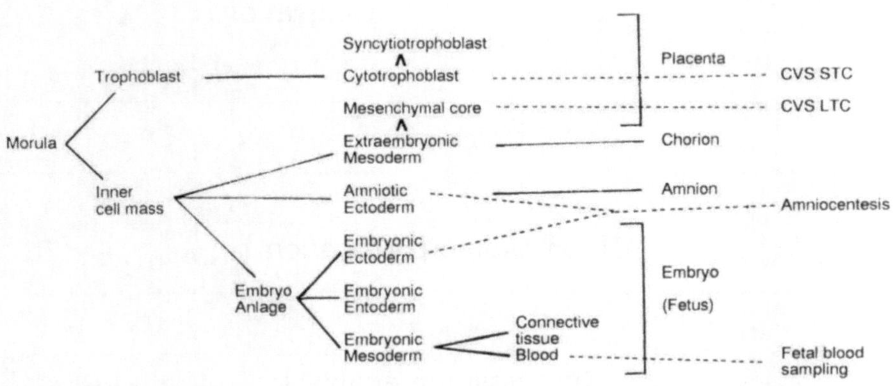

Fig. 2. Diagram presenting a model of lineage differentiation in early embryonic development and the tissues analyzed by various cytogenetic techniques (Wegner, 1995).

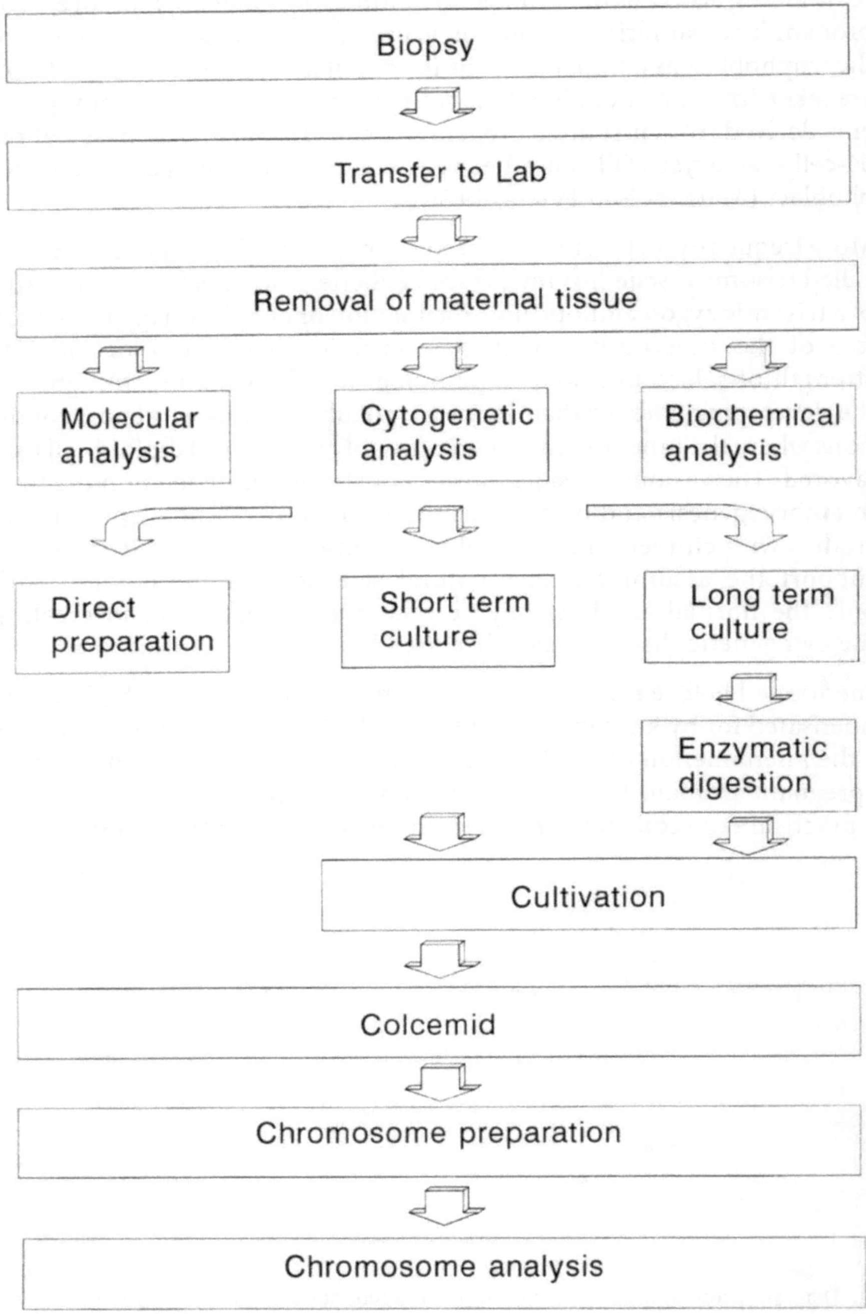

Fig. 3. Flow sheet demonstrating the processing of chorionic villi samples

A further topic of concern, the assumption of limb defects induced by CVS is still a matter of discussion. Proposals are issued recommending tissue sampling not before the 11th week of pregnancy. While the outcome of studies performing CVS before the 10th week of pregnancy are discussed controversially it is widely agreed that CVS is safe thereafter.

For molecular or biochemical analysis sample sizes of 10 - 20 mg tissue are requested. When applicable, CVS is the method of choice for such investigations.

Today women opting for an invasive prenatal chromosome analysis have the choice between CVS and amniocentesis (see Chapter 12) notwithstanding fetal blood sampling at later times. Therefore a short comparison of the three techniques seems worthwhile and is given in the introduction to the chapters dealing with prenatal diagnosis (see Chapter 11).

The following protocol for STC is a modified technique initially described by Simoni et al., (1983). The enzymatic digestion protocol for LTC is based widely on a protocol provided by Mikkelsen (pers. communication). The flow chart (Figure 3) shows how analysis of chorionic villi might take place.

Subprotocol 1
Setting-up a Short Term Culture

Materials

Equipment

- Glass pipettes (150 mm) [Roth E 326.1]
- Plastic petri dishes 94 mm (PD)[Greiner # 633171]
- Plastic petri dish 60 mm [Greiner # 628160]
- Plastic petri dish 35 mm [Greiner # 627160]
- Invertoscope [Zeiss]
- Screw cap tubes [Recker # 1795-G, sterile]

Solutions

Transport medium:
- Streptomycin [Grünenthal #757753B]
- Penicillin [Grünenthal # E744114]
- Dulbecco's MEM/Ham's F12 [Gibco/BRL # 21331-020]

- L-glutamine (200 mM) [Seromed # K0282]
- Hepes buffer (1M) [Biochrom # L1613]
- Liquemin (25000 I.E.) [Roche # 47197]

Dissolve Streptomycin (1g) in 5 ml aqua dest. (sterile).

Dissolve Penicillin (1 000 000 U) in 5 ml aqua dest. (sterile).

Prepare transport medium
- 500 ml medium Dulbecco's MEM/Ham's F12
- 7.5 ml L-glutamine
- 10 ml Hepes buffer
- 7.5 ml Liquemin
- 0.25 ml Streptomycin
- 0.25 ml Penicillin
  Aliquot in 7 ml portions in screw cap tubes.

Washing solutions
- Nystatin (500 000 U) [Sigma # N6261]
- Penicillin (1 000 000 U) [Grünenthal # E744114]
- Eagle's MEM [Gibco/BRL # 072-01500]
- Hepes buffer (1M) [Biochrom # L1613]
- Fetal calf serum, FCS [Gibco/BRL # 10270-106]

Prepare antibiotic solution
- Dissolve Nystatin in 1 ml 0.9% (w/v) NaCl (sterile)
- Dissolve Penicillin in 10 ml 0.9% (w/v) NaCl (sterile)
- Mix 1 ml Nystatin and 1 ml Penicillin with 98 ml NaCl (0.9% (w/v); sterile).

Prepare washing medium
- Eagle's MEM medium containing 2% Hepes buffer

Prepare culture medium
- Eagle's MEM + 5% FCS

Procedure

1. Fill 10 ml of washing medium into a sterile 94 mm plastic petri dish and add 50 µl of the antibiotic solution.
2. Fill 10 ml of washing medium into a second sterile 94 mm plastic petri dish and add 25 µl of the antibiotic solution.
3. Fill a third petri dish with washing medium.
4. Pour the chorionic villi sample received in transport medium into the first petri dish and note blood and detritus contamination in a protocol; rinse tissue by rocking the dish.
5. Transfer chorionic villi (CV) in small drops with the aid of a Pasteur pipette into the second petri dish, rinse thoroughly.
6. Transfer CV into the third dish and wash villi again so that no blood clots are sticking to the tissue.
7. Suck up each piece of CV separately and transfer the tissue in a drop to a 60 mm dish
8. Dissipate each tissue piece with a fine forceps, check for maternal contamination in an inverted microscope and remove all obscure material; mark CV which are not completely free of contaminating tissue for short term culture.
9. Transfer 5-6 mg CV in a 35 mm petri dish and add 2 ml culture medium.
10. Incubate CV over night in an incubator (37°C; 5% CO_2) for short term culture.
11. Process remaining CV immediately for long term culture (preferably >12 mg) or place villi in a petri dish (60 mm) with 3 ml Eagle's MEM and keep it overnight in an incubator before processing for LTC.

Subprotocol 2
Chromosome Preparation from Short Term Culture

Materials

Equipment

- Cleaned slides (Superfrost) [Menzel # 507894]
- Glass pipettes (150 mm) [Roth E 326.1]
- Filter paper stripes [Schleicher and Schuell # 311645]
- Automatic preparation device:
  Appareil ECT No 91181, Type ECT 85 No 71, Type A/Watts 60/Volts 220/ HZ 50 (Productions TOULEMONDE, Rue Pascal, 93120 LA COURNEUVE)

Solutions

- Colcemid (10 µg/ml) [Gibco/BRL # 15210-016]
- Sodium citrate 1% (w/v) [Merck # 6404.1000E]
- Acetic acid [Merck # 1.00063]
- Deionized water
- Fixative (3:1 (v/v) methanol/acetic acid)

Procedure

1. Add 20 µl Colcemid to STC and incubate 1.5 h (37°C).
2. For automatic slide preparation switch on the machine (temperature) and place the ethanol cleaned slides on the heated plate (40°C).
3. Aspirate medium with Pasteur pipette, add 3 ml sodium citrate and incubate for 15 minutes at room temperature.
4. Prepare hydration series (see 7.) and fixative (store at -20°C); aspirate sodium citrate, add 3 ml fixative and incubate for 10 minutes at 4°C.
5. Change the fixative and incubate for 10 minutes at 4°C in fresh fixative.
6. Prepare 60% acetic acid.
7. Remove fixative and perform hydration with an ethanol series as follows

abs. EtOH, 3 ml, 30 sec
70 % EtOH, 3 ml, 30 sec
50 % EtOH, 3 ml, 30 sec

8. Rinse in deionized water for 30 seconds
9. Rinse in deionized water for 2 minutes.
10. Aspirate deionized water, dry dish and villi gently with filter paper.
11. Add 16-18 drops of 60 % acetic acid and keep the inclined dish for 3 minutes at room temperature.
12. Place on each cleaned slide one fixed villi and run apparatus for 3 minutes.
13. Remove slides, blot any remaining fluid by tapping edge of slide, discard surplus material and allow remaining acetic acid to evaporate.
14. Stain cells for 10 min in Giemsa solution (10 %).

Note: Fixative should be prepared freshly before starting a next series.

Subprotocol 3
Setting-up a Long Term Culture by Physical Maceration

Materials

Equipment

- Glass coverslips 10X30 mm [Menzel, special manufacturing]
- Plastic petri dish 60 mm [Greiner # 628160]
- Plastic petri dish 150 mm [Falcon # 3025]
- Scalpel blades [Feather # 5504633]
- Ambitubes [Nunc # 156758]
- Fine forceps
- Tissue culture flasks (50 ml) [Falcon # 3013]

Solutions and medium

- Amniomax [Gibco/BRL # 17001-082]
- Dulbecco's MEM/Ham's F12 [Gibco/BRL # 21331-020]
- Nutrient mix Ham F-12 [Gibco/BRL # 21765-011]

- Dulbecco's MEM [Gibco/BRL # 31885-015]
- L-glutamine (200 mM) [Biochrom # K0282]
- FCS
- Ultroser [Gibco/BRL # 15950-017]
- Trypsin/EDTA solution, 0,5 g/l / 0,2 g/l [Gibco/BRL # 45300-019]

Prepare DHU culture medium:

- 50 ml Nutrient mix Ham F-12
- 50 ml Dulbecco's MEM
- 2,5 ml L-glutamine
- 10 ml FCS
- 2 ml Ultroser
- 150 µl Penicillin and Streptomycin-mix (see transport medium)

Procedure

1. Put 2 sterile glass coverslips side by side in a petri dish (60 mm).
2. Place 1-2 mg cleaned villi on each coverslip and aspirate surplus liquid.
3. Mince CV with scalpel blades on the cover slip until all the tissue is a fine mush of tiny fragments, collect and spread the material on two coverslips.
4. Put each coverslip in an ambitube upside down to sandwich the tissue in between the coverslip and bottom.
5. Press the coverslip lightly against the bottom of the tube and add one pipette amniomax taking care not to allow the coverslip to float off.
6. To the second culture add one pipette DHU medium.
7. Incubate the cultures overnight at 37°C and 5% CO_2.
8. Close the tubes next day.
9. After 3 or more days cell growth will be seen at the edge of the tissue pieces.

10. Feed the cultures after 7 days.

11. Cells in colonies should be trypsinized to obtain an evenly distributed cell lawn. Prepare metaphases a few days later when culture is growing exponentially.

Subprotocol 4
Setting-up a Long Term Culture by Enzymatic Dissociation

Materials

Equipment

- Snap cap tubes (14 ml) [Falcon # 2001]
- Tissue culture flasks (50 ml) [Falcon # 3013]
- Ambitubes [Nunc # 156758]
- Centrifuge [Heraeus Sepatech, Rotor # 3360/BS4402/A]

Solutions

- Trypsin/EDTA [Gibco/BRL # 45300-019]
- Amniomax [Gibco/BRL # 17001-082]
- Collagenase Type IV solution [Sigma # C5138] 4000 U/ml stock solution (sterile filtrated)

Dissolve in 10 ml DHU:

- 78.125 mg collagenase Type IV (512 U/mg)
- 2.0 mg DNase [Sigma # DN25]
- Make aliquots of 200 µl and store at - 20°C

Note: Collagenase batches vary in activity. In the present example the weight of the Collagenase powder relates to the activity of 512 U/mg. The actual weight has to be calculated for every new batch of the enzyme.

Procedure

1. Transfer CV unequivocally free of maternal tissue from medium into a snap cap tube containing 2 ml trypsin/EDTA. Take care not to transfer medium into the enzyme solution, in particular no FCS-containing medium which will inactivate the enzymatic activity.
2. Incubate for 60 min in a 37°C water bath and agitate gently every 15 minutes.
3. After incubation spin for 10 min at 1000 rpm.
4. Transfer sediment into a second tube containing 2 ml working solution of collagenase, mix thoroughly and incubate for 90 min in a water bath (37°C).
5. After incubation fill up to 10 ml with Eagle's medium and spin at 1000 rpm for 10 min.
6. Remove supernatant until 1 ml and resuspend the sediment.
7. Transfer 2/3 of the suspension into the first culture flask and fill up with 3 ml amniomax (A-culture).
8. Transfer the remaining suspension into an ambitube filled up with 1ml DHU (B-culture).
9. Renew the medium after two days and note the growth of cells, e.g. evenly distributed cell lawn or - in case of a few number of surviving cells - the number of colonies.
10. After appropriate cell growth harvest cells and prepare metaphases.

Subprotocol 5
Cell Harvest and Chromosome Preparation

Materials

Equipment

- Snap cap tube (13 ml) [Greiner # 172101]
- Glass slides [Menzel # 507894]

Solutions

- Colcemid (10μg/ml) [Gibco/BRL # 15210-016]
- Trypsin/EDTA [Gibco/BRL # 45300-019]
- Amniomax [Gibco/BRL # 17001-082]
- Hypotonic solution (KCl 0.075M)[Merck # 4936.1000]
- Cold fixative (3:1 (v/v) methanol/acetic acid)

Procedure

Pay attention that all steps are done under sterile conditions. Cell harvest must be done under a sterile working bench.

1. When cell density and mitotic index is appropriate add 100 μl Colcemid and incubate for 1.5 h.
2. Collect the medium of the culture flask in a snap cap tube.
3. Rinse the culture with 2 ml trypsin and also collect this solution in the tube.
4. Add 6 drops of trypsin/EDTA to the culture and wait until 2/3 of cells are detached.
5. Add 3 ml amniomax and transfer the cell suspension into a second snap cap tube.
6. Add 3 ml amniomax to the culture flask and incubate the culture.
7. Spin snap cap tubes at 800 rpm for 10 min.
8. Discard supernatant and resuspend sediment gently.
9. Add 5 ml prewarmed (37°C) hypotonic solution, resuspend the pellet gently and incubate 20 min in a water bath (37°C).
10. Spin at 800 rpm for 10 min, discard supernatant and resuspend sediment.
11. Add 5 ml fixative (-20°C) dropwise (slowly) to the cells and resuspend gently.
12. Incubate 15 min at 4°C and spin at 800 rpm for 10 min.
13. Remove supernatant, resuspend sediment and add 5 ml fresh fixative.

14. Incubate 10 min at 4°C, spin, discard supernatant, resuspend pellet and add fresh fixative (5 ml).
15. Incubate for 30 min at -20°C.
16. Centrifuge, remove supernatant and adjust density of cell suspension with fresh fixative to get ready for preparing slides.
17. Drop cell suspension on cooled (-20°C) ethanol-cleaned and moistened slides. Air-dry the slides.

Troubleshooting

In the introduction problems have already been mentioned which might arise by confined placenta mosaicism (CPM). CPM, with mosaic or non-mosaic aberrations restricted to the extraembryonic tissue while the fetal karyotype is normal, presents a situation known as false positive case. The experienced cytogeneticist can recognize and cope easily with most of these false positives (see Example 1). On the other hand examples of false negative findings in STC - we became aware of two cases in the course of more than 2000 CVS - can pose serious problems when fast decisions are necessary (example 2) or findings are unexpected (example 3).

Example 1 A CVS-STC carried out due to maternal anxiety resulted in a mosaic with cells showing a normal karyotype and cells with two Robertsonian translocation chromosomes t(13;14). The Robertsonian translocation was inherited from the father. After genetic counselling to provide information about the high probability that the fetus will not possess cells with the double trisomy 13 and 14 the mother asked for an amniotic fluid cell analysis. This investigation resulted in a mosaic ratio very similar to the CVS finding. Further intensive discussions with the mother led to her decision to opt for a fetal blood analysis. Eventually no abnormal cells were detected and the mother carried the pregnancy to term. A completely normal boy was delivered.

Recommendations providing standards for CVS, mirroring the German experience with CVS, has been published (Vogel, 1995; Guidelines of the Professional Association "Medical Genetics", 1997). Therein, the necessity that the cytogeneticist who is engaged in CVS needs special experience/ training is underlined.

Example 2

A prenatal diagnosis was requested in the 26th week of pregnancy because of serious fetal distress. Ultrasonically, the fetus did not represent any malformation. CVS-STC revealed a normal male karyotype in all analyzed metaphases. Two days later the child's condition became life threatening, labour was induced, a boy was delivered and taken under neonatal intensive care. The analysis of cells from CVS-LTC showed a completely unexpected result: trisomy 21, 47,XY,+21, in all metaphases. Knowing the infant's karyotype mild clinical symptoms of Down syndrome were found pointing out the problem of clinical diagnosis in preterm infants even in cases of Down syndrome. A lymphocyte culture confirmed the pathological result obtained from CVS-LTC.

Example 3

Prenatal diagnosis was indicated due to maternal age. Structural analysis of metaphases from CVS-STC (~ 400 bands/haploid genome) resulted in a

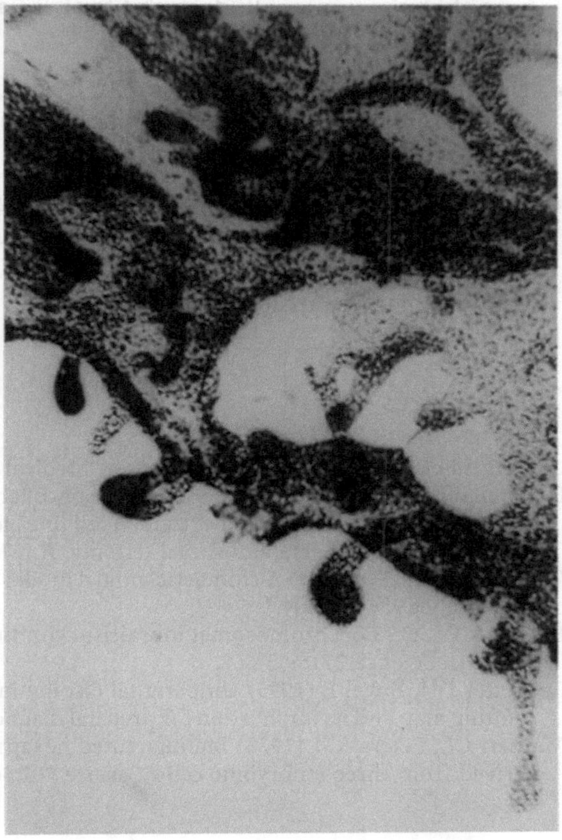

Fig. 4. Microphotograph of chorionic villi showing the typical club-like ramifications.

normal karyotype. Numerical analysis of CVS-LTC also showed normal results. After delivery the newborn showed malformations, e.g. heart defect, microcephaly, microphthalmia and hydrocephalus internus. In lymphocyte metaphases a structural aberrant chromosome 6 was seen. Reanalysis of CVS-STC confirmed the normal karyotype in these cells. However, the metaphases from LTC possessed the aberrant chromosome 6 which had been missed in the course of the numerical analysis. Since that time the strategy of CVS chromosome analysis was remodelled and structural analysis is performed on cells from LTC.

This first case of a non-mosaic discrepant structural chromosomal aberration between CVS-STC and CVS-LTC has been published in detail elsewhere (Wegner et al., 1996).

Technical tips

- Sometimes it is difficult to recognize and/or remove all maternal tissue. Chorionic villi typically show club-like ramifications (Figure 4). Make sure that chorionic villi destined for LTC are unequivocally free of maternal tissue.
- During slide preparation by hand or by machine do not allow the fixative to evaporate completely. This will lead to a high background of detritus hampering chromosome analysis.
- In case of very poor metaphase spreading after STC which prevents any chromosome counting you should flame-dry the slides. This might provoke spreading of chromosomes but will give poor chromosome quality usually insufficient for banding.

References

Guidelines of the Professional Association "Medical Genetics" (1997) Med. Genetik 9

Hahnemann JM, Vejerslev LO (1997) European collaborative research on mosaicism in CVS (EUCROMIC) - fetal and extrafetal cell lineages in 192 gestations with CVS mosaicism involving single autosomal trisomy. Am J Med Genet 70:179-187.

James RM, West JD (1994) A chimaeric animal model for confined placental mosaicism. Hum Genet 93: 603-604.

Kalousek DK (1983) Chromosomal mosaicism confined to the placenta in human conceptions. Science 221:665-667.

Ledbetter DH, Engel E (1995) Uniparental disomy in humans: development of an imprinting map and its implication for prenatal diagnosis. Hum Mol Genet 4: 1757-1764.

Markert CL, Peters RM (1978) Manufactured hexaparental mice show that adults are derived from three embryonic cells. Science 202:56-58.

Simoni G, Brambati B, Danesino C, Rosella F, Terzoli GL, Ferrari M, Fraccaro M (1983) Efficient direct chromosome analysis and enzyme determinations from chorionic villi samples in the first trimester of pregnancy. Hum Genet 63:349-357.

Vogel W (1995) Recommendations for prenatal diagnosis from chorionic villi by the Scientific Advisory Board of the German Collaborative Study . In: Stengel-Rutkowski S (ed) Early prenatal diagnostics. Dr. Kovac Verlag, Hamburg.

Wegner RD (1993) Chorionic villi analysis. In Obe G (ed) Advances in mutagenesis research 4. Springer Verlag Berlin, Heidelberg, New York.

Wegner RD (1995) Cytogenetic reliability of CVS: the German collaborative study in comparison to other multi-center studies. In: Stengel-Rutkowski S (ed) Early prenatal diagnostics. Dr. Kovac Verlag, Hamburg.

Wegner RD. Schröck E, Obladen M, Becker R, Stumm M, Sperling K (1996) Partial trisomy/monosomy 6q in fetal cells and CVS long term culture not present in CVS short term culture. Prenat Diagn 16:741-748.

Part IV

Special Applications

Chapter 14

Diagnosis of Chromosomal Instability Syndromes

ROLF-DIETER WEGNER AND MARKUS STUMM

Introduction

Some monogenic diseases show genetic defects which are also expressed at the cellular or at the chromosomal level (Table 1). Characteristic cytogenetic abnormalities allow their unequivocal identification. For diagnosis a routine chromosome analysis is sufficient in a number of disorders while others require either an exposure to mutagens prior to the cytogenetic investigation or a cell cycle analysis by fluorescence activated cell sorting (FACS).

Several of the underlying genes of the above mentioned disorders are already localized or even cloned (Table 1). However, a fast and simple molecular analysis is frequently hampered due to genetic heterogeneity and/or the large size of the genes. Therefore, cytogenetic or FACS analysis is still commonly requested as the primary laboratory investigation.

The specific group of classical chromosomal instability syndromes includes three genetic disorders namely Bloom's syndrome (BS), Fanconi's anemia (FA) and Ataxia telangiectasia (AT). In the last decade, cytogenetic studies show that the Nijmegen breakage syndrome (NBS) has to be added to this group.

For decades, the ultimate diagnosis of chromosomal instability syndromes had been based on cytogenetic means. Thus, Bloom's syndrome (MIM 210900) can be unequivocally confirmed by the highly increased rate of sister chromatid exchanges (SCEs) in lymphocyte cultures (Chaganti et al., 1974). A detailed protocol for SCE analysis is given elsewhere in this manual (Chapter 2). In the present chapter, protocols are provided to confirm cytogenetically the clinical diagnosis of FA, AT and NBS. Testing for AT

Correspondence to Rolf-Dieter Wegner, Charité Campus Virchow-Klinikum, Institut für Humangenetik, Augustenburger Platz 1, Berlin, 13353, Germany (*phone* +49-30-450-66288-123; *fax* +49-30-450-66904; *e-mail* rolf-dieter.wegner@charite.de), Markus Stumm, Universitätsklinikum, Institut für Humangenetik, Leipziger Str. 44, Magdeburg, 39120, Germany

Table 1. Genetic disorders exhibiting characteristic chromosomal peculiarities. The extent of genetic heterogeneity, diagnostic chromosomal features, gene localization, and the type of analysis, ie a routine blood analysis vs. a specialized investigation are listed. References to each genetic entity are ordered to refer to gene localization, to gene structure, and to a review or a specific clinical paper. IR: Ionizing radiation, RDS: Radio-resistant DNA-Synthesis, SCE: Sister chromatid exchange.

Disorder Gene	Cytogenetic abnormality	Chromosomal Localization	Specialized diagnostics	References
Ataxia telangiectasia ATM	Hypersensitivity to IR, RDS	11q23.1	+	1, 2, 3, 4
Bloom's syndrome BLM	Increased rate of SCEs	15q26.1	+	5, 6, 7
Fanconi's anemia	Hypersensitivity to alkylating agents			
FACA		16q24.3	+	8, 9, 10,11
FACB		?	+	12
FACC		9q22.3	+	12
FACD		3p22-26	+	12, 13
FACE		?	+	14, 15
FACF to FACH				16
ICF	Heterochromatin anomalies (stickiness, condensation)	20q11-q13[a]	-	17, 18
Nijmegen breakage syndrome NBS	Hypersensitivity to IR, RDS	8q22	+	19, 20, 21
Robert's syndrome	Heterochromatin repulsion	?	-	22, 23
Werner's syndrome WRN	Chromosomal variegation	8p12	- fibroblasts	24, 25, 26, 27

References: 1 Gatti et al. (1988); 2 Savitsky et al.(1995).; 3 Uziel et al. (1996); 4 Shiloh Y (1995); 5 German et al. (1994); 6 Ellis et al., 1995; 7 German, 1993; 8 Duckworth-Rysiecki et al., 1985; 9 Pronk et al., 1995; 10 Lo Ten Foe et al., 1996; 11 The Fanconi anaemia/Breast cancer consortium, 1996; 12 Strathdee et al., 1992; 13 Whitney et al., 1995; 14 Joenje et al., 1995; 15 Wegner et al., 1996; 16 Joenje et al., 1997; 17 Hulten 1978; 18 Smeets et al.,1994; 19 Hustinx et al., 1979; 20 Saar et al., ; 21 Wegner et al. 1999; 22 Freeman et al. (1974); 23 van den Berg and Francke, 1993; 24 Salk et al. (1981); 25 Goto et al., 1992; 26 Schellenberg et al., 1992; 27 Yu et al., 1996, [a] Wijmenga et al. (1998)

or NBS can be done by identical approaches since the cellular features of both disorders are almost identical despite a very distinct clinical presentation.

The blood culture technique and the analysis of chromosomal breakage rates are described in the first part of this contribution since these protocols are applicable for all the disorders. Subsequently, descriptions of the mutagenic treatments specific for FA or AT/NBS are presented.

Materials

Tissue culture

Blood culture medium

- HAM's F-12 [Biochrom, F0815] or RPMI 1640 Medium [Biochrom, F1215]
- Fetal calf serum - FCS - (15 %) [Biochrom, S0115]
- Glutamine L (2 mM) [Biochrom, K0282]
- Antibiotics (Penicillin 100 I.E./ml; Streptomycin 100 μg/ml) [Biochrom, A2212]
- Phytohaemagglutinin - PHA (1 %) [Biochrom, M5030]

Radio-resistant DNA-Synthesis (RDS)

- Black plastic box with rack [Roth, 2285.1]
- Dipping jar with a small volume and an opening slit just sufficient to let two slides pass.
- Methyl-^{3}H-thymidine, aqueous solution (50-90 Ci/mmol) [DuPont, NET-027Z] or Methyl,1',2',-^{3}H-thymidine, ethanol : water (7:3) (20 Ci/mmol) [DuPont, NET-027E]
- Bleomycin [Mack, 10216]
- Nuclear Research Emulsion K2 in gel form [Ilford]
- Developer D19 [Kodak, 1464593]
- Kodak Fixer [Kodak, 1971746]

Procedure

Blood culture

1. Add 0,4 ml peripheral blood to 4,6 ml blood culture medium contained in a sterile tube. Set up triplicate cultures for each individual to be tested (Figure 1). Analyze one untreated culture and two exposed cultures to determine the spontaneous breakage rate and the induced breakage rates, respectively. Apart from the index patient, blood cultures of a normal individual (negative control, eg a tested person from the lab staff) and if available of a positive control (eg index patient in a familial case) should be initiated and handled simultaneously.
2. Incubate for a total of 72 hours at 37°C. The timing of mutagenic treatment depends on the used clastogen (see below "Working with mutagenic agents", "TRE treatment", "Mitomycin C (MMC)" and Figure 1).
3. Add Colcemid (0,06 μg/ml) two hours before harvest.
4. Prepare air-dried slides as described in Chapter 5.
5. It should be kept in mind that cells from patients with chromosomal instability syndromes may show a very poor PHA-stimulation response. Thus, EBV-transformation of lymphocytes to obtain LCLs (see Chapter 6) is advisable for further studies but will not overcome the problems of time-consuming work or culture failure when an immediate result is needed.

Chromosome breakage analysis

Working with mutagenic agents

Mitomycin C (MMC), Diepoxybutane (DEB), Trenimon (TRE) and Bleomycin (BLE) are mutagens and potential carcinogens. Precautions should be taken when handling these compounds. A proper disposal of waste is mandatory. It is essential to follow closely general safety guidelines and to adhere to local policy advice.

Note:

- All work with these agents should be done under a class II biological safety cabinet.
- The investigator should wear an apron and gloves. In the case of handling powder use an inhalation protection mask.

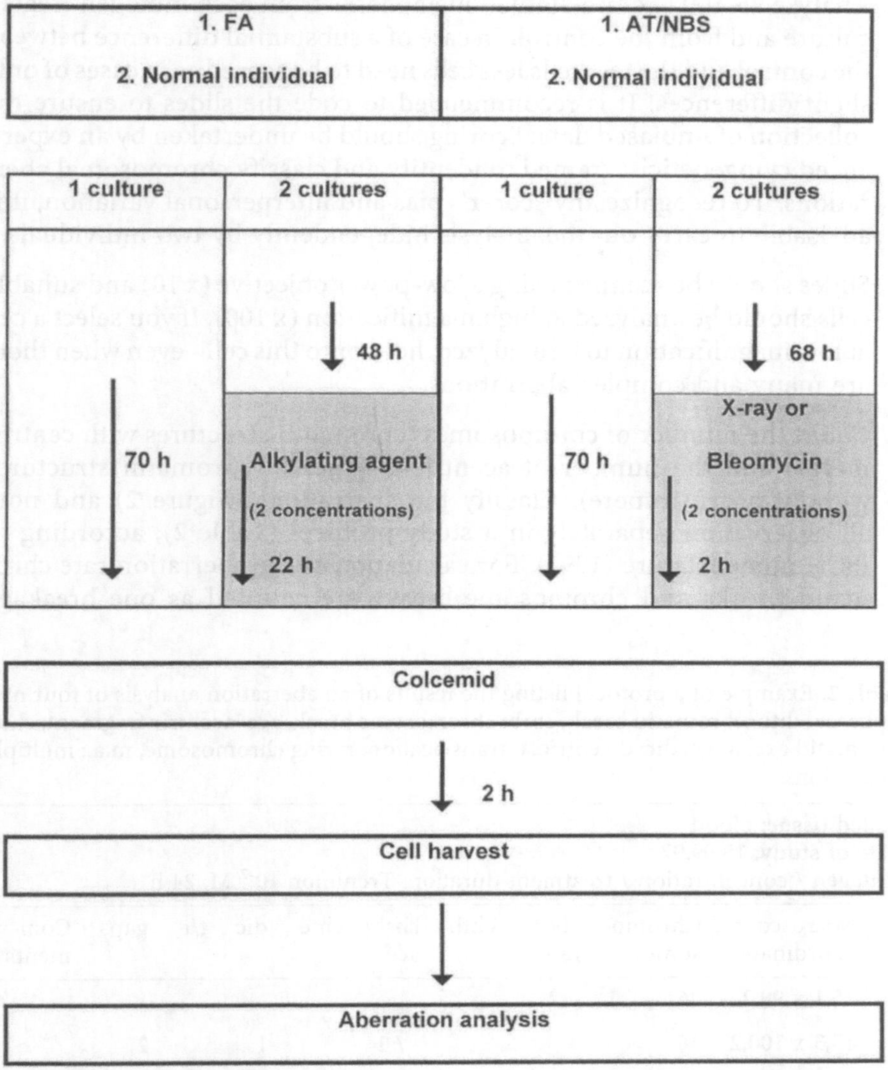

Fig. 1. Flow-chart showing the steps in the cytogenetic diagnosis of Fanconi's anemia (FA), Ataxia telangiectasia (AT) and Nijmegen Breakage Syndrome (NBS). Cultures of a normal individual should be treated simultaneously as reference.

- In all experiments disposable plastic items should be used which have to be disposed as mutagenic waste.
- Medium and rinsing solutions containing mutagens should also be collected and disposed as mutagenic waste.

Aberration analysis

1. Analyze 25-100 Giemsa-stained metaphases from each mutagen treated culture and from the control. In case of a substantial difference between the control and the test cells less cells need to be scored as in cases of only slight differences. It is recommended to code the slides to ensure the collection of unbiased data. Scoring should be undertaken by an experienced cytogeneticist trained to identify and classify chromosomal aberrations. To recognize any scorer's bias and interpersonal variation, it is advisable to carry out the analysis independently by two individuals.

2. Slides should be scanned using a low-power objective (x 10) and suitable cells should be analyzed at high magnification (x 100). If you select a cell at low magnification to be analyzed, hold onto this cell - even when there are many and complex aberrations.

3. Count the number of chromosomes (chromatin structures with centromeres) and the number of acentric fragments (chromatin structures without a centromere). Classify the aberrations (Figure 2) and note all observation separately in a study protocol (Table 2), according to ISCN nomenclature (1995). For calculation of the aberration rate chromatid breaks and chromosome breaks are counted as one breakage

Table 2. Example of a protocol listing the results of an aberration analysis of four metaphases. chtb: chromatid break, chrb: chromosome break, ace: acentric fragment, chte: chromatid exchange, dic: dicentric, t: translocation, r: ring chromosome, m.a.: multiple aberrations.

Tested tissue: blood
Date of study: 18.09.97
Mutagen / concentration / treatment duration: Trenimon 10^{-7} M, 24 h

n	Microscope coordinates	Chromosomes	aberrant	chtb	chsb/ ace	chte	dic	t/r	gaps	Comments
1	15,1 x 99,2	46	-							
2	15,3 x 100,2	46	+		- / 1		1		2	
3	16,9 x 102,6	45	+	2		1				
4	17,5 x 101,7	?	+	> 15	> 10	4	1	2		m. a.
¶	...									
50	...									
Σ										
	Breaks/affected cells:					Breaks/cell:				

event. Chromatid exchanges are counted as one breakage event per involved chromosome. Aberrations which include the formation of an acentric fragment, eg dicentrics, and ring chromosomes, are counted as two breakage events and corresponding acentric fragments are not taken into consideration. Cells with too many aberrations (>15) are noted in a special category "multiple aberrations" but will not be included in the calculation of the breakage rate.

4. Decode the slides and calculate the aberration frequency as breaks per cell (Table 2).
5. For statistical data analysis the χ^2-test can be used. However, a clear-cut difference in the breakage rates between exposed control cells and exposed patient's cells is expected (Tables 3 and 4) and must be found to confirm the clinical diagnosis.

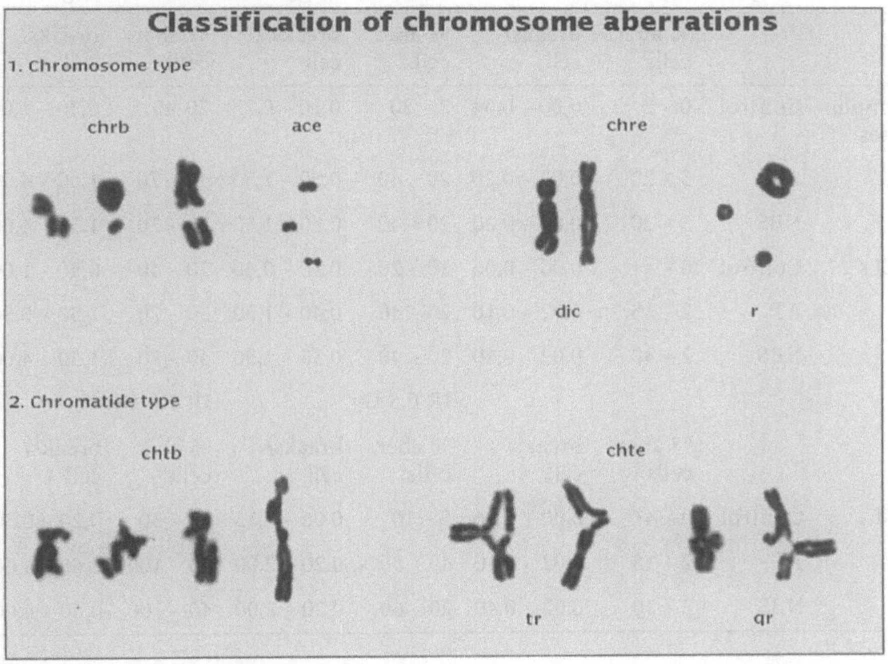

Fig. 2. Cut-outs showing chromosomal aberrations induced in lymphocytes of a patient with Fanconi anemia by Trenimon treatment for the last 24 hrs of the 72-hour culture time. chrb: chromosome break, ace: acentric fragment, chre: chromosome exchange, dic: dicentric, r: ring chromosome, chtb: chromatid break, chte: chromatid exchange, tr: triradial, qr: quadriradial

Table 3. Mean range of spontaneous and Trenimon-induced chromosomal damage in peripheral blood lymphocytes of some patients with Fanconi's anemia diagnosed in Berlin.

Tissue	Diagnosis	Untreated		Tre 10^{-8}M		Tre 10^{-7}M	
		% aber. cells	breaks/ cell	% aber. cells	breaks/ cell	% aber. cells	breaks/ cell
Lymphocytes	Control	0 - 2	0,00 - 0,04	0 - 16	0,00 - 0,20	25 - 80	0,30 - 1,00
	FA	5 - 50	0,10 - 0,80	35 - 100	0,50 - 2,00	50 - 100	2,00 - 6,00

Table 4. Mean range of spontaneous, irradiation- and Bleomycin-induced chromosomal damage in lymphocytes and LCLs of some patients with Ataxia-telangiectasia (AT) or Nijmegen Breakage Syndrome (NBS) diagnosed in Berlin. Bleomycin treatment took place in G_2-phase, irradiation in G_1-phase. BLE: Bleomycin, IR: Ionizing radiation, γ-rays.

Tissue	Diagnosis	Untreated		BLE 1 µg/ml		BLE 10 µg/ml	
		% aber. cells	breaks/ cell	% aber. cells	breaks/ cell	% aber. cells	breaks/ cell
Lymphocytes	Control	0 - 2	0,00 - 0,04	2 - 20	0,10 - 0,25	20-40	0,30 - 1,00
	AT	5 - 20	0,02 - 0,20	20 - 40	0,50 - 1,50	40 - 70	1,50 - 4,00
	NBS	5 - 20	0,02 - 0,20	20 - 40	0,50 - 1,50	40 - 70	1,50 - 4,00
LCLs	Control	0 - 4	0,00 - 0,06	10 - 20	0,10 - 0,40	20 - 40	0,40 - 1,00
	AT	2 - 15	0,02 - 0,10	20 - 40	0,40 - 1,50	50 - 70	1,50 - 5,50
	NBS	2 - 30	0,02 - 0,40	20 - 40	0,35 - 1,50	40 - 70	1,50 - 4,00
				IR 0,5 Gy		IR 1,0 Gy	
		% aber. cells	breaks/ cell	% aber. cells	breaks/ cell	% aber. cells	breaks/ cell
LCLs	Control	0 - 4	0,00 - 0,06	5 - 10	0,06 - 0,15	10 - 30	0,20 - 0,50
	AT	2 - 15	0,02 - 0,10	40 - 80	0,20 - 2,00	50 - 100	0,60 - 4,00
	NBS	2 - 30	0,02 - 0,40	20 - 60	0,20 - 2,00	40 - 100	0,60 - 4,00

Diagnosis of Fanconi anemia (FA)

FA (MIM 227645-227660) is inherited in an autosomal recessive mode with a prevalence of about 1:80.000 live births. Patients exhibit short stature, radial ray defects, skin hyperpigmentation, pancytopenia and predisposition to malignancies. Clinical diagnosis of FA is often equivocal because of the overlap of the FA phenotype with other diseases.

Cytogenetic hallmarks of FA cells are an increased spontaneous chromosome breakage and, as a specific feature, hypersensitivity to cross-linking agents such as DEB, MMC and TRE. A comprehensive review of the recent findings in FA has been given by Joenje (1996).

The following protocols allow to reveal chromosomal hypersensitivity of FA cells by exposure to the cross-linkers TRE and MMC, respectively. The preferred tissue for diagnosis is peripheral blood.

Results of chromosome mutation tests can be obtained at the earliest within one to two weeks.

Trenimon (TRE)

Trenimon or triaziquone is a trifunctional benzoquinone derivative. It was used formerly as a synthetic anticancer drug (for review see Obe and Beek, 1979; Rademacher and Obe, 1996). TRE is provided as a dark violet crystalline substance. It is soluble in water and in organic solvents such as ethanol.

TRE solution

1. Prepare a stock solution of 10^{-5} M TRE by dissolving it in sterile distilled water. Use a precision-scale since only a few crystals are needed. Dissolving Trenimon requires about 30 min until all crystals have disappeared.
2. Dilute the stock solution in complete medium to concentrations of 10^{-7}M and 10^{-8}M. Prepare sufficient medium for the treatment of all cultures (5 ml per tube).
3. Warm the medium to 37°C.

TRE treatment

1. After 48 hrs cultivation centrifuge the blood cultures for 10 minutes at 100 x g.
2. Remove the supernatant and add 5 ml warm Trenimon-containing medium to the pellet.
3. Resuspend the pellet carefully to obtain a single cell suspension.
4. Incubate cultures for an additional 22 hrs.
5. Add Colcemid (0,06 µg/ml) to each culture for further 2 hrs and prepare air-dried metaphases following standard protocols (see Chapter 5).

6. Analyze Giemsa-stained metaphases for chromosome aberrations (see "Aberration analysis").

Prenatal diagnosis

Prenatal diagnosis on amnion fluid cells or chorionic villi cells should be accompanied whenever possible by molecular analysis. If only a cytogenetic analysis is possible fibroblasts of the index patient should be available and his cells must be treated in parallel to the prenatal tissue. TRE concentration of 10^{-8} M and 10^{-9} M are recommended. Complementing the breakage analysis flow cytometry can be applied (see Chapter 15).

Personal experience and troubleshooting

- There is a wide interindividual range of chromosome damage induced by the Trenimon treatment as found by our group (Table 3). Typical aberrations induced by alkylating agents are shown in Figure 2.
- A few FA patients do not show any spontaneous chromosomal instability. Therefore, in case of a strong clinical suspicion of FA, hypersensitivity to crosslinking agents should be tested.
- FA testing should be offered to siblings of affected probands to take care of presymptomatic patients.
- A critical step in the treatment protocols is slide preparation. Metaphases need to be well spread, but not broken.
- TRE solution is very unstable and should not be stored for more than 24 hours. Do not use the TRE solution when it has lost its light purple color.
- LCLs can be treated with the same protocol as for lymphocytes.
- Fibroblasts are much more sensitive to TRE than lymphocytes. Therefore, exposure of fibroblasts requires a 10fold reduction of the TRE concentration.

Mitomycin C (MMC)

MMC is a bifunctional alkylating agent inducing cross-links in DNA which results in DNA strand breaks and chromosomal breakage (for review see Friedberg et al., 1995; Vos, 1995). MMC is used as an anti-tumor drug. MMC [SERVA, 29805.01] is provided as lyophilized substrate in 48 mg sodium chloride which serves as diluent.

1. Prepare a stock solution of 200 ng/µl by dissolving the lyophilized substrate in sterile distilled water.
2. Dilute the stock solution in complete medium to concentrations of 50 and 100 ng/ml. Prepare sufficient medium for the treatment of all cultures (5 ml per tube).

3. Prewarm the medium to 37°C.
4. For MMC treatment and all other steps in chromosome preparation and breakage analysis follow the protocol described for TRE treatment.

Personal experience and troubleshooting

- MMC in solution is very unstable and should not be stored for more than 24 hours at a temperature of 4°C. For longer storage freeze stock solutions immediately after preparation in a deep freezer at -°20°C.
- LCLs can be treated with the same protocol as for lymphocytes.

Diagnosis of Ataxia telangiectasia (AT) and of Nijmegen breakage syndrome (NBS)

AT (MIM 208900/208905) is an autosomal recessive disorder known to occur in various populations. Its estimated prevalence is between 1:40.000 and 1:300.000 in newborns. Clinical hallmarks are ataxia, telangiectasia, neurological deterioration, immunodeficiency, predisposition to cancer, premature aging. Despite earlier assumptions of four complementation groups, molecular studies demonstrated the existence of various mutations in only one single gene, the ATM gene, which is responsible for AT (Savitsky et al., 1995). This gene is localized in chromosome band 11q23.1 (Gatti et al., 1988). It spans 150 kb of genomic DNA and encodes a transcript of 13 kb representing 66 exons (Uziel et al., 1996) For a thorough presentation of the findings in AT see Shiloh (1995).

NBS is a rare disease and shows an autosomal recessive mode of inheritance. Clinically, patients presented with microcephaly, microgenia, growth retardation, immunodeficiency, tumor predisposition (see Wegner et al., 1999). The underlying gene was localized to 8q21 (Saar et al., 1997) and has been cloned very recently (Carney et al., 1998; Varon et al., 1998).

Cytogenetically, AT and NBS present almost identical abnormalities such as spontaneous chromosomal instability involving in particular chromosomes 7 and 14, cellular and chromosomal hypersensitivity to irradiation or radiomimetica and radio-resistant DNA-synthesis (RDS) (for review see Wegner, 1991). Thus, the following protocols are applicable to both AT and NBS.

Ionizing irradiation

Ionizing irradiation induces a variety of DNA lesions, including single and double strand breaks (for review see Friedberg et al., 1995; Vos, 1995). It produces chromosomal aberrations when used at any stages of the cell cycle, however, $G2_2$-treatment yields the most reliable results in the diagnosis of AT and NBS.

1. Set up blood cultures (see Procedure "Blood culture")
2. After 68 hrs incubation start irradiation. Avoid cooling of the cells for example by transferring the tubes in an insulated box to the irradiation source.
3. Expose cultures to 0,5 and 1,0 Gy ionizing radiation (X-rays or γ-rays) with a dose rate of about 0,5 Gy/min. Our source has been a therapeutical cobalt γ-ray device (Gammatron 3, Siemens) 1,17-1,33 MeV, with plexi glass filter, and a 5 mm gaze layer.
4. Immediately after treatment incubate cultures for an additional 4 hrs until harvest, the last two hours in the presence of Colcemid. For culture harvest, chromosome preparation and breakage analysis follow instructions given in the according sections.
5. Fibroblasts and LCLs can be treated with the same protocol as for lymphocytes.

Prenatal diagnosis

Prenatal diagnosis should be accompanied whenever possible by molecular analysis. If amnion fluid cells or chorionic villi cells are tested cytogenetically it is recommended to treat fibroblasts of the index patient in parallel. Complementing the breakage analysis flow cytometry can be applied (see Chapter 15).

Bleomycin (BLE)

BLE belongs to the group of radiomimetic glycopeptide antibiotics and induces DNA damage comparable to ionizing radiation. Thus, it provokes DNA single- and double-strand-breaks and inhibits the replicative DNA synthesis (D'Andrea and Haseltine, 1978). Therefore, BLE can be used instead of ionizing radiation for radiosensitivity testing. BLE is sold under the name Bleomycinum [Mack, 10216]. The injection ampoule contains 7 to 11 mg Bleomycin sulfate with a biological activity of 15 mg BLE (15 I. E.).

1. Dissolve the lyophilized BLE in 3 ml 0,9% NaCl solution. The resulting stock solution (5 mg/ml) must be used immediately to prepare the working dilutions or must be stored at -20°C.
2. Dilute the stock solution in medium to concentrations of 1 µg/ml and 10 µg/ml. Prepare sufficient medium for the treatment of all cultures (5ml per tube).
3. Warm the medium to 37°C.

BLE exposure

1. Incubate blood cultures for 68 hrs at 37°C.
2. Centrifuge for 10 min at 100 x g.
3. Remove the supernatant and add 5ml BLE-containing medium to the pellet.
4. Resuspend the pellet by pipetting carefully and mix well with the medium.
5. Incubate cultures for additional 4 hrs until harvest, during the last 2 hrs in the presence of Colcemid. For culture harvest, chromosome preparation and breakage analysis follow instructions given in the section Tissue Culture.

Personal experience and troubleshooting

- A few AT or NBS patients do not show any spontaneous chromosomal instability.
- Fibroblasts are much more sensitive to BLE than lymphocytes. Therefore, it is recommended to expose fibroblasts to 0,1 μg/ml and 1 μg/ml BLE.
- LCLs can be treated with the same protocol as lymphocytes.
- G_2-phase treatment is preferred compared to treatment in G_1 leading to the most distinct breakage rates between control and radiosensitive cells. The frequency of induced aberrations may be more than 10 fold higher in AT and NBS cells, than in controls (Table 4).
- AT and NBS cells show chromosome type damage even after G_2 treatment, whereas control cells just show chromatid type damage.
- BLE in solution is very unstable and treatment of cells should be initiated immediately after preparing the solution. Deep frozen at -20°C the solution shows an unchanged activity for several months. Do not refreeze thawed Bleomycin solution since there is an unpredictable loss of activity.

Radio-resistant DNA-synthesis (RDS)

RDS is a typical cellular property of AT and NBS cells. Here we describe a method to perform a RDS-test on a single cell level. In comparison to extraction methods followed by scintillation counting this test enables quantitative autoradiographic detection of DNA synthesis in individual cells. As a measure for the rate of incorporated ^{3}H-labelled thymidine the number of silver grains in the photo emulsion on top of a nuclei is counted under the light microscopy.

1. Prepare a stock solution of BLE as described above.
2. Dilute the stock solution in medium to concentrations of 200 μg/ml and 300 μg/ml. Prepare sufficient medium for the treatment of all cultures (5 ml per tube).
3. Warm the medium to 37°C.

BLE exposure

1. Set up and incubate blood cultures for 72 hrs as described above.
2. Centrifuge for 10 minutes at 100 x g.
3. Remove the supernatant and add 5 ml BLE-containing medium to the pellet.
4. Resuspend the pellet by pipetting carefully and mix well with the medium.
5. Incubate cultures 15 min in BLE containing medium at 37°C.
6. Add ^3H-thymidine (0,5-1 μCi/ml // 50 Ci/mmol) and incubate for further 15 min at 37°C.
7. Harvest cells without Colcemid treatment as described and prepare air-dried slides following instructions given in Chapter 7.
8. Incubate the air dried slides two times for 5 min in 5% TCA (4°C) to reduce the radioactive background.
9. Rinse for 1 min each in 80% and 100% ethanol.

Autoradiography

Slides for autoradiography are prepared by a dipping method which results in a thin autoradiographic film on top of the slide.

Preparation of film emulsion

1. Open the container with the gelled film emulsion in a darkroom equipped with a yellow-brown safelight illumination.
2. Transfer a sufficient volume of emulsion into a small clean beaker, which is used only for autoradiography.
3. Place the beaker in a 42°C water bath until the emulsion is molten (~ 30 min).
4. Dilute emulsion 1:1 with prewarmed (42°C) aqua dest, mix well and transfer enough emulsion into the dipping jar contained in the 42°C-waterbath.

Coating the slides with the emulsion

1. Dip two test slides back to back into the emulsion to test the homogeneity of the molten solution. Slides should be uniformly coated with a thin yellow film when held against the yellow-brown safelight.
2. Dip two experimental slides back to back into emulsion. Keep them for 5 seconds in a vertical position.
3. Withdraw slides slowly and steadily. Keep them vertical for further 5 seconds to drain off surplus emulsion.
4. Wipe the back of the slides free of emulsion and place them face up on an even ice cooled surface in an horizontal position for 5-10 minutes. Take care of keeping the slides exactly in the horizontal position to obtain an even layer of emulsion all over the slide. This is an essential prerequisite for quantitative autoradiography.
5. After gelling, transfer the slides to a rack and dry them light protected for 1-2 hours at room temperature.
6. Place the air-dried slides in light protected plastic boxes containing a drying agent (e. g. $CaCl_2$ wrapped in paper tissue), wrap each box in a black plastic bag and place it in a refrigerator (4°C).
7. It is recommended to develop test slides at different time intervals to evaluate the best exposure time (nuclei of untreated cells should show 30-50 grains on its surface). Using ^{3}H-thymidine as specified above an incubation time of 2 to 3 days is common.

Developing the emulsion

1. Dip slides for 2 minutes at room temperature in Kodak D19 developer.
2. Rinse in 2% acetic acid.
3. Fix for 5-10 minutes until the film is clear.
4. Rinse again in 2% acetic acid.
5. Wash slides for 10 minutes in running tap water, followed by three washes with distilled water.

Staining and interpretation

Slides can be stained after drying but better results are obtained by direct staining after the wash in distilled water.

1. Stain slides for 10 minutes in a 2% Giemsa solution. Nuclei should show only a faint staining to allow an easy analysis of the silver grains located over the nuclei.

2. The labeling index, eg the proportion of cells in S-phase, will be given as a percentage of the number of labelled cells related to the total number of cells.
3. The replicational activity of an individual cell is determined by counting the number of silver grains over the nucleus subtracted by the background rate, eg the number of grains found in an area of identical size over the cytoplasm. The DNA-synthesis rate is stated as the mean of silver grains per nucleus found in 50 - 100 labelled cells.

Troubleshooting

- Ilford K2 emulsion should not be stored longer than six months at 4°C. Ilford emulsions should not be melted more than once as this results in a high background.
- Tritium emits electrons with low energy travelling no more than 2 μm through the emulsion. Therefore, quantitative autoradiography requires an emulsion thickness of more than 2 μm all over the slide. This aim can be reached commonly by following precisely the present protocol. However, it is always advisable to check the emulsion thickness of each dipping series by day-light inspection of a control slide.
- Drying should be slow and gentle in a cold air stream to prevent curling of the emulsion.

References

Carney JP, Maser RS, Olivares H, Davis EM, Le Beau M, Yates III JR, Hays L, Morgan WF, Petrini JH (1998) The hMre11/hRad50 protein complex and Nijmegen breakage syndrome: linkage of double-strand break repair to the cellular DNA damage response. Cell 93: 477-486.

Chaganti RS, Schonberg S, GermanJ (1974) A manifold increase in sister chromatid exchanges in Bloom's syndrome lymphocytes. Proc Natl Acad Sci 71: 4508-4512.

D'Andrea A and Haseltine WA (1978) Sequence specific cleavage of DNA by the antitumor antibiotics neocarzinostatin and bleomycin. Proc Natl Acad Sci USA 75, 8: 3608-3612.

Duckworth-Rysiecki G, Cornish K, Clarke CA and Buchwald M (1985) Identification of two complemenation groups in Fanconi anemia. Somat Cell Mol Genet 11: 35-41

Ellis NA, Groden J, Ye T-Z, Straughen J, Lennon DJ, Ciocci S, Proytcheva M, German J (1995) The Bloom's syndrome gene product is homologous to RecQ helicases. Cell 83: 655-666.

Freeman MVR, Williams DW, Schimke RN, Temtamy A, Vachier E, German J (1974) The Roberts syndrome. Clin Genet 5: 1-16.

Friedberg EC, Walker GC and Siede W (eds) (1995) DNA Repair and Mutagenesis. ASM Press, Washington DC.

Gatti RA, Berkel I, Boder E, Braedt G, Charmley P et al. (1988) Localization of the ataxia-telangiectasia gene to chromosome 11q22-23. Nature 336: 577-580.
German J (1993) Bloom syndrome: a mendelian prototype of somatic mutational disease. Medicine 72: 393-406.
German J, Roe AM, Leppert MF, Ellis NA (1994) Bloom syndrome: An analysis of consanguineous families assigns the locus mutated to chromosome band 15q26.1. Proc Natl Acad Sci 91: 6669-6673.
Goto M, Rubenstein M, Weber J, Woods K, Drayna D (1992) Genetic linkage of Werner's syndrome to five markers on chromosome 8 Nature 335: 735-738.
Hulten M (1978) Selective somatic pairing and fragility at 1q12 in a boy with common variable immunodeficiency: anew syndrome. Clin Genet 14: 294.
Hustinx TWJ, Scheres JMJC, Weemaes CMR, ter Haar BGA, Janssen AH (1979) Karyotype instability with multiple 7/14 and 7/7 rearrangements. Hum Genet 49: 199-208
ISCN (1995) An International System for Human Cytogenetic Nomenclature. Mitelman (ed) S Karger, Basel.
Joenje H (1996) Fanconi anemia complementation groups in Germany and The Netherlands. Hum Genet 97: 280-282.
Joenje H, Lo Ten Foe JR, Oostra AB, van Berkel CGM, Rooimans MA, Schroeder-Kurth T, Wegner RD, Gille JJP, Buchwald M and Arwert F (1995) Classification of Fanconi anemia patients by complementation analysis: evidence for a fifth genetic subtype. Blood 86: 2156-2160
Joenje H, Oostra AB, Wijker M, di Summa FM, van Berkel CGM, Rooimans MA, Ebell W, van Weel M, Pronk JC, Buchwald M, Arwert F (1997) Evidence for at least eight Fanconi anemia genes. Am J Hum Genet 61: 940-944.
Lo Ten Foe JR, Rooimans M, Bosnoyan-Collins L, Alon N, Wijker M, Parker L, Lightfoot J, Carreau M, Callen DF, Savoia A, Cheng NC, van Berkel CGM, Strunk MHP, Gille JJP, Pals G, Kruyt FAE, Pronk JC, Arwert F, Buchwald M, Joenje H (1996). Expression cloning of a cDNA for the major Fanconi anaemia gene, FAA. Nature Genet 14: 320-323.
Obe G, Beek B (1979) Trenimon: Biochemical, physiological and genetic effects on cells and organism. Mut Res 65: 21-70.
Pronk JC, Gibson RA, Savoia A, Wijker M, Morgan NV, Melchionda S, Ford D, Temtamy S, Ortega JJ, Jansen S, Havenga C, Cohn RJ, deRavel TJ, Roberts I, Westerveld A, Easton DF, Joenje H, Mathe CG, Arwert F (1995) Localisation of the Fanconi anaemia complementation group A gene to chromosome 16q24.3. Nature Genet 11: 338-340.
Rademacher P, Obe G (1995) Trenimon: structure and reactivity of a versatile chemical agent. Mut Res 340: 37-49.
Rogers AW (ed) (1979) Techniques of autoradiography. Elsevier, Amsterdam, New York, Oxford.
Saar K, Chrzanowska KH, Stumm M, Jung M, Nürnberg G, Wienker TF, Seemanová E, Wegner RD, Reis A, Sperling K. The gene for Ataxia-Telangiectasia-Variant (Nijmegen Breakage Syndrome) maps to a 1 cM interval on chromosome 8q21. Am J Hum Genet 60: 605-610.
Salk D, Au K, Höhn H, Martin GM (1981) Effects of radical-scavenging enzymes and reduced oxygen exposure on growth and chromosome abnormalities of Werner syndrome cultured skin fibroblasts Hum Genet 57: 269-275.
Savitsky K, Bar-Shira A, Gilad S, Rotmann G, Ziv Y, Vanagaite L, Tagle DA, Smith S, Uziel T, Sfez S, Ashkenazi M, Pecker I, Frydman M, Harnik R, Patanjali SR, Simmons A, Clines GA, Sartiel A, Gatti RA, Chessa L, Sanal O, Lavin MF, Jaspers NGJ, Taylor AMR,

Arlett CF, Miki T, Weissman SM, Lovett M, Collins FS, Shiloh Y (1995) A single ataxia-telangiectasia gene with a product similar to PI-3 kinase. Science 268: 1749-1753.

Schellenberg GD, Martin GM, Wijman EM, Nakura J, Miki T, Ogihara T (1992) Homozygosity mapping and Werner's syndrome. Lancet 339: 1002

Shiloh Y (1995) Ataxia-telangiectasia: closer to unraveling the mystery. Eur J Hum Genet 3: 116-138.

Smeets DFCM, Moog U, Weemaes CMR, Vaes-Peeters G, Merkx GFM, Niehof JP, Hamers G (1994) ICF syndrome: a new case and review of the literature. Hum Genet 94: 240-246.

Strathdee CA, AMV Duncan, Buchwald M (1992) Evidence for at least four Fanconi anaemia genes including FACC on chromosome 9. Nature Genet 1: 196-198.

The Fanconi anaemia/Breast cancer consortium. Positional cloning of the Fanconi anaemia group A gene (1996). Nature Genet 14:324-328.

Uziel T, Savitsky K, Platzer M, Ziv Y, Helbitz T, Nehls M, Boehm T, Rosenthal A, Shiloh Y and Rotman G (1996) Genomic organization of the ATM gene. Genomics 33, 317-320.

van den Berg DJ, Francke U (1993) Roberts syndrome: A review of 100 cases and a new rating system for severity. Am J Med Genet 47: 1104-1123.

Varon R, Vissinga C, Platzer M, Cerosaletti KM, Chrzanowska C, Saar K, Beckmann G, Seemanova E, Cooper PR, Nowak NJ, Stumm M, Weemaes CMR, Gatti R, Wilson RK, Digweed M, Rosenthal A, Sperling K, Concannon P, Reis A (1998) Nibrin, a novel DNA double-strand break repair protein, is mutated in Nijmegen breakage syndrome. Cell 93: 467-476.

Vos JMH (ed) (1995) DNA Repair Mechanisms: Impact on human diseases and cancer. Springer-Verlag, Heidelberg.

Wegner R-D (1991) Chromosomal instability syndromes in man. In: Obe G (ed) Advances in mutagenesis research 3. Springer: 81-117

Wegner R-D, Chrzanowska K, Sperling K, Stumm M (1999) Ataxia Telangiectasia Variants (Nijmegen Breakage Syndrome) In: Ochs HD, Puck J, Smith E (eds) Primary immunodeficiency diseases, a molecular and genetic approach. Oxford University Press, New York.

Wegner R-D, Henrichs I, Joenje H, Schroeder-Kurth T (1996) Fanconi anemia complementation group E: clinical and cytogenetic data of the first patient. Clin Genet 50: 479-482.

Whitney M, Thayer M, Reifsteck C, Olson S, Smith L, Jakobs PM, Leach R, Naylor S, Joenje H, Grompe M (1995) Microcell mediated chromosome transfer maps the Fanconi anemia group D gene to chromosome 3p. Nature Genet 11: 341-343.

Wijmenga C, van den Heuvel LP, Strengman E, Luyten JA, van der Burgt IJ, de Groot R, Smeets DF, Draaisma JM, van Dongen JJ, de Abreu RA, Pearson PL, Sandkuigl LA, Weemaes CM (1998) Localisation of the ICF syndrome to chromosome 20 by homozygosity mapping. Am J Hum Genet 63: 803-809.

Yu C-E, Oshima J, Fu Y-H, Wijsman EM, Hisama F, Alisch R, Matthews S, Nakura J, Miki T, Quais S, Martin GM, Mulligan J, Schellenberg GD (1996) Positional cloning of the Werner's syndrome gene. Science 272: 258-262.

Chapter 15

Flow Cytometric Testing for Syndromes with Chromosomal Instability, Aplastic Anemia and Related Hematological Disorders

DETLEV SCHINDLER AND HOLGER HOEHN

Introduction

Traditionally, the evaluation of metaphase chromosomes for breaks and rearrangements has evolved as the diagnostic tool for the identification or confirmation of syndromes associated with chromosomal instability. However, many of these syndromes exhibit poor cell proliferation, and often metaphases are not abundant. Moreover, the analysis of metaphases at the microscope is cumbersome and time-consuming, and only limited numbers (50 – 100) can be scored. We have developed a simple, fast, and, at least in part, automated procedure which complements or, in certain cases, may substitute microscopic analyses with regards to accuracy and reproducibility (Rabinovitch et al 1988, Kubbies et al 1989, Poot et al 1994). This alternative assay using flow cytometry employs cell cycle analysis of peripheral blood lymphocytes activated *in vitro* (Figure 1). The cell cycle test has been successfully applied to the diagnosis of Fanconi anemia (FA) (Kubbies et al 1985, Seyschab et al 1993, Seyschab et al 1995), severe forms of non-Fanconi type aplastic anemia (non-FA) and related hematological disorders (Tschampa et al, submitted), ataxia telangiectasia (A-T) (Seyschab et al 1992), as well as the Nijmegen breakage syndrome (NBS) (Barbi et al 1991).

Correspondence to Detlev Schindler, University of Wuerzburg, Biozentrum, Dept. of Human Genetics, Am Hubland, Wuerzburg, 97074, Germany (*phone* +49-931-888-4088; *fax* +49-931-888-4069; *e-mail* schindler@biozentrum.uni-wuerzburg.de), Holger Hoehn, University of Wuerzburg, Biozentrum, Dept. of Human Genetics, Am Hubland, Wuerzburg, 97074, Germany

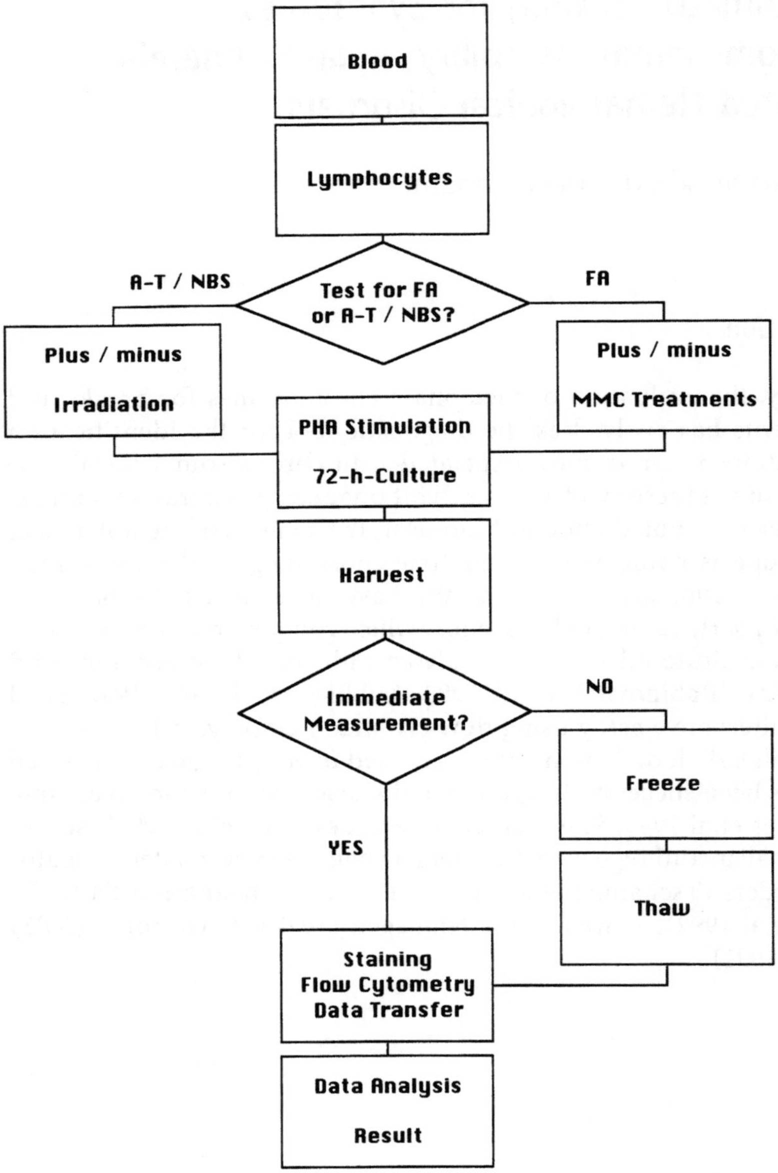

Fig. 1. Flow chart showing the processing, culture and further handling of lymphocytes for flow cytometry and cell cycle analysis.

Materials

Reagents

- Ficoll-Paque (density, 1.077 g/ml; Pharmacia, Uppsala, Sweden).
- Phytohemagglutinin, purified (PHA HA16; Murex Diagnostica, Burgwedel, Germany; order no. F400310). Dissolve at 2 mg/5 ml in dH_2O as stock. Store in aliquots at − 20 °C and avoid repeated freeze-thawing!
- 5-Bromo-2'-deoxyuridine (BrdUrd; Sigma, Deisenhofen, Germany; order no. B 5002). Dissolve at $1x10^{-2}$ M in PBS as stock. Store sterile at 4 °C protected from light.
- 2-Deoxycytidine (DC; Sigma; D 3897). Dissolve at $1x10^{-2}$ M in PBS as stock. Store sterile at 4 °C protected from light.
- 1-Monothioglycerol (ATG; Sigma; M 1753). Dissolve at $1x10^{-2}$ M in PBS as stock. Store sterile at 4 °C protected from light.
- Mitomycin C (MMC; mitomycin 2; medac, Hamburg, Germany). Dissolve vial contents in 4 ml of dH_2O (0.5 mg/ml). Dilute 1:9 (10-fold) with dH_2O (50,000 ng/ml. Further dilute 1:4 (5-fold) to make a 10,000 ng/ml stock. Store at 4 °C protected from light. Do not freeze stocks!
- Hoechst 33258 (Hoechst; Sigma B 2883). Dissolve at 2 mg/ml in dH_2O as stock.
- Ethidium bromide (EB; Sigma E 8751). Dissolve at 150 μg/ml in dH_2O as stock. Store the staining stock solutions in light-protected containers. Protect from microbial contamination and refrigerate. Do not keep longer than 1 year!
- Dimethylsulfoxide (DMSO; Merck-Schuchardt, Darmstadt, Germany; order no. 802912). Add 10 % (v/v) to RPMI 1640 cell culture medium, filter sterile, and add 10 % (v/v) fetal bovine serum (FBS).

Equipment

- Source of ionizing radiation, either X-ray or ^{60}Co γ-ray-source, with dosimetry.
- Flow cytometer, e.g., PAS II (Partec, Muenster, Germany), equipped with a mercury-arc lamp system and appropriate filter settings.
- Personal computer, IBM or clone, MS-DOS or WINDOWS based.
- Special software: MULTI 2D and MCYCLE (Phoenix Flow Systems, San Diego, USA).

Procedure

Lymphocyte isolation

Blood processing

1. Draw blood using 0.1 ml (corresponding to 500 IU) of heparin per each 5 ml volume. Avoid use of heparin preparations containing preservatives toxic to cells. Transfer to laboratory at rt within maximum 48 h.
2. Centrifuge at 200xg for 10 min at rt. Obtain 2 ml of autologous plasma and replace with Hank's balanced salt solution (HBSS).
3. Further dilute blood sample with an equal volume of HBSS.

Density gradient centrifugation

4. Overlay 3 ml of Ficoll Paque with 4 ml of diluted blood at 4 °C.
5. Centrifuge at 400xg for 30 min at 4 °C.
6. Discard upper phase. Carefully collect the interphase using a Pasteur pipette. Transfer interphase into 6 ml of HBSS.

Washes

7. Spin at 200xg for 10 min at rt. Discard supernatant. Resuspend pellet in 6 ml of HBSS.
8. Again spin at 200xg for 10 min at rt. Discard supernatant and suspend the cell pellet in 3 ml of RPMI 1640 medium.

Cell counts

9. Remove 10 μl of this suspension. Add 10 μl of 1 % acetic acid to this aliquot in order to distroy residual erythrocytes. Use 10 μl of this aliquot and add 10 μl of 0.2 % Trypan blue. Determine cell viability; dye exclusion should occur in >95 % of the cells. Count mononuclear cells excluding the dye in a hemocytometer (Fuchs-Rosenthal).
10. Adjust cell number to 10^6/ml in RPMI 1640.

Cell culture

Set-up and treatments

1. Set up 4 flasks for each test, ie, 2 duplicate flasks without treatments and 2 duplicate flasks with appropriate treatments causing chromosome breakage in the condition under question, such as MMC in case of FA and ionizing irradiation for A-T and NBS (see Chapter 14).
2. Use 0.5 ml of the cell suspension (corresponding to $5x10^5$ cells) for transfer into 3.56 ml of RPMI 1640 per 25 cm^2/50 ml tissue culture flask. Keep the flasks upright.

3. At this point, expose 2 duplicate flasks to ionizing radiation at 1.5 Gy each when testing for A-T or NBS.

4. Complete cultures by adding from the stocks (per flask):

Table 1.

Autologous plasma	0.05 ml	(final concentration 1 %)
Fetal bovine serum (FBS)	0.75 ml	(15 %)
BrdUrd	0.05 ml	(10^{-4} M)
DC	0.05 ml	(10^{-4} M)
ATG	0.01 ml	($2x10^{-5}$ M)
PHA	0.023 ml	(1.2 μg/ml)

to make cultures of 5 ml volume.

5. Add 5 μl of the MMC stock to make a final concentration of 10 ng/ml when testing for FA.

6. Culture flasks must be wrapped in aluminum foil in order to avoid exposure to light during culture and handling.

7. Incubation of cultures is for 72 h at 37.5 °C. Incubators are set at 5% CO_2 in air (v/v); reduction of atmospheric oxygen to 5% (v/v) by replacement with nitrogen assures optimal growth in case of FA cell cultures. Incubation

Cell harvest

- Use indirect, subdued, or red light in the working area throughout harvest. Precautions

1. Transfer cell cultures to 15 ml conical centrifuge tubes and pellet for 10 min at 200xg at rt. Freezing

2. Resuspend pellet in 1 ml freezing medium (RPMI 1640: 80%; FBS: 10%; DMSO: 10%, by volume).

3. Freeze samples at – 20 °C. Can be stored frozen for up to 2 years.

Cell staining

Hoechst staining

1. Prepare staining buffer:

Table 2.

Tris-HCl (1 M, pH 7.4)	10 ml	(final concentration 0.1 M)
NaCl (1.54 M)	10 ml	(0.154 M)
$MgCl_2$ (50 mM)	1 ml	(0.5 mM)
$CaCl_2$ (0.1 M)	1 ml	(1 mM)
Nonidet P-40 (10 %)	1 ml	(0.1 %)
Bovine serum albumin (BSA)	0.2 mg	(0.2 % w/v)
dH_2O	77 ml	(ad 100 ml)

2. Prepare Hoechst staining solution: Add 60 μl of the Hoechst stock to 100 ml of staining buffer (Hoechst final concentration 1.2 μg/ml).
3. Thaw cell samples in waterbath at rt. Centrifuge for 8 min at 200xg and rt. Resuspend pellet in Hoechst staining solution at 400.000 cells/ml. Incubate for 15 min at 4 °C in the dark.

EB staining

4. Add 0.01 volume of the EB stock to give a final concentration of 1.5 μg/ml and continue incubation for another 15 min at 4 °C. Stained cells can be kept at 4 °C and protected from light for several hours.

Flow cytometry

Instrument adjustment

1. Samples are measured in a flow cytometer designed for two parameter measurements. The prefered light source for the UV-excitable dye Hoechst 33258 is a mercury arc lamp, but excitation can also be by a UV-laser. Filter combinations appropriate for the two dyes, Hoechst and EB, must be used.
2. Internal standardization of the instrument is with chicken erythrocytes or a commercial bead standard.
3. Adjust flow rate to 100-300 cells per second. A minimum of 50,000 cells should be measured.
4. Two parameter data are stored in a dedicated computer and analyzed by curve fitting procedures, using appropriate software. We use software

developed by P.S. Rabinovitch, University of Washington, Seattle, WA 98195 (e.g., MULTI 2D-AV, MCYCLE-AV, MPLUS-AV).

The experimental procedure is displayed schematically in Figure 1.

Results

Evaluation of flow cytometric data

Figure 2 shows a 72 h cell harvest of peripheral blood mononuclear cells from a healthy individual. Panels A to D illustrate the electronic framing of first (A), second (B), third (C) and fourth (D) cell cycle of the same specimen. Panels E through H show the corresponding conversion to single parameter histograms that is achieved by rotation and collapse of the two parameter single cell cycle data onto the X-axis. The resulting single parameter histograms are analyzed by curve fitting. As a result of these steps, the distribution of cells throughout four consecutive cell cycles can be enumerated.

Flow cytometric diagnosis of Fanconi anemia (FA)

Figure 3 shows two parameter cytograms of 72 h lymphocyte harvests (without MMC treatments) from a control donor and a patient with FA (upper panels). Note that the first and second cycle G2 phases show massive accumulations of cells in the FA patient. The lower panels convert the signal-density data into three-dimensional cytograms by showing the number of cells present in a given cell cycle compartment by height (vertical axis). The resulting peaks denoting the G2 phase compartments of the first and second cell cycles are much more prominent in the FA compared to the control donor.

Figure 4 illustrates the results of the flow cytometric evaluation of 20 FA patients (solid squares), 47 healthy donors (solid circles) and 17 patients with acquired (non-FA) aplastic anemia (open circles). Plotted is the non-cycling (G0-G1) cell fraction, i.e., the number of cells that failed to respond to the mitogen within the 72 h incubation period, vs. the sum of all G2 phases divided by the growth fraction (GF). The growth fraction is the total cell population measured by flow cytometry minus the non-cycling (G0–G1) cell fraction. The figure shows a clear separation between FA and non-FA donors in cultures devoid of MMC. The ratio ΣG2/GF for normal controls was shown to be 0.22 ± 0.03 (mean±SD, range 0.13 – 0.27) in contrast to

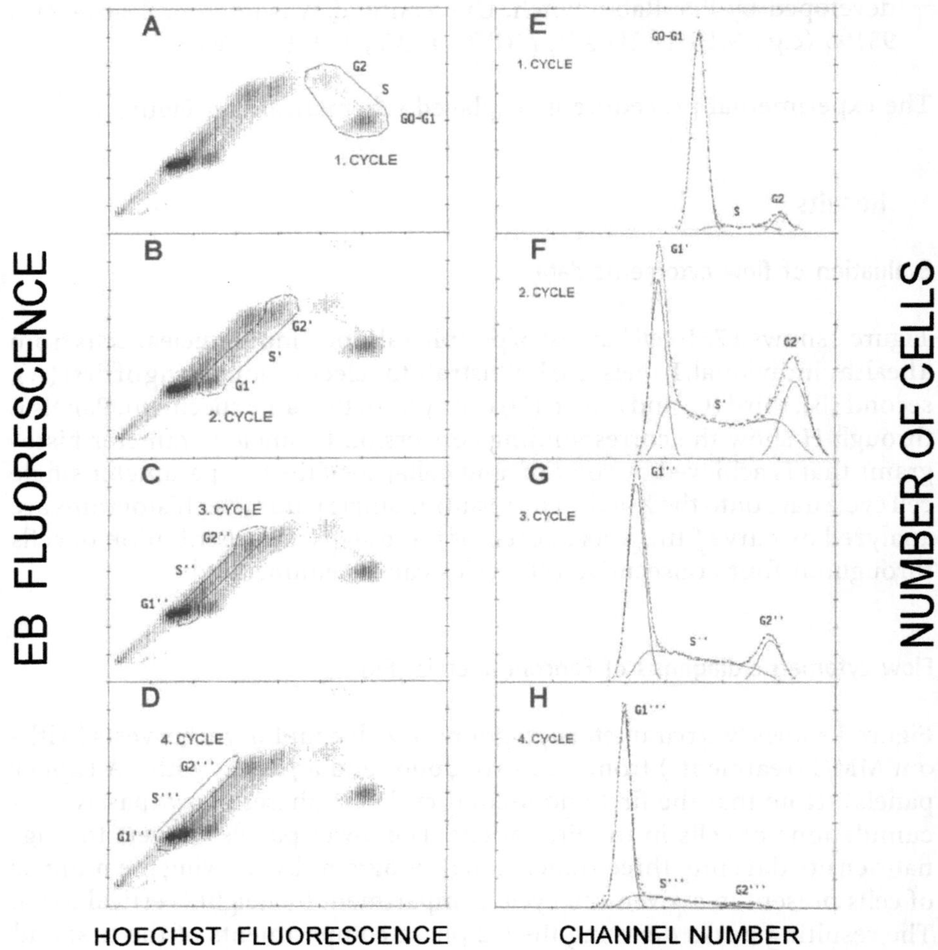

Fig. 2. Examplified bivariate flow histogram demonstrating a 72 h cell harvest of peripheral blood mononuclear cells from a healthy individual. The consecutive cell cycles framed in A-D are isolated and deconvoluted in E-G.

0.53±0.08 for FA cells (range of the latter, 0.38 – 0.64). The normal control and the non-FA (aplastic anemia) clusters are distinct by their different G0-G1 cell fractions which amount to 19.9±7.3 % (mean±SD) for the controls in contrast to 57.8±18.5 % (mean±SD) for non-FA aplastic anemia. There is some overlap between the control and the non-FA (aplastic anemia) clusters, indicating that the diagnosis of non-FA aplastic anemia cannot solely rely on flow cytometry. Likewise, poor lymphocyte responsiveness to PHA stimulation was observed in the myelodysplastic syndrome.

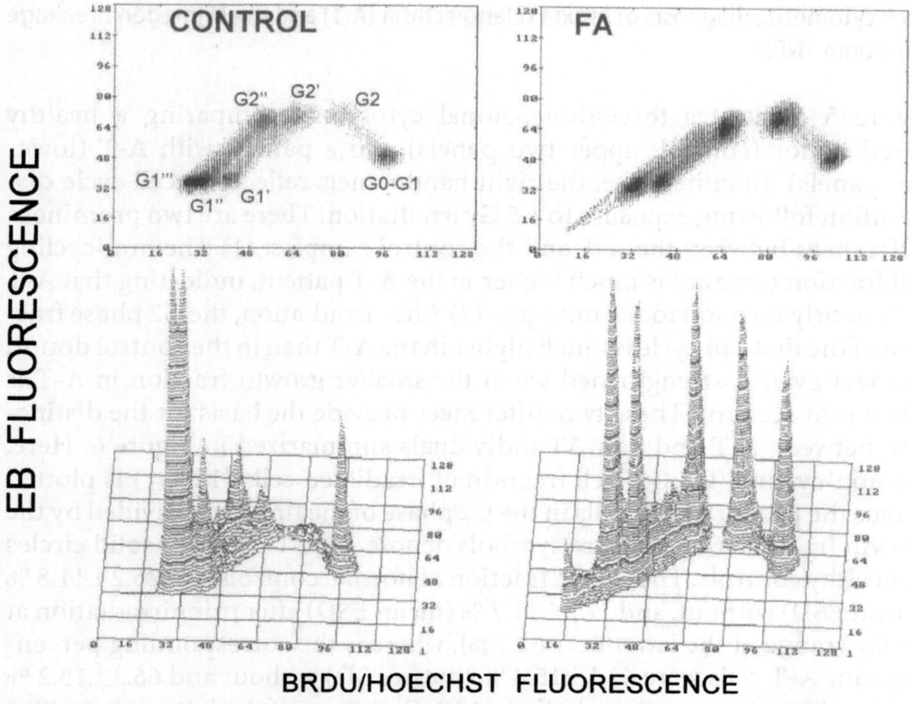

Fig. 3. Normal control (left) and FA cells (right) following 72 h cultures without MMC are shown in signal-density (upper panel) and signal-height (lower panel) representations of bivariate flow histograms.

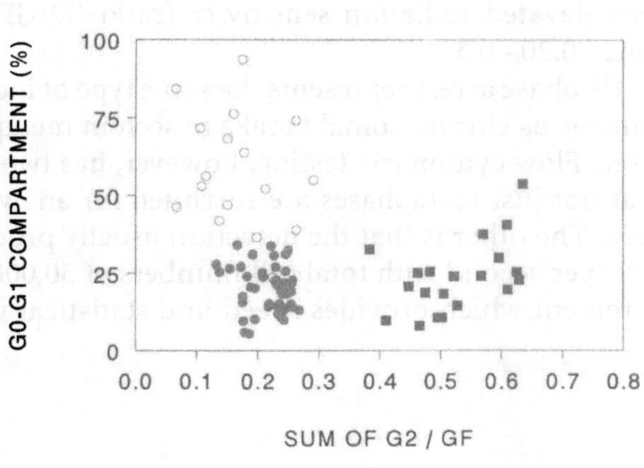

Fig. 4. Scattergram of the ratio sum of G2 phases/GF vs. G0-G1 values in 72 h harvests of lymphocyte cultures without MMC treatments: FA cells (solid squares) are distinct from those of patients with aquired anemia (open circles), and normal controls (solid circles).

Flow cytometric diagnosis of Ataxia telangiectasia (A-T) and the Nijmegen breakage syndrome (NBS)

Figure 5 illustrates three-dimensional cytograms comparing a healthy blood donor (control, upper two panels) and a patient with A-T (lower two panels). In either case, the right hand panels reflect the cell cycle distribution following exposure to 1.5 Gy irradiation. There are two prominent differences between the A-T and the control samples: (1) The non-cycling cell fraction (arrows) is much higher in the A-T patient, indicating that A-T cells poorly respond to the mitogen. (2) After irradiation, the G2 phase fraction of the first cell cycle is much higher in the A-T than in the control donor; this fact even is strengthened when the smaller growth fraction in A-T is taken into account. These two differences provide the basis for the distinction between A-T and non-AT individuals summarized in Figure 6. Here, the non-cycling (G0-G1) cell fraction of irradiated cells (1.5 Gy) is plotted versus the proportion of cells in the G2 phase of the first cycle divided by the growth fraction. Solid square symbols denote 16 A-T patients, solid circles 32 healthy controls. The G0-G1 fraction of normal controls was 25.2±11.8 % (mean±SD) without, and 27.2±11.7 % (mean±SD) after prior irradiation at 1.5 Gy (range of the latter 6.5 – 52.7%), whereas the corresponding percentages for A-T cells were 62.1±15.4 % (mean±SD) without and 66.3±15.2 % (mean±SD) after prior irradiation at 1.5 Gy (range of the latter, 43.7 – 88.5 %). The ratio G2 (of the 1st cell cycle)/GF following prior irradiation of the cells at 1.5 Gy was shown to be 0.08±0.03 (mean±SD) for normal controls (range 0.02 – 0.16), in contrast to 0.45±0.10 (mean±SD) for A-T cells (range of the latter, 0.26 – 0.60). The four solid triangles denote patients with NBS that is clinically distinct from A-T but also shows poor mitogen response and elevated radiation sensitivity (ratio G2/GF, 0.26±0.08, mean ± SD, range 0.20 – 0.37).

G2 phase arrest represents the same type of indicator of unrepaired DNA damage as chromosomal breakage seen in metaphases in cytogenetic analyses. Flow cytometric testing, however, has two major advantages: One is that not just metaphases are recruited for analysis but also all interphase cells. The other is that the detection usually proceeds at rates of about 200 cells per second with total cell numbers of 50,000 to 100,000 cells per measurement which provides speed and statistical validity.

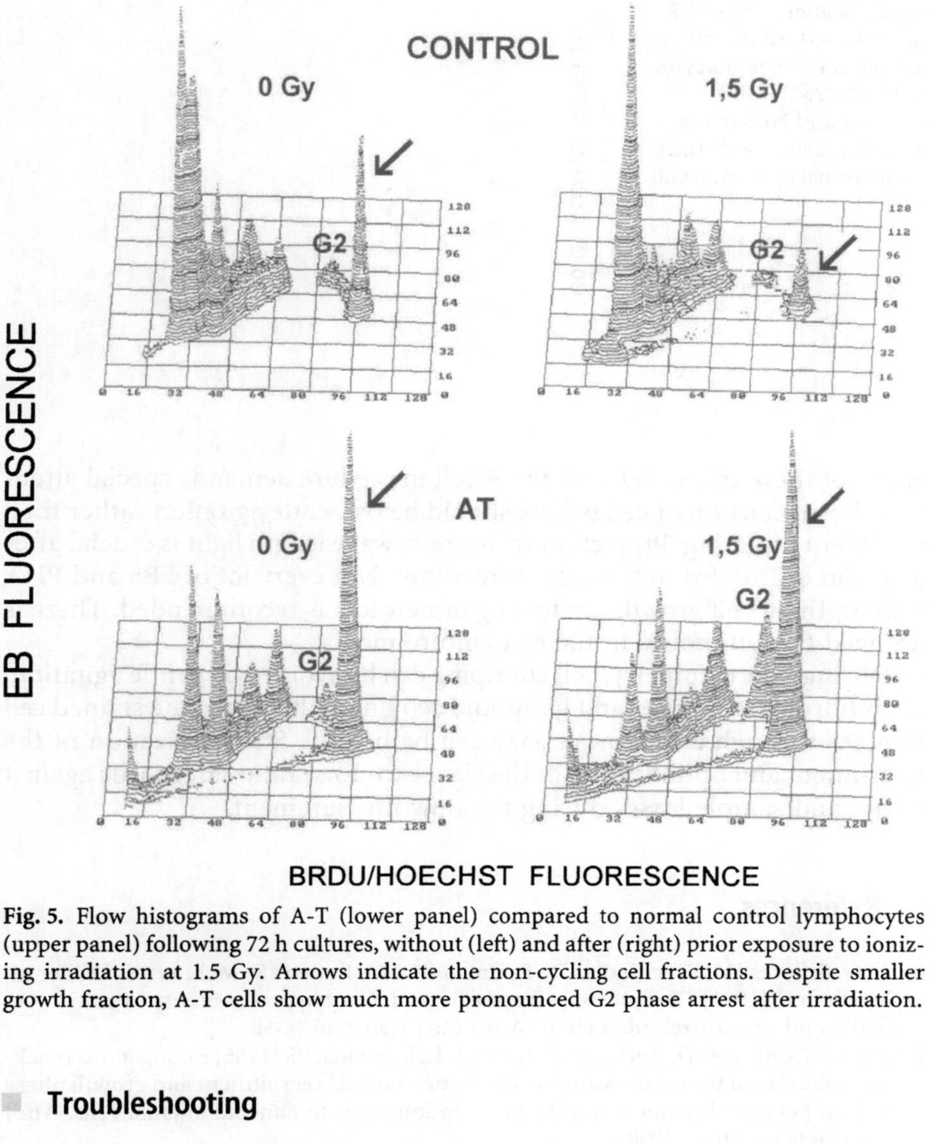

Fig. 5. Flow histograms of A-T (lower panel) compared to normal control lymphocytes (upper panel) following 72 h cultures, without (left) and after (right) prior exposure to ionizing irradiation at 1.5 Gy. Arrows indicate the non-cycling cell fractions. Despite smaller growth fraction, A-T cells show much more pronounced G2 phase arrest after irradiation.

Troubleshooting

We recommend standardization of blood collection by using commercial heparinized tubes, shipment and storage at room temperature, and culture set-up within 48 h after venipuncture. EDTA and citrate blood samples yield very poor cell growth and should be avoided.

Patients affected by aplastic anemia (FA and non-FA types) have low numbers of mononuclear cells in their peripheral blood samples. The iso-

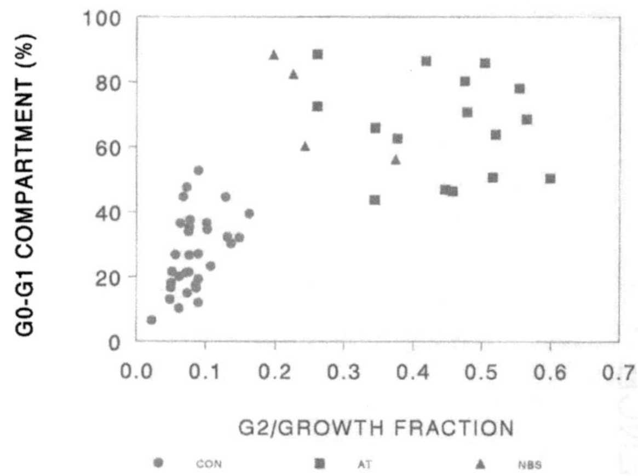

Fig. 6. Scattergram of G2/GF vs. G0-G1 values after γ-irradiation of lymphocytes at 1.5 Gy: A-T (solid squares) and NBS (solid triangles) cells are distinct from normal controls (solid circles).

lation of these sparse cells via the Ficoll procedure demands special attention. Resuspension of cell pellets should be by gentle agitation rather than by violent pipetting. Protection from short wave length light is crucial after addition of BrdUrd to the culture medium. Not every lot of FBS and PHA yield optimal cell growth. Pretesting of new lots is recommended. There is no need for antibiotics in the cell culture media.

During flow cytometry, cell clumping can be avoided by gentle agitation. With fibroblast cultures and lymphoid cell lines, filtering of the stained cell samples through a fine mesh gaze can be helpful. Standardization of the instrument and optimization of the signal-to-noise ratio safeguards against errors and sample losses during the flow measurement.

References

Barbi G, Scheres JMJC, Schindler D, Taalman RDFM, Rodens K, Mehnert K, Müller M, Seyschab H (1991) Chromosome instability and X-ray hypersensitivity in a microcephalic and growth-retarded child. Am J Med Genet 40:44-50

Kubbies M, Schindler D, Hoehn H, Schinzel A, Rabinovitch PS (1985) Endogenous blockage and delay of the chromosome cycle despite normal recruitment and growth phase explain poor proliferation and frequent endomitosis in Fanconi anemia cells. Am J Hum Genet 37:1022-1030

Kubbies M, Hoehn H, Schindler D, Chen Y, Rabinovitch PS (1989) Cell cycle analysis via BrdU-Hoechst flow cytometry - Principles and applications. In: Yen A (ed) Flow cytometry: advanced research and clinical applications, vol 2. CRC Press, Boca Raton, pp 6-28

Poot M, Hoehn H, Kubbies M, Grossmann A, Chen YC, Rabinovitch PS (1994) In vitro cell cycle analysis using continuous BrdUrd labeling and bivariate Hoechst 33258/Ethidium bromide flowcytometry. Meth Cell Biol 41:327-340

Rabinovitch PS, Kubbies M, Chen YC, Schindler D, Hoehn H (1988) BrdU-Hoechst flow cytometry: a unique tool for quantitative cell cycle analysis. Exp Cell Res 174:309-318

Seyschab H, Schindler D, Friedl R, Barbi G, Boltshauser E, Fryns JP, Hanefeld F, Korinthenberg R, Krägeloh-Mann I, Scheres JMJC, Schinzel A, Seemanová E, Tommerup N, Hoehn H (1992) Simultaneous measurement, using flow cytometry, of radiosensitivity and defective mitogen response in ataxia telangiectasia and related syndromes. Eur J Pediat 151:756-760

Seyschab H, Sun Y, Friedl R, Schindler D, Hoehn H (1993) G2 phase cell cycle disturbance as a manifestation of genetic cell damage. Hum Genet 92:61-68

Seyschab H, Friedl R, Sun Y, Schindler D, Hoehn H, Hentze S, Schroeder-Kurth T (1995) Comparative evaluation of diepoxybutane sensitivity and cell cycle blockage in the diagnosis of Fanconi anemia. Blood 85:2233-2237

Tschampa H, Friedl R, Fuehrer M, Haubitz I, Kubbies M, Hoehn H, Schindler D High resolution cell cycle analysis of lymphocyte proliferation in aplastic anemia and other hematological disorders. Pediat Hematol Oncol (submitted)

Chapter 16

Cell Fusion

MARKUS STUMM AND ROLF-DIETER WEGNER

Introduction

Cell fusion is used in somatic cell genetics to obtain intra- or interspecies heterokaryons or hybrid cells. Thus, inter-species cell fusion has been employed for such different approaches as the production of monoclonal antibodies by hybridoma cells made from man and mouse cells (for review see Westerwoudt, 1985) or the construction of human chromosome libraries in human-rodent-hybrids (Cowell, 1992). Furthermore, cell fusion studies are applied for diagnostic and research purposes to study the dominant or recessive nature of specific cell functions and to examine genetic heterogeneity in various human disorders (Bryant et al., 1979; Jaspers and Bootsma, 1982; Sperling, 1982; Zakrzewski and Sperling, 1982; Chen et al., 1984; Jaspers et al., 1988; Wegner et al., 1988; Chen and Kidson; 1993; Joenje et al., 1995; Schuffenhauer et al., 1995, Stumm et al., 1997).

Moreover, when an interphase cell is fused with a mitotic cell the interphase chromatin can be forced to condense into chromosomes. This technique, called premature chromosome condensation (PCC) (Johnson and Rao, 1970) has been intensively used to study interactions between cells fused in different phases of the cell cycle.

The principle of cell fusion is simple. The cell membranes of different parental cells are brought in close contact to each other and are briefly dissolved using either a virus (e. g. Sendai-virus) or chemical agents (e. g. polyethylene glycol). During the recovery period the cell membranes reform and create multinucleate cells. The cell fusion method using inactivated Sendai-

Correspondence to Markus Stumm, Universitätsklinikum, Institut für Humangenetik, Leipziger Str. 44, Magdeburg, 39120, Germany (*phone* +49-391-67-15344; *fax* +49-391-67-15066; *e-mail* Markus.Stumm@Medizin.Uni-Magdeburg.de), Rolf-Dieter Wegner, Charité Campus Virchow-Klinikum, Institut für Humangenetik, Augustenburger Platz 1, Berlin, 13353, Germany virus is complex (Marshall and Dave, 1978) and has meanwhile been replaced by the quick and easy polyethylene glycol (PEG) mediated cell fusion (Davidson and Gerald, 1976). To determine the parental origin of the nuclei in di- or multikaryons labelling is required. When such cells pass the next mitoses and the different genomes are enclosed in one nuclear membrane a true hybrid cell has been formed.

In this chapter the production of heterokaryons/hybrid cells and two different techniques to label parental cells are described. With these procedures heterokaryons/hybrids can be examined a few hours after fusion without the necessity of selection. For the sake of simplicity we will use the term heterokaryon also for cells in the first mitosis after fusion when no nuclear membrane is present but a true hybrid cell has not been formed.

Our recent studies focused on the nature of inheritance of the genetic defect in chromosomal instability syndromes as the ICF-syndrome (Schuffenhauer et al., 1995), ataxia-telangiectasia and Nijmegen breakage syndrome (Stumm et al., 1997). A flow chart (Figure 1) shows the main steps of fusion procedures. Experience gathered during these investigations are presented here focusing in particular on technical details.

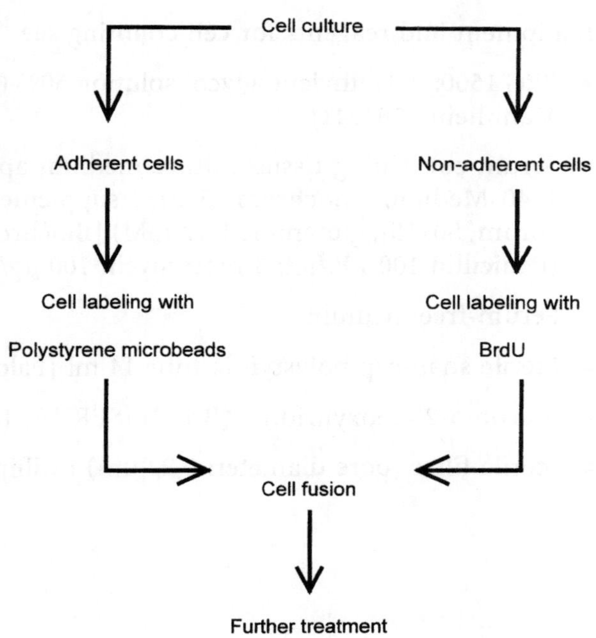

Fig. 1. Flow chart of the cell fusion procedures

Materials

Adherent cells

- Carboxylated polystyrene-microbeads-beads of different size (Solution 10%; Size 0,73 μM and 1,01 μM) [SERVA, on request]

For cell counting

- hemocytometer with coverslip
- 0,4% trypan blue
- PEG 1500: Polyethylene-glycol solution 50% (w/v) in Hepes [Boehringer Mannheim; 783641]
- Dulbecco's MEM Medium (Minimal Essential Medium with Earl's salts) [Biochrom, F0405] supplemented with 15% fetal calf serum - FCS [Biochrom, S0115], glutamine L (2 mM) [Biochrom, K0282] and antibiotics (Penicillin 100 I.E./ml; Streptomycin 100 μg/ml) [Biochrom, A2212]
- Identical medium without FCS

Non-adherent cells

Equipment and reagents for cell counting see "Adherent cells".

- PEG 1500: Polyethylene-glycol solution 50% (w/v) in Hepes [Boehringer Mannheim; 783641]
- Serum-containing tissue culture medium appropriate for LCLs: RPMI 1640 Medium [Biochrom, F1215] supplemented with 15% FCS [Biochrom, S0115], glutamine L (2 mM) [Biochrom, K0282] and antibiotics (Penicillin 100 I.E./ml; Streptomycin 100 μg/ml) [Biochrom, A2212]
- Serum-free medium
- Sterile snap-cap polystyrole tube 14 ml [Falcon, 2001]
- 5-bromo-2'-deoxyuridine (BrdU) [SERVA, 15240.02]
- Sterile filter (pore diameter: 0,22 μm) [Millipore, SLGV025BS]

Procedure

Fusion of adherent cells

This method describes the production of heterokaryons from two parental cell lines growing as monolayer cultures (e. g. human fibroblasts). The cells are fused by using polyethylene glycol (PEG). For identification of the parental nuclei in the resulting heterodikaryons carboxylated polystyrene-microbeads of different sizes are used.

Polystyrene microbeads suspensions

1. Carboxylated polystyrene-microbeads (10%)
2. Prepare 0,05% stock solutions in PBS [Biochrom, L1823] and autoclave.
3. Dilute 1:200 to prepare ready for use suspension (0,00025%) with microbeads of each.

We received the best labelling results with microbeads of sizes 0,73 μM and 1,01 μM. Smaller beads cause high rates of cross-contamination of the other parental cell line and larger ones show a small uptake rate leading to insufficient labelling. The chosen sizes of beads are clearly distinguishable under the microscope.

Preparation of parental cell lines

1. Grow parental cells (500 000 / 5 ml medium) in an appropriate number of tissue culture flasks (25 cm^2) to confluence and check that the cells are free of mycoplasm contamination (see Chapter 1). The number of parental cell lines and flasks depends on the planned experiments.
2. Trypsinize confluent cultures of two parental cell lines and subculture in a ratio 1:2.
3. Keep each parental cell line in medium containing microbeads of different size for three days, to allow efficient uptake of microbeads.
4. Rinse the cultures thoroughly twice with serum-free medium to remove all non-incorporated microbeads.
5. Trypsinize cultures and count the cells by aid of a hemocytometer. A short description for cell counting is given in the appendices of the "Current protocols in human genetics" Vol. 2 (Dracopoli et al., 1994). Test for viability using a selective labelling method, e. g. trypan blue staining. Calculate the adequate number of viable cells for cell fusion. Transfer equal numbers ($1x10^6$) of the parental cells preloaded with different sizes of microbeads into 60 mm^2 petri dishes.

6. Incubate cells in petri dishes for a further 24 hours.

Before starting cell fusion prewarm all reagents to 37°C!

Cell fusion

1. Check cell density of the culture dishes. For cell fusion 70-80% of the petri dish's bottom should be covered. Lower or higher cell densities will decrease the maximal number of dikaryons. If required, adjust cell density to reach optimal fusion conditions.
2. Suck off culture medium and rinse the dishes twice with prewarmed (37°C) serum-free medium. Pay attention to aspirate medium completely because FCS reduces the activity of the cell fusion solution.
3. Add 1 ml prewarmed PEG solution (37°C), distribute solution evenly over cells and incubate for 0,5-1,5 minutes. The incubation time is a critical step in fusion, therefore expose all areas of a plate as quickly as possible and control the treatment time precisely to seconds. Optimal PEG concentrations and incubation times may vary with the used cell material. The parameters for optimal fusion conditions have to be tested. In our laboratory each fusion experiment includes the test of triplicate dishes, treated for 30, 60 and 90 seconds. When the number of cells is limited use a 60-second treatment which might work in the majority of cases.
4. Immediately, add carefully 5 ml of serum-free medium (37°C), swirl gently to dilute PEG with medium and aspirate the fluid. The first wash must be done very cautiously because cell membranes are very fragile by the PEG treatment.
5. Wash twice with serum-free medium to remove PEG completely.
6. Feed cells with 5 ml culture medium (37°C) and incubate for 24 to 48 hours. Cells will show a delayed growth after fusion.
7. Trypsinize cells and seed at lower densities in complete medium before running any experiments or before harvesting.

Fusion behavior varies from cell line to cell line. Providing that two parental cell lines show identical membrane properties with respect to the cell fusion process, a ratio of 50% heterodikaryons and 25 % homokaryons each of the parental cell lines has to be expected. In fact, in most experiments homodikaryons of one parental cell line predominate.

PCC will also occur after cell fusion. The frequencies of cells showing PCC depend on the mitotic activity of the cultures.

Fusion of non-adherent cells

This procedure is used for cell fusion, when one or both of the parental cell lines are growing in suspension (e. g. lymphocytes, lymphoblastoid cell lines - LCLs). LCLs do not take up Latex beads and so a different labelling procedure is necessary (e. g. BrdU) to allow subsequent identification of parental cell origin in heterodikaryons.

Preparation of BrdU-solution

Differentially stained sister chromatids of mitotic chromosomes are obtained by incorporating the thymidine-analogue BrdU into replicating DNA for one cell cycle followed by one cell cycle in standard medium and subsequent staining. For details of the procedure see Chapter 2. BrdU is a mutagenic agent and requires special handling and disposal.

1. Prepare stock solution 0,5 mg/ml BrdU in aqua dest and sterilize by passing the solution through a 0,22 µm pore filter. BrdU-solution is light-sensitive, therefore, keep the stock solution in the dark at 4°C. Under these conditions the solution can be stored for at least six months.
2. For cell labelling prepare medium containing 10 µg/ml BrdU.

Preparation of parental cell lines

1. Make sure before any experiment that the cell lines are free of mycoplasma contamination (see Chapter 1).
2. Incubate one parental cell type (500 000 cells/ml) in BrdU-containing medium (10µg/ml) for 24 hours before fusion.

Cell fusion

1. Count parental cell lines and mix equal numbers (2 x 10^6) of BrdU-labeled and unlabelled cells in a snap-cap polystyrole tube. The cells can be counted in a hemocytometer. Test for viability using a selective labelling method, e. g. trypan blue staining. Calculate the adequate number of viable cells for fusion.
2. Centrifuge 10 minutes at 100 x g (room temperature) and aspirate the medium. BrdU-containing medium has to be discharged as mutagenic waste.
3. Wash twice with 10 ml prewarmed (37°C) serum-free medium, centrifuge for 10 minutes each at 100 x g. Wash thoroughly to remove serum completely before fusion.
4. Aspirate all medium and tilt tube carefully to remove all medium drops.
5. Gently break up the pellet by flicking the tube with a finger.

6. Carefully add 1 ml prewarmed (37°C) 50 % PEG solution to the pellet. Incubate 30-90 seconds while gently swirling the tube. If cell numbers are not limited, test for best fusion conditions, otherwise use 60 seconds. Agitate the tube to keep the cells in contact with the fusing agent. The mechanism of cell fusion is not entirely understood. We observed quite different fusion properties between various cell lines and, therefore, recommend to test the fusion conditions in a pilot experiment.
7. Add 10 ml of prewarmed (37°C) serum-free medium and centrifuge without any delay at 100 x g.
8. Aspirate medium and wash twice with serum-free medium.
9. Resuspend cell pellet in complete medium and transfer the cells to a tissue culture flask. Add culture medium and incubate for 24 to 48 hours, before harvesting cells. Any treatment must be done during this time interval to make sure that chromosomes show differentially stained sister chromatid.
10. Differential staining of sister chromatids follows the Fluorescense-Plus-Giema (FPG) technique described in Chapter 2.

However, the fusion ratio of heterodikaryons is low in human cells and many slides have to be screened to get a sufficient amount of cells for reliable results. Using this cell fusion technique it is also possible to get proliferating somatic cell hybrids.

Personal experience

- Whenever possible test for the optimal PEG concentration and incubation time. This is time saving since you have the maximal number of heterodikaryons per slide and you avoid screening a considerable higher number of slides.
- The absolute number of cells and the 1:1 ratio of the parental cells are critical for a successful experiment.
- For optimal results it is essential that the cells are in the logarithmic phase of growth and are free of mycoplasma.
- Figures 2 and 3 show heterokaryons of adherent cells labeled with Latex beads and of non-adherent cells labeled with BrdU.

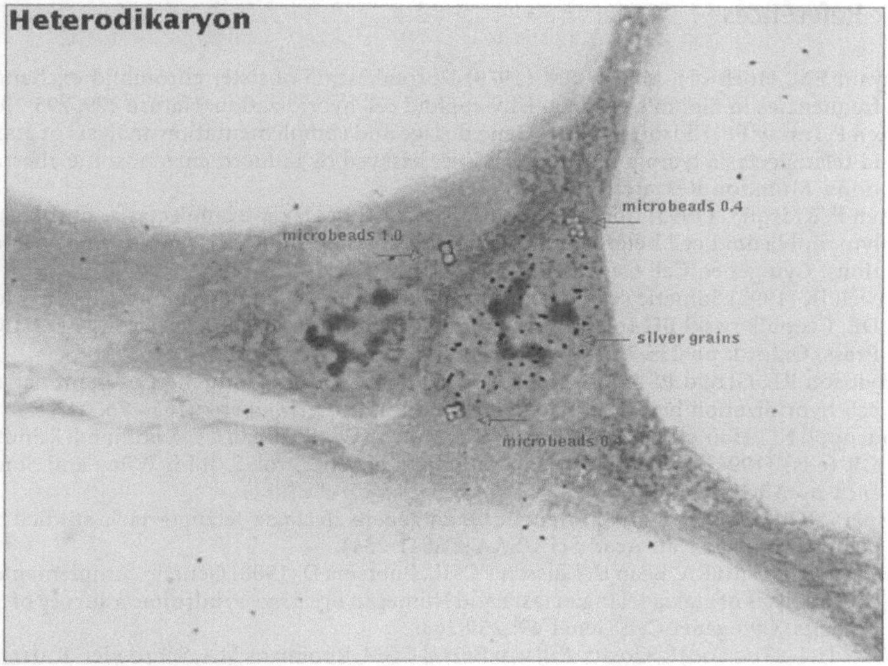

Fig. 2. Heterodikaryon obtained after fusion of two fibroblast cells labeled with microbeads of different size. One nuclei is in S-phase, as shown by the presence of silver grains after ^{3}H-Thymidine incorporation and autoradiography.

Fig. 3. Metaphase of a heterodikaryon obtained after fusion of two LCL cell lines. After differentially staining (FPG-staining) of sister chromatids the incorporation of BrdU in one parental cell line can be visualized. Thus, the parental origin of every chromosome can be traced back.

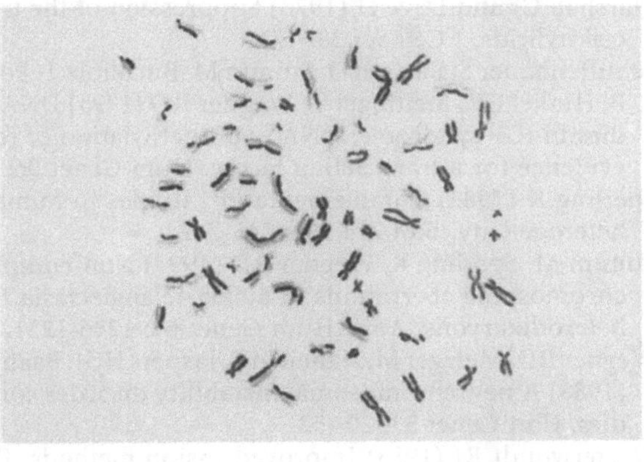

References

Bryant EM, Hoehn H, Martin GM (1979) Normalization of sister chromatid exchange frequencies in Bloom's syndrome by euploid cell hybridization. Nature 279: 795-796.

Chen P, Imray FP, Kidson C (1984) Gene dosage and complementation analysis of ataxia-telangiectasia lymphoblastoid cell lines assayed by induced chromosome aberrations. Mutation Research 129: 165-172.

Chen P, Kidson C (1993) Complementation analysis in ataxia-telangiectasia: fibroblast-lymphoblastoid cell heterokaryon assay by radiation-induced chromosome aberrations. Cytogenet. Cell Genet 64: 9-11.

Cowell JK (1992) Somatic cell hybrids in the analysis of the human genome. In: Rooney DE, Czepulkowski BH (eds) Human Cytogenetics: A Practical Approach, vol. 2. IRL Press, Oxford, pp 235-252.

Davidson RL, Gerald PS (1976) Improved techniques for the induction of mammalian cell hybridization by polyethylene glycol. Somatic Cell Genet 2: 165-176.

Dracopoli NC, Haines JL, Korf BR, Moir DT, Morton CC, Seidman CE, Seidman JG, Smith DR (eds) (1994) Current protocols in human genetics, vol. 2. John Wiley and Sons, Inc., pp A3G8-A3G11

Jaspers NGJ, Bootsma D (1982) Genetic heterogeneity in ataxia-telangiectasia studied by cell fusion. Proc Natl Acad Sci USA 79: 2641-2644.

Jaspers NGJ, Gatti RA, Baan C, Linssen PCML, Bootsma D (1988) Genetic complementation analysis of ataxia telangiectasia and Nijmegen breakage syndrome: a survey of 50 patients. Cytogenet Cell Genet 49: 259-263.

Joenje H, Lo Ten Foe JR, Oostra AB, van Berkel CGM, Rooimans MA, Schroeder-Kurth T, Wegner RD, Gille JJP, Buchwald M, Arwert F (1995) Classification of Fanconi anemia patients by complementation analysis: evidence for a fifth genetic subtype. Blood 86: 2156-2160.

Johnson RT, Rao PN (1970) Mammalian cell fusion: Induction of premature chromosome condensation in interphase nuclei. Nature 226: 717-722.

Marshall CJ and Dave H (1978) Suppression of the transformed phenotype in somatic cell hybrids. J Cell Sci 33: 171.

Schuffenhauer S, Bartsch O, Stumm M, Buchholz T, Petropoulou T, Kraft S, Belohradsky B, Hinkel GK, Meitinger T, Wegner RD (1995) DNA, FISH and complementation studies in ICF syndrome: DNA hypomethylation of repetitive and single copy loci and evidence for a trans acting factor. Hum Genet 96: 562-571.

Sperling K (1982) Complementation studies in human genetic disorders: analysis for heterogeneity. Biol Zbl 101: 745-762.

Stumm M, Sperling K, Wegner RD (1997) Non-complementation of radiation induced chromosome aberrations in ataxia-telangiectasia / ataxia-telangiectasia-variant cell heterodikaryons. Am J Hum Genet 60: 1246-1251.

Wegner RD, Metzger M, Hanefeld F, Jaspers NGJ, Baan C, Magdorf K, Kunze J, Sperling K (1988) A new chromosomal instability disorder confirmed by complementation studies. Clin Genet 33: 20 -32.

Westerwoudt RJ (1985) Improved fusion methods. IV. Technical aspects. J Immunol Methods: 181-196.

Zakrezewski S, Sperling K (1982) Genetic heterogeneity of Fanconi's anemia demonstrated by somatic cell hybrids. Hum. Genet. 56: 81-84.

Chapter 17

Origin of Trisomies

GESA SCHWANITZ AND THOMAS EGGERMANN

Introduction

Trisomy is the most commonly identified chromosome abnormality in humans, occurring in at least 4% of all clinically recognized pregnancies. Some trisomies are compatible with livebirth as approximately 0.3% of all newborn infants have an aneuploidy of the gonosomes or a trisomy of the autosomes 13, 18, or 21. However, most trisomic fetuses do not survive to term. They are associated either with clinically recognized spontaneous abortions, where they account for approximately 60% of all such fetuses, or with intrauterine fetal death, or they terminate as subclinical spontaneous abortions.

Despite the high frequency of trisomies, their aetiology remains imperfectly understood. There is no consensus regarding a causative role for any of the many factors that have been investigated, with the single important exception of maternal age.

In large parts, this is due to the difficulties inherent in obtaining the relevant meiotic material, i.e. the testicular and ovarian tissues in which the nondisjunctional event occurs. Thus, most studies on human nondisjunction have focused on determining the parent and meiotic stage of origin of the additional chromosome in the fetus/child.

Studies beginning in the 1960s made use of blood group polymorphisms to evaluate the origin of sex chromosome aneuploidy. During the 1970s and 1980s, chromosome heteromorphisms were used to determine the origin of autosomal trisomies, especially trisomy 21. However, this approach has also had limitations because these marker systems are less informative and, in

Gesa Schwanitz, Institut für Humangenetik, Wilhemstr. 31, Bonn, 53111, Germany, *Correspondence to* Thomas Eggermann, Technical University of Aachen, Institute of Human Genetics, Pauwelsstr. 30, Aachen, 52074, Germany (*phone* +49-241-8088008; *fax* +49-241-8888580; *e-mail* TEGGERMANN@POST.KLINIKUM.RWTM-Aachen.de) the case of cytogenetic heteromorphisms, they showed reduced reliability (Antonarakis et al., 1992). The identification of DNA polymorphisms has eliminated these problems. Highly polymorphic markers are now available for all human chromosomes and, for some chromosomes, centromere-specific polymorphisms have been described as well. These DNA markers consist of Restriction Fragment Length Polymorphisms (RFLPs) and Short Tandem Repeat markers (microsatellites; short sequence repeats=STRs). RFLPs were developed in the 1980s: the analyses include digestion of genomic DNA with restriction endonucleases, gel electrophoresis, Southern-blotting, DNA-probe labelling and hybridization, a very time- and material- (including genomic DNA) consuming procedure. Another disadvantage is the relatively low informative value with an averaged polymorphism information content of 0.375 (PIC-value). Therefore, STRs as another class of DNA-markers had increasing significance in investigations on the human genome since 1988. Firstly, they are highly informative (PIC-value›0.6) and easy and fast to handle. Furthermore, only minimal amounts of genomic DNA are necessary for molecular genetic analyses. STRs have been described as an abundant class of DNA polymorphisms in the human genome, consisting of highly repetitive short DNA sequences. They can be typed by using PCR and single-copy primers flanking the repeats. In 1991, Petersen and co-workers were the first who described the application of these markers to determine the parental origin of the extra chromosome in families with trisomy 21. All recently published studies on the origin of trisomies are based on these marker systems, sometimes completed by RFLPs, when centromeric STR markers are not available.

By following the transmission of informative polymorphic markers spanning the whole chromosome in a trisomic patient and his parents, it is possible to determine both the parent and the meiotic stage of origin of every supernumerary chromosome. One or more additional chromosomes can result from an error in a paternal or maternal meiotic division, or from an error at an early mitotic division of the zygote.

To distinguish between these possibilities the physical localisation of the marker on the chromosome is important: Heterozygosity for a **centromere marker** indicates that nondisjunction occurred at the first meiotic division whereas homozygosity indicates a mistake at the second meiotic division. In contrast, **noncentromeric markers** provide direct information only on the parent who contributed the additional chromosome. On the basis of centromere markers alone, it is not possible to distinguish between a second meiotic division error and a mitotic error occurring in an early cell division of a normal zygote. However, centromere markers used in conjunction with a large number of more distally situated markers allow such a distinction to be made. Mitotic errors are associated with complete homozygosity for the duplicated chromosomes, whereas errors at meiosis II are associated with homozygosity for proximal markers but heterozygosity for more distally located markers is caused by crossing over events.

Furthermore, the use of such markers allows to ask new questions regarding the origin of human trisomies. One of the most important of these is whether or not abnormal genetic recombination is associated with nondisjunction in humans, as supposed for trisomy 21 (Sherman et al., 1994). So far, molecular techniques have been used to determine parent and stage of nondisjunction in the most frequent autosomal trisomies in man: The vast majority of autosomal trisomies results from errors during maternal meiosis. In trisomy for chromosomes 13, 15, 16, 21 and 22, the predominant stage of nondisjunction is maternal meiosis I (MI) (Hassold et al., 1991; Antonarakis et al., 1993; Zaragoza et al., 1994). In contrast, an excess of maternal meiosis II (MII) compared with MI in nondisjunction resulting in trisomy 18 could be demonstrated (Fisher et al., 1995; Eggermann et al., 1996a). However, paternally derived additional acrocentric chromosomes result predominantly from nondisjunction at MII. Postzygotic mitotic errors account for some 5% of non-mosaic autosomal trisomies. Here, the parental origin of the additional chromosome is equally likely to be maternal or paternal and there is no increased parental age effect seen in trisomies of postzygotic mitotic origin.

Apart from studies on the aetiology of trisomies, similar investigations have been published about the origin of structural aberrations (Eggermann et al., 1996b).

For many chromosome disturbances, only little numbers of patients are available for each aberration because of the rareness of the abnormality. Therefore, retrospective studies using archival tissues seem reasonable. In this respect, the most important sources for DNA extraction are chromosomal spreads from routine cytogenetic diagnosis, and formalin fixed and paraffin embedded tissues from histological investigations. DNA recovered from these samples is often degraded thus rendering study by conventional Southern blot analysis difficult. For this assay PCR provides a fast and simple method where no high molecular weight DNA is required.

The procedure described here may generally be applicable for retrospective diagnosis of chromosomal aberrations when pathological specimens are available. They can be used also for the detection of uniparental disomies.

In this context, the use of the **D**egenerate **O**ligonucleotide **P**rimed polymerase chain reaction (DOP-PCR) for a general amplification of human genomic DNA is suggested. By using DOP-PCR prior to locus-specific genotyping, one can increase the amount of typable DNA by the tenfold.

Materials

Molecular biology grade reagents should be utilized whenever possible. Distilled water and solutions should be autoclaved. To avoid contamination in the PCR, we usually use sterile tips and tubes. According to our experience, we were well satisfied with oligonucleotide primers from MWG-Biotech, Munich, Germany, or PERKIN ELMER, Applied Biosystems. Good quality Proteinase K and Pronase E can be obtained from Boehringer Mannheim. PCR process and the polymerase that drive PCR are the subjects of patents now held by Hoffmann-LaRoche. If you purchase reagents from one of the companies licensed by Hoffmann-LaRoche to sell Taq-Polymerase then you are essentially free to use PCR for whatever non-profit-making research you wish. The situation is different when there is some commercial application.

Study population

To raise a study population for getting statistically significant information, the retrospective inclusion of patients of whom only archival specimens are available should be strived, enlarged by a multicentric collaboration. For each patient, pre- or postnatal diagnosis of trisomy should be performed by chromosome analysis. When this diagnostic purpose has not been successful or has not been carried out, Fluorescence *in situ* hybridization (see Part V, Chapter 18) or molecular analysis can be used for diagnosis (see Comments, Personal experience).

Procedure

DNA extraction

Two types of DNA sources can be distinguished: Fresh tissues consist of blood, amniotic fluid, chorionic villi, fibroblasts and muscle biopsies. Archival specimens are chromosomal spreads from routine diagnosis, paraffin embedded tissues and suspensions of cultured cells stored in fixative. Of course, the former are preferred because of the higher integrity and the larger amount of the extracted DNA. But the retrospectivity of the type of study suggested here makes it probable that the majority of specimens consist of archival tissues.

Fresh tissues

Fresh tissues are a widely used source for DNA-extraction. Therefore, many protocols have been published. Here, we suggest two strategies for isolating DNA from blood and other tissues, particularly fetal cells. Whereas in the case of adults large amounts of blood are available, the amount of fetal cells like chorionic villi and amniotic cells is limited. Thus, we propose the application of phenol/chloroform/isoamylalcohol extraction after Proteinase K digestion to obtain the maximal yield of highly purified DNA as described by Guay-Woodford et al. (1995).

In adults and living patients, venous blood samples, anticoagulated with EDTA, can be drawn. DNA is extracted by salting out with saturated NaCl solution according to the protocol of Miller et al. (1988).

Archival specimens: paraffin embedded tissues

Probands' DNA can be extracted from unstained or haematoxylin and eosin-stained formalin fixed and paraffin embedded specimens from different tissues (eg liver, lung, kidney, spleen, muscle, gut).

1. Cut the paraffined tissues in extremely thin slices (10 µm) by a microtome or a scalpel. Put three or four slices in a 1.5 ml-tube.
2. Deparaffine and rehydrate the specimen by resuspension in 500 µl xylene and washing twice in xylene, twice in absolute ethanol and finally in 70% ethanol.
3. Resuspend the final pellet in 100 µl digestion buffer (0.2 M TrisHCl pH 8.0, 10 mM EDTA) containing 10% SDS and 2.5 µl Proteinase K (20 mg/ml).
4. Incubate the digest at 50°C for 24 to 48h (If the digest is incomplete, add the same amounts of SDS and Proteinase).
5. Purify the digest by two extractions with 200 µl phenol (buffered with 0.25 M TrisHCl pH 8.0) and chloroform/isoamylalcohol (24:1) each and by a last extraction with chloroform/isoamylalcohol.
6. Estimate the final volume and add 1/10 vol of 3 M NaAc (pH 4.8) and 2 vol of ice cold 100% ethanol.
7. Invert the preparation several times and precipitate DNA overnight at -20°C or for 30 min on dry ice.
8. Recover the precipitate by centrifugation for 30 min at 13.000 rpm. Discard the supernatant and resuspend the pellet in 100 µl TE^{-4} (10 mM TrisHCl pH 8.0, 0.1 mM EDTA).

Archival specimens: chromosomal spreads

DNA from the probands can be isolated from routinely fixed metaphase spreads on glass slides using a modification of the protocol described by Jonveaux (1991). Metaphase spreads (native, G-or Q-banded) from different cell systems (lymphocytes, fibroblasts, amniotic fluid cells and chorionic villi) that have been stored at room temperature for approximately 6 months to 14 years can be used.

1. Remove coverslip (If the slide is mounted, put it in xylene until the coverslip can be removed gently).
2. Treat the chromosomal spreads directly with Proteinase K (0.5 mg/ml) in 50 µl extraction buffer (100 mM NaCl, 50 mM TrisHCl pH 7.5, 1 mM EDTA, 0.5% SDS) under a coverslip sealed with rubber cement at 55°C for 2 h.
3. Remove the coverslip gently and rinse the slide briefly with 100 µl extraction buffer alone.
4. Purify the preparations by two extractions in phenol/chloroform/isoamylalcohol and one extraction in chloroform/iosamylalcohol.
5. Precipitate DNA by addition of 1/10 vol 3 M NaAc (pH 4.8) and 2 vol icecold absolute ethanol at -70°C for at least 12 h.
6. Add 1 µl glassmilk (Dianova Cat. 1801-484) to 300 µl DNA/ethanol mixture. Incubate for 30 min at room temperature and wash subsequently with absolute ethanol.
7. Redissolve DNA with 50 µl TE^{-4}.

Because of the small amount of cells per slide the isolated DNA will not be detectable by standard DNA-estimation procedures. Therefore, we propose to use 2 µl of DNA solution directly in the following analyses.

Archival specimens: cell suspensions

If a collaborating laboratory has not the capacity to extract DNA from freshly drawn blood and a fast mailing can not be guaranteed, the cells should be stored in fixative after cultivation. DNA can be obtained by centrifugation of the suspension, followed by Proteinase K digest and phenol/chloroform/isoamylalcohol purification as described above.

Polymerase chain reaction (PCR)

Nowadays, PCR is one of the most widely used basic biological techniques due to its remarkable speed, sensitivity and flexibility. Therefore, a great number of applications of PCR is described. In this chapter, we can describe PCR in respect to our special applications. For basic protocols and trouble-shooting, we refer to special PCR literature (Erlich (ed.), 1989; Pherson et al. (eds.), 1995).

Before performing PCR, you have to make some considerations about contamination. Particularly the analysis of minimal amounts of genomic DNA extracted from archival tissues and the sensitivity of PCR makes accurate working necessary. To prevent contamination of samples by DNA and previously amplified products, considerable thought and care should be taken to the actual logistics of the experiments. A set of micropipettors should be devoted exclusively to preparation of the reaction mixtures at a designated PCR workstation set apart from the rest of the laboratory. Negative controls should be included in each experiment as an assay for overt contamination.

DOP-PCR

When only small amounts of genomic DNA are obtained, DOP-PCR (Telenius et al., 1992a, b) prior to locus-specific typing of STRs is used. DOP-PCR employs oligonucleotides of partially degenerate sequences; this allows priming from multiple, regularly dispersed sites within a genome.

1. Prepare a 50 µl reaction mixture containing 5 µl patients' DNA solution, 200 µM each dNTP (Gibco BRL), 2mM $MgCl_2$, 50 mM KCl, 10 mM TrisHCl (pH 8.3), 0.01% gelatine, 1.5 µM universal primer (sequence 5'CCG ACT CGA GNN NNN NAT GTG G3') and 2.5 U Taq-polymerase.
2. Cycle in thermal cycler as follows: 93°C for 10 min initially, followed by: 5 cycles of
   - 94°C for 1 min
   - 30°C for 1.5 min
   - 30-72°C transition for 3 min
   - 72°C for 3 min, followed by: 30 cycles of
   - 94°C for 1 min
   - 62°C for 1 min
   - 72°C for 3 min with an addition of 1 sec/cycle final extension for 10 min hold at 4°C

DOP-PCR products can be purified with different PCR product purification systems, for example "Qiaquick PCR purification system" (Quiagen Cat. 28104).

STR typing

The great increase in studies on the origin of trisomies during recent years has essentially resulted from great improvements in reference genetic linkage maps, which at present mainly consist of STRs. For example, the previously published Généthon genetic linkage map (Dib et al., 1996) includes 5264 STRs spanning the whole human genome with an average interval size of 1.6 cM.

The meaningfulness of the type of studies described here depends on the choice of polymorphic markers. For getting meaningful results, the following conditions should be fulfilled: Firstly, a panel of markers spanning the whole chromosome should be used. Thereby, the distinction between meiotic or mitotic error as well as the detection of recombinational events become possible. Secondly, several centromeric or pericentromeric polymorphisms are necessary to distinguish between meiosis I and meiosis II nondisjunction. In case of pericentromeric markers, an error rate of stage determination proportional to the genetic distance between the most proximal informative STRs has to be taken in consideration (see Results).

If DNA from archival tissues is analysed, it should be noted that this DNA is often fragmented. Therefore, only STR markers with a PCR product length of up to 200 bp should be typed to achieve reliable results.

Furthermore, PCR conditions for the amplification of this type of DNA and DOP-PCR preamplified DNA should be modified as follows. Samples are processed through an increased number of cycles (up to 40). To avoid primer dimerization and pre-PCR mis-priming, hot start PCR is performed: The reaction tubes are opened in order to add Taq-polymerase and are reclosed as they sit in the thermal cycler at 94°C. To achieve stronger allelic bands and to reduce unspecific bands, variations like asymmetric PCR and nested PCR should be tested.

STR PCR products are separated on denaturing acrylamide gels. The detection can be performed by different techniques, for example silver staining or radioactive labelling followed by autoradiography. Each method has its own advantages and problems, which is why the authors would refer to specialist PCR literature.

Results

The use of STRs for studying the origin of chromosomal disturbances is a fast and reliable technique. It allows us to perform molecular investigations in cases where only minimal amounts of genomic DNA are available.

For each trisomy, at least two informative markers should be used to determine the parent of origin in each family. Information includes different parental alleles and the presence of three alleles in the patient. The allelic status in the proband can consist of three different alleles or two alleles with different intensities (Figure 1). When three alleles are present in the patient and the parents have different alleles, it is unambiguously possible to determine that two alleles are from one parent and one from the second parent (Figure 1a). When only two alleles are present in a patient, one allele is normally amplified to a greater intensity (Figure 1b, c). In this case, the parent of origin is determined to be the one contributing the more intense allele. But this status is difficult to prove because there is only a limited proportional correlation between template DNA and successful amplification (Figure 1b, c). For example, smaller DNA fragments are amplified faster than larger templates. Therefore, differences in band intensities should be confirmed by densitometric analysis and compared to intensities of normal samples with the same allelic status or by cytogenetic analysis.

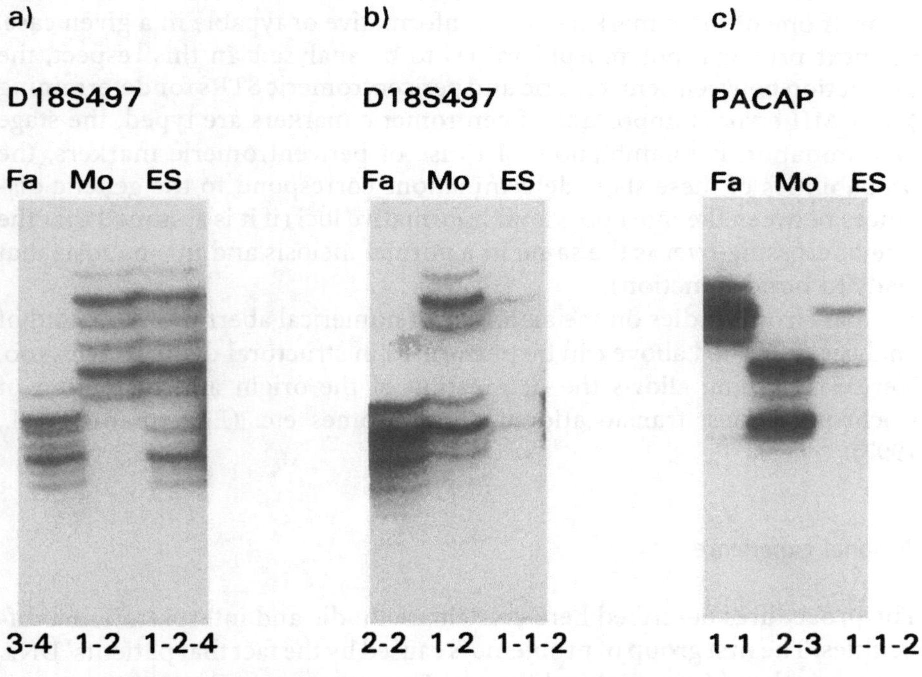

Fig. 1. a) Examples for STR typing in trisomy 18 families. Note the weak differences between the intensities of proband's alleles in b) and c) which make an interpretation difficult. a) Trisomy 18 in the proband (ES) is demonstrated by typing the STR D18S497. The patient inherited two chromosomes 18 (alleles 1 and 2) from his mother, the third allele (4) is the paternal one. b) By typing the same marker the maternal origin of the extra chromosome 18 is proven by one maternal allele (allele 1) which is more intense than the paternal allele (2). c) In this case the extra chromosome is paternal in origin (allele 1). (Fa father, Mo mother, ES Edwards syndrome patient)

Subsequently, the categorization of the trisomic offspring for each analysed locus at which the parent who contributed for example an extra chromosome 18 is heterozygous allows the determination of the stage on nondisjunction. By this characterization, one can delineate the cell division of origin of trisomy by typing at least five informative markers for each case: if parental heterozygosity is retained in all analysed loci, an error at MI can be concluded. If parental heterozygosity is reduced to homozygosity in all informative markers, one concludes a MII or postzygotic mitotic error. In cases where both reduction and nonreduction are observed, a postzygotic mitotic nondisjunction is excluded, and one must determine the cell division of origin by typing the most proximal markers or the centromeric markers on the chromosome, respectively. If parental heterozygosity in this region appears to be nonreduced in the offspring, it is assumed that the error has occurred at MI, whereas if it appears to be reduced, one concludes a MII error. If one of these markers is not informative or typable in a given case, the next proximal polymorphism has to be analysed. In this respect, the distinction between centromeric and pericentromeric STRs for determining MI or MII error is important: If centromeric markers are typed, the stage determination is unambiguous. In case of pericentromeric markers, the probabilities of these stage determinations correspond to the genetic distances between the most proximal informative loci (if it is assumed that the rate of crossing-over is the same in a normal meiosis and in a meiosis that leads to nondisjunction).

Apart from studies on the aetiology of numerical aberration the kind of analysis described above can be performed in structural disturbances, too. Here STR typing allows the delineation of the origin and formation of isochromosomes, translocational chromosomes etc. (Eggermann et al., 1997).

Personal experience

The procedures described here contain methodic and interpretational difficulties. The first group of problems is caused by the fact that patients' DNA is often isolated from archival tissues. These tissues can become contaminated by genomic DNA during storage or preparation. Wrong results caused by these contaminations are excluded by the simultaneous STR analysis of parental DNA and patients' DNA which detects contamination directly. Another problem is the contamination of fetal tissues with maternal cells, particularly in chorionic villous cells and placental specimens which can change the intensities of allelic band in STR typing. Furthermore, differences in intensities of alleles are difficult to prove when the trisomic proband shows only two alleles (see Results; Figure 1b, c).

The interpretational difficulties include undetected recombinations events. Theoretically, crossing-over can remain undetected though a close-meshed STR typing for the chromosome of question has been performed. Therefore, cases with a MI error can be misclassified as MII or MII/postzygotic mitosis and vice versa.

Furthermore, it should be noted, that a reduction to homozygosity at all analysed loci, a finding corresponding to a postzygotic mitotic nondisjunction, also be consistent with a failure of recombination at MI, followed by nondisjunction at MII. If there is no polymorphic centromeric marker for a chromosome, it is not possible to distinguish with certainty between MI and MII errors. Errors that occur postzygotically show reduction to homozygosity at all informative loci.

Finally, it should be mentioned, that for diagnostic purposes only three different bands in patients DNA are a reliable clue for a trisomy (Figure 1a). Different band intensities are an indication for trisomy (Figure 1b, c), but this has to be proven by cytogenetic or molecular cytogenetic analysis.

References

Antonarakis SE, Petersen MB, McInnis MG, Adelsberger PA, Schinzel AA, Binkert F, Pangalos C, Raoul O, Slaugenhaupt SA, Hafez M, Cohen MM, Roulson D, Schwartz S, Mikkelsen M, Tranebjaerg L, Greenberg F, Hoar DI, Rudd NL, Warren AC, Metaxotou C, Bartsocas C, Chakravarti A (1992) The meiotic stage of nondisjunction in trisomy 21: determination by using DNA polymorphisms. Am J Hum Genet 50: 544-550

Antonarakis SE, Avramopoulos D, Blouin J-L, Talbot Jr CC, Schinzel AA (1993) Mitotic errors in somatic cells cause trisomy 21 in about 4.5% of cases and are not associated with advance maternal age. Nature Genet 3: 146-150

Dib C, Fauré S, Fizames C, Samson D, Drouot N, Vignal A, Millasseau P, Marc S, Hazan J, Seboun E, Lathrop M, Gyapay G, Morissette J, Weissenbach J (1996) A comprehensive genetic map of the human genome based on 5264 microsatellites. Nature 380: 152-154

Eggermann T, Nöthen MM, Eiben B, Hofmann D, Hinkel K, Fimmers R, Schwanitz G (1996a) Trisomy of human chromosome 18: molecular studies on parental origin and cell stage of nondisjunction. Hum Genet 97:218-223

Eggermann T, Engels H, Moskalonek B, Nöthen MM, Müller-Navia J, Schleiermacher E, Schwanitz G, Stengel-Rutkowski S (1996b) Tetrasomy 18p de novo: identification by FISH with library and microdissection probes and analysis of parental origin and mode of formation by short sequence repeat typing. Hum Genet 97: 568-572

Eggermann T, Engels H, Heidrich-Kaul, C, Moderau I, Schwanitz G (1997) Molecular investigations of the parental origin of a de novo unbalanced translocation 13/18. Hum Genet 99:521-522

Erlich HA (ed.) (1989) PCR technology. Stockton Press, New York

Fisher JM, Harvey JF, Morton NE, Jacobs PA (1995) Trisomy 18: studies of parent and cell division of origin and the effect of aberrant recombination on nondisjunction. Am J Hum Genet 56: 669-675

Guay-Woodford LM, Muecher G, Hopkins SD, Avner ED, Germino GG, Guillot AP, Herrin J, Hollemann R, Irons DA, Primack W, Thomson PD, Waldo FB, Lunt PW, Zerres K (1995) The severe perinatal form of autosomal recessive polycystic kidney disease (ARPKD) maps to chromosome 6p21.1-p12: implications for genetic counselling. Am J Hum Genet 56: 1101-1107

Hassold T, Pettay D, Freeman SB, Grantham M, Takaesu N (1991) Molecular studies of nondisjunction in trisomy 16. J Med Genet 28: 159-162

Miller SA, Dykes DD, Polesky HF (1988) A simple salting out procedure for extracting DNA from human nucleated cells. Nucleic Acids Res 16: 1215

Petersen MB, Schinzel AA, Binkert F, Tranebjaerg L, Mikkelsen M, Collins FA, Econoumou EP, Antonarakis SE (1991) Use of short sequence repeat DNA polymorphisms after PCR amplification to detect the parental origin of the additional chromosome 21 in Down syndrome. Am J Hum Genet 48: 65-71

McPherson MJ, Hames BD, Taylor GR (eds.) (1995) PCR 2: A practical approach. Oxford University Press, Oxford

Sherman SL, Petersen MB, Freeman SB, Herswey J, Pettay D, Taft L, Frantzen M, Mikkelsen M, Hassold TJ (1994) Nondisjunction of chromosome 21 in maternal meiosis I: evidence for a maternal age-dependent mechanism involving reduced recombination. Hum Mol Genet 3: 1529-1535

Telenius H, Carter NP, Bebb CE, Nordenskjöld M, Ponder BAJ, Tunnacliffe A (1992a) Degenerate Oligonucleotide-primed PCR: general amplification of target DNA by single degenerate primer. Genomics 13: 718-725

Telenius H, Pelmer AH, Tunnacliffe A, Carter NP, Behmel A, Ferguson-Smith MA, Nordenskjöld M, Pfragner R, Ponder BAJ (1992b) Cytogenetic analysis by chromosome painting using DOP-PCR amplified flow-sorted chromosome. Genes Chrom Cancer 4: 257-263

Zaragoza MV, Jacobs PA, James RS, Rogan P, Sherman S, Hassold T (1994) Nondisjunction of human acrocentric chromosome: studies on 432 trisomic fetuses and liveborns. Hum Genet 94: 411-417

Part V

Molecular Cytogenetics

Chapter 18

Fluorescence *in Situ* Hybridization

GESA SCHWANITZ AND REGINE SCHUBERT

Introduction

In situ hybridization (ISH) combines cytogenetic and molecular genetic techniques. The principle of ISH is the interaction of a labelled single-stranded DNA or RNA probe with a complementary single stranded target DNA sequence. The presence of the target DNA is proven in metaphase chromosomes or in interphase nuclei which are fixed on slides hence *in situ* (Gall and Pardue, 1969; John et al., 1969).

Using specific DNA probes, different target sequences can be analysed. These probes can be labelled either radioactively or – more often-non-radioactively. Direct non-radioactive *in situ* hybridization is most generally performed with fluorescent dyes which have been linked to the probe prior to labelling. Indirect methods include non-fluorescence labelling of probe DNA, e.g. with marker molecules such as biotin or digoxigenin. Subsequent identification of the probe's location is performed by molecules that possess high affinity to the marker molecules and are labelled with fluorochromes.

The most widely used method of **fluorescence in situ hybridization** (FISH) employs non-radioactive labelled probes which proved to be very sensitive (Bauman et al., 1980; Lichter et al., 1992).

Special fields for application of this method are the identification and characterization of numerical and structural chromosome aberrations in clinical cytogenetics, the analysis of typical chromosome changes in different types of tumors, the identification of microdeletions and microduplications, gene mapping and the analysis of karyotype evolution (Figure 1-7).

At first this technique was limited to *metaphase analysis*, meanwhile it has gained importance in *interphase diagnostics* (Figure 8, 9) as well (Cre-

Correspondence to Gesa Schwanitz, Institut für Humangenetik, Wilhelmstr. 31, Bonn, 53111, Germany (*phone* +49-228-287-2338; *fax* +49-228-287-2380; *e-mail* cytogen@-humgen.uni-bonn.de), Regine Schubert, Institut für Humangenetik, Wilhelmstr. 31, Bonn, 53111, Germany

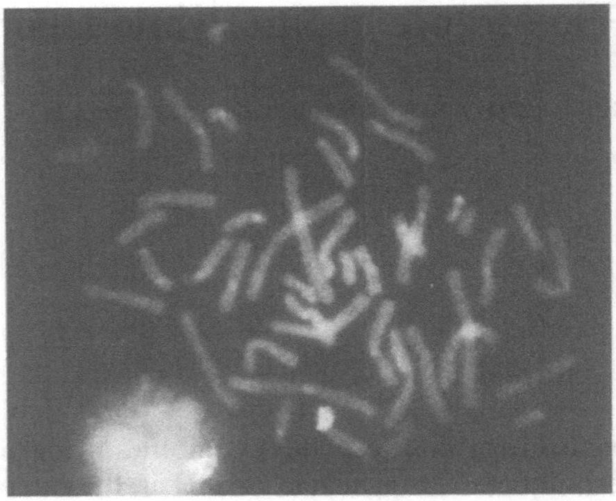

Fig. 1. Identification of an Y/autosome translocation (karyotype: 46,XX,t(Y;10)(q12;q24). Positive hybridization signals at the der (10) with the repetitive probe YpH3.4, detecting Yq12 (probe DYZ1, origin SatIII, E. Rubin Livermore).

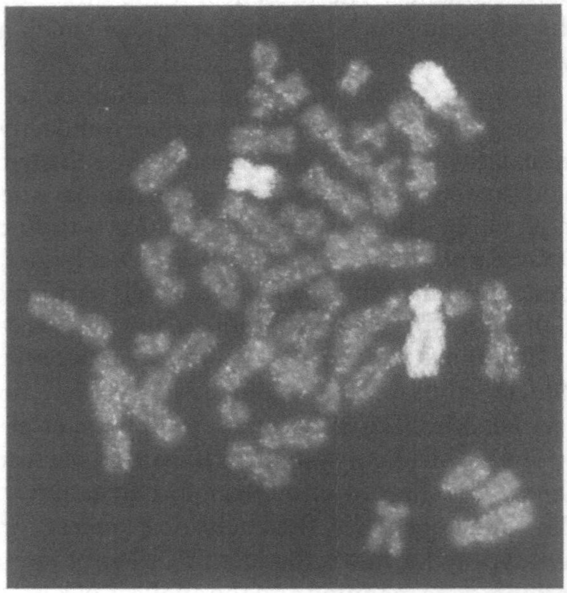

Fig. 2. Metaphase with reciprocal translocation 4/10 hybridized with a whole chromosome 4 paint probe giving positive signals in the normal chromosome 4, in 4pter to q21 in the first translocation chromosome and in 4q21 to 4qter in the other translocation chromosome (probe pBs – 4, plasmid – vector, Collins et al., 1991).

mer et al., 1988; Cremer et al., 1990; Jauch et al., 1990). Both types of investigation have a number of advantages and disadvantages, therefore they are combined now providing us with new approaches (see Applications).

With FISH different types of sequences of the human genome are detectable. *Whole chromosome paint probes* are used, which usually detect the euchromatic regions of an entire chromosome or chromosome arm (Figure

Fig. 3. Microdeletion analysis in 22q11.2 (di George-/VCF-syndrome) by hybridization with the probe D22S75. Only one of the chromosomes 22 gives a positive signal in the region. The distally (22q13.3) located probe D22S39 serves as a control (Cosmid probe, digoxigenin labelled, Oncor).

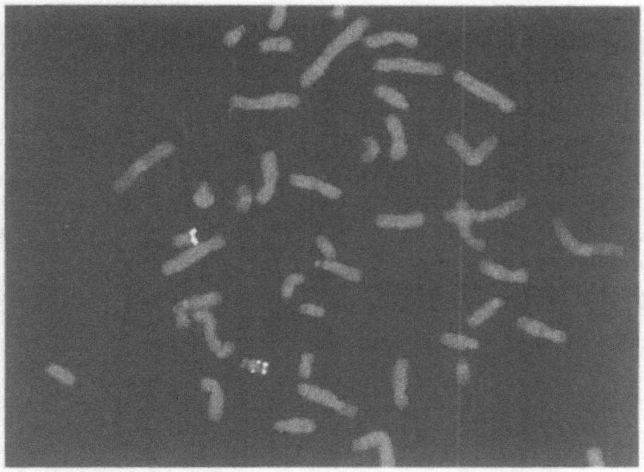

Fig. 4. The whole chromosome paint probe X gives a positive signal over the whole length of the X chromosome and in the pseudoautosomal region (PAR) of the Y – chromosome (probe wcpX, AGS).

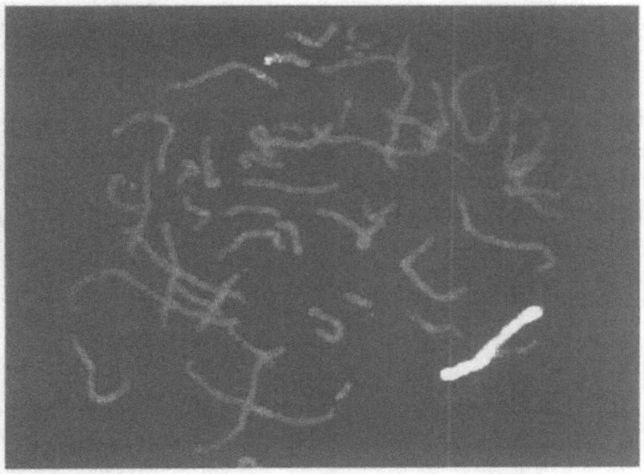

Fig. 5. CISS – hybridization with human whole chromosome paint 2 (HSA) on chromosome preparations of the chimpanzee (PTR) with positive signals on chromosomes 12 and 13, thus demonstrating the karyotype phylogeny (probe pBS – 2, plasmid – vector, Collins et al., 1991).

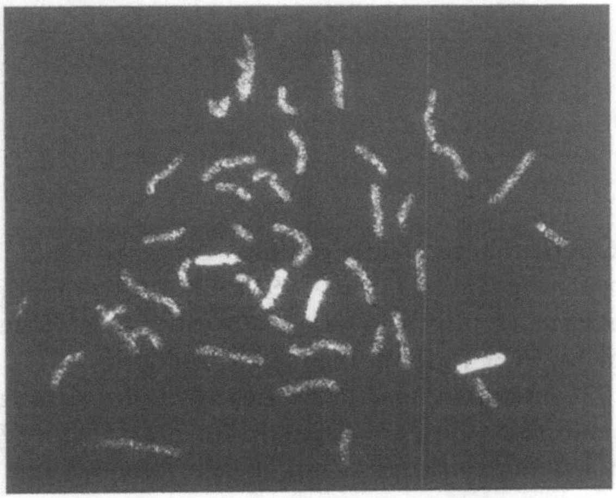

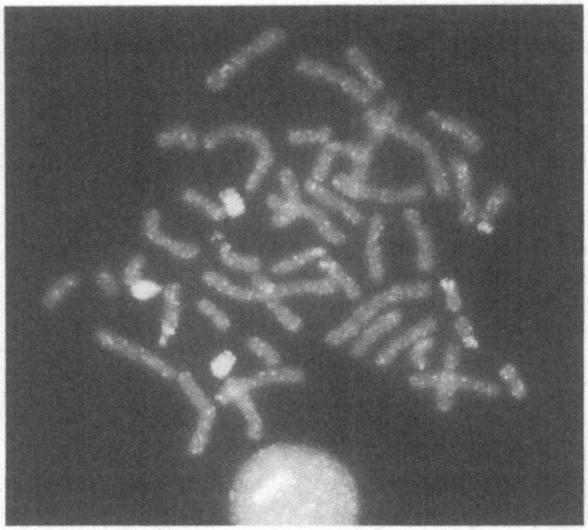

Fig. 6. Metaphase of a patient with trisomy 21. Hybridization with whole chromosome paint 21 gives not only positive signals on the target chromosomes but also on the short arm regions of all other acrocentric chromosomes (probe 1066 - 21, AGS).

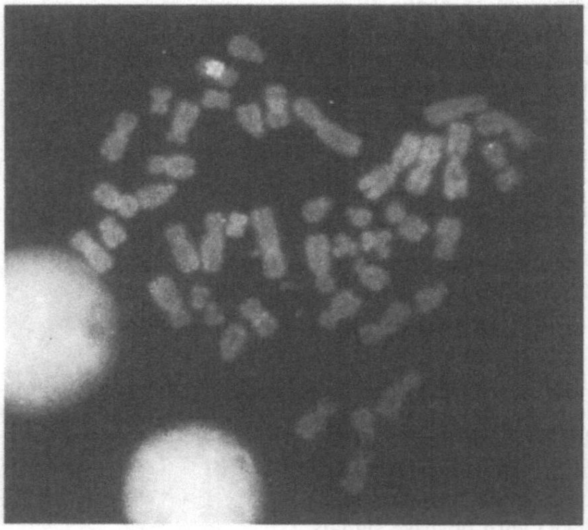

Fig. 7. Dicentric Y-chromosome with one inactivated centromere giving 2 signals with Y centromere probe (DYZ3, probe pDP97, ATCC).

2, 4, 6). These probes contain clones of DNA libraries of the specific chromosome (Collins et al., 1991). They are available commercially, but it is also easy to amplify and label the probes. Amplification is performed in vectors or by PCR (polymerase chain reaction).

Centromere specific probes which are highly repetitive sequences (alpha – satellite probes) are also used for FISH (Figure 7, 8). Within this group cross-hybridization may occur in homologous regions. For instance, chromo-

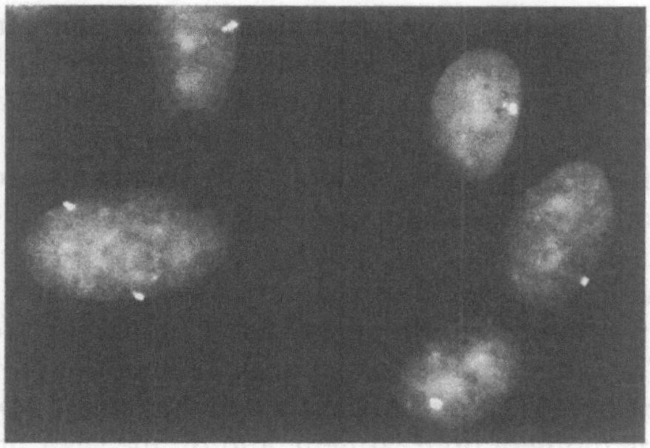

Fig. 8. Analysis of a mosaic 45,X/46,XX by interphase diagnostics. Nuclei from cells of the amniotic membranes are hybridized with an X centromere probe (DXZ1, probe 5060, Oncor).

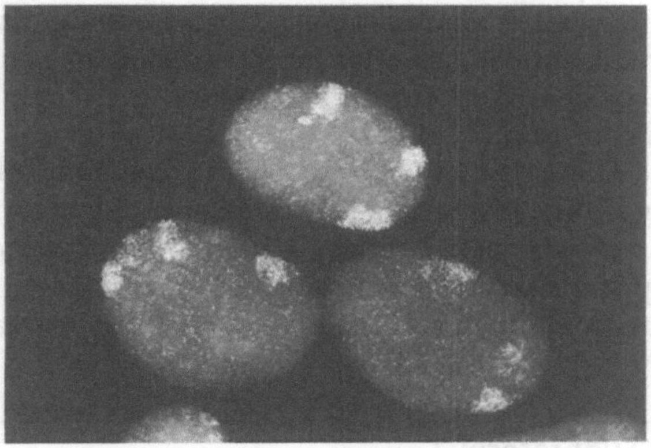

Fig. 9. Interphase nuclei from amniotic fluid cells of the reciprocal translocation 4/10 demonstrated in Fig. 2. After hybridization with whole chromosome paint 4 the 3 domains of the normal chromosome 4 and the 2 translocation chromosomes are recognizable.

somes 13/21 and 14/22 contain homologous sequences in their centromere regions and therefore only combined centromere probes are available for them so far.

Other groups of DNA probes used for FISH include sequences from highly repetitive *constitutive heterochromatic regions* such as Yq12 (Figure 1).

Finally, there is an increasing number of small euchromatic probes (*unique sequence* or *single copy probes*), which are mainly used for gene mapping, analysis of microdeletion syndromes and in tumor genetics (Figure 3). There are also probes for the most distal chromosome specific sequences (*subtelomeric probes*) which are useful e.g. for the identification of cryptic telomeric translocations or deletions.

Hybridization can be performed simultaneously with several probes labelled with different fluorochromes (or other labels), thus permitting simultaneous detection of different target sequences.

The sensitivity of the hybridization is influenced by a number of factors. One parameter is the investigated cell type. Cells from lymphocyte cultures proved to be especially suitable for a wide variety of probes. Cells from amniotic fluid and fibroblast cultures also give good results in certain cases after specific pretreatment. More frequently problems arise using cells after chorionic villus sampling (CVS), biopsies of solid tumors and tissue sections.

Rapid results after *interphase analyses* are mainly obtained on tissue samples of uncultivated amniocytes, mucosa cells, cells of the hair bulb or different solid tissues (fresh or formalin fixed and paraffin embedded). The latter require very specific methods of pretreatment. The effectivity of the hybridization here again depends on the type of the investigated cells, on the mode of pretreatment, and on the DNA probe that is applied. Centromere probes give the best results, e.g. the highest correspondence between the expected and observed number of hybridization signals. Small euchromatic probes are often more difficult to handle, while paint-probes are usually inappropriate because of overlapping domains of homologous chromosomes.

The sensitivity of *in situ hybridization* is furthermore influenced by the age of the investigated chromosome preparations and the mode of storage. Freshly prepared slides (2-5 days) usually lead to the best results. But by appropriate conservation (70% ethanol, 4°C) good hybridization results are possible to achieve even months up to years later.

Subprotocol 1
Preparation of Slides

For DNA – DNA *in situ hybridization* high quality slides of interphase or chromosome preparations are essential. Cytoplasm and other cellular structures may cause strong background fluorescence.

Good slides show metaphases and/or interphase nuclei clearly separate from each other, but not too far apart. In metaphase spreads the chromosomes should have no or only few overlaps. Hybridization on formalin fixed, paraffin embedded tissue sections is also feasible. Here a pretreatment is neccessary to remove the paraffin.

Note: It is always important to mark a good hybridization area, e.g. with a sketch on the frosted side of the slide.

Slides with cellular and chromosomal preparations can be hybridized after air drying (stored at 60°C over night or 2 days at room temperature, minimum). Two days after preparation they must be stored in 70% ethanol at 4°C to preserve the DNA from aging effects.

Unspecific hybridization of the probe with other sequences than the target sequence is to be avoided by RNase/pepsin pretreatment of the slides. We only perform such a pretreatment on older slides of lymphocyte preparations or on slides of fibroblast long term cultures and amniocytes because of the large amount of cytoplasm in these cells.

Materials

- Tris (Boehringer Mannheim, Nr.: 708976)
- Tris/HCl 10mM: 30 g Tris and 2,5 l aqua dest, adjust with HCl to pH 8,0
- RNase A (Boehringer, No 109169)
- RNase stock solution: 20 mg RNase/ ml 10 mM Tris/HCl, pH 7,5 + 15 mM NaCl DNase free: cook for 15 min, cool at room temperature, store at −20°C.
- RNase working solution: 1:200 dilution of the stock solution in 2xSSC, 100μg/ml 2 x SSC
- Pepsin (Sigma, No P−6887)
- Pepsin stock solution (10% pepsin)
- Pepsin working solution 1: 0,005%: 99 ml H_2O + 1 ml 1M HCl, prewarm to 37°C, adjust to pH 2,0, add to 50μl stock solution just before use
- Pepsin working solution 2: 0,5% in physiological saline
- 2 x SSC: 50 ml 20 x SSC + 450 ml aqua dest
- 20 x SSC: 350,6 g NaCl + 176,5 g $Na_3C_6H_5O_7$ x $2H_2O$ (trinatriumcitrate – dihydrate) (0,3 M) with aqua dest fill up to a total volume of 2,0 l; adjust to pH 7,0 with 10M solution of NaOH, store at room temperature
- 1 x PBS (Sigma, Nr.: 1000-3): 120 mM NaCl + 7 mM Na_2HPO_4 + 3 mM NaH_2PO_4 + 2,7 mM KCl
- 1 x PBS−$MgCl_2$: 50 ml 1 M $MgCl_2$ + 950 ml 1 x PBS
- Formaldehyde (37%) (Merck, Nr. 104 000)
- 1% Formaldehyde: 2,7 ml formaldehyde (35 – 37%) +97,3 ml PBS-$MgCl_2$
- Fixative: methanol:acetic acid (3:1)

Procedure

RNase/pepsin treatment

Note: All washing steps are performed in a coplin jar on a shaker.

1. Equilibrate slides for a short time in 2 x SSC at room temperature.
2. For digestion cover slides with 200µl RNase working solution and a coverslip (24 x 60 mm) and incubate at 37°C for 1 h in a moist chamber.
3. Remove coverslip and wash 3 x 5 min in 2 x SSC at room temperature.
4. Incubate the slides in a coplin jar with pepsin working solution 1 at 37°C for 5 min.
5. Wash 2 x 5 min in 1 x PBS.
6. Wash in 1 x PBS–$MgCl_2$ for 5 min.
7. Incubate in 1% formaldehyde in 1 x PBS–$MgCl_2$ for 10 min at room temperature.
8. Wash in 1 x PBS for 5 min.
9. Dehydrate in a series of ethanol (70%, 90%, 100%) for 3 min each.
10. Air dry slides.

Preparation of slides from formalin fixed, paraffin embedded tissue sections

The method presented here is a modification of the technique published by Cowles et al., 1995.

For best results the sections of formalin fixed, paraffin embedded tissues should be 30 - 50µm thick. From 1 - 2 of these sections a suspension of isolated nuclei is prepared and then dropped onto a slide.

A higher concentration of cells is achieved by centrifugation at 1000 rpm for 3 min between the different steps (take off supernatant).

1. The paraffin is removed by washing the sections for 3 x 5 min in 100% xylene in an Eppendorf cap at room temperature.
2. Rehydrate tissue in 100%, 90%, 70%, 50% ethanol for 10 min each at room temperature.
3. Rinse tissue in aqua dest over night.

4. A suspension of isolated nuclei is prepared by incubation of the tissue in 0,5% pepsin working solution 2 at 37°C for 30 min. Vortex every 5 min to promote cell segregation.
5. Place 2 – 3 drops of concentrated suspension on a clean, grease-free slide and examine under a microscope to determine the concentration of nuclei.
6. Air dry the slides.
7. To minimize cell loss, bake over night at 65°C.
8. Refix in freshly mixed fixative for 5 min and air dry.

Subprotocol 2
Labelling of DNA Probes

A fair amount of labelled DNA probes is commercially available (e.g. Oncor, AGS, ATCC, Vysis). A cheaper alternative for FISH is to amplify DNA in vectors and lable it by nick translation.

Generally for FISH double stranded DNA probes are used. Usually they are directly or indirectly labelled by nick translation which results in an optimal probe size for penetration into the target DNA (100-500 nucleotides). The direct labelling employs the integration of a fluorescence labelled nucleotide into the DNA. The indirect method is based on the attachment of a hapten (e.g. biotin, digoxigenin) to nucleotides of the probe. After the hybridization a labelled binding protein (e.g. avidin, streptavidin) is detected by a specific antibody (Höfler, 1990). Biotin and digoxigenin labelling combined with fluorescence detection are the most widely used procedures because of the high sensitivity and the commercial availibility of the reagents.

Nonisotopic *in situ hybridization* is the preferred method because of the disadvantages of radioisotopes.

Preparation of DNA probes

Using gene technology it is possible to clone each DNA sequence. As cloning vectors plasmids, bacteriophages, cosmids and YACs (yeast artificial chromosomes) are appropriate, depending on the size of the DNA fragment which is to be incorporated.

An enrichment of the vector with its insert is performed first by incorporating the fragment and then cloning it in host cells with high doubling rate.

Vector sequences in a pool of probes can intensify the background. Using single probes an isolation of the insert is not necessary. To reduce the rate of background signals, which are caused by repetitive sequences of complex probes (e.g. whole chromosome paint probes), it is neccessary to add unlabelled blocking DNA or cot −1-*DNA (chromosomal in situ suppression hybridization = CISS hybridization*, Lichter et al., 1988).

Labelling by nick translation

One method for labelling DNA probes is the *nick translation* (Rigby et al., 1977). In this procedure two enzymes are used: DNase I and DNA polymerase I (from E. coli).

In the presence of Mg^{2+} the DNase I induces single strand breaks ("nicks") in the DNA. DNA polymerase I detects the free 3' hydroxy groups of the single strand nicks and uses them as starting points for its exonuclease activity to remove nucleotides in 5' - 3' direction. At the same time its polymerase activity incorporates labelled desoxyribonucleosid triphosphates (dNTPs).

For easy handling we use a nick translation kit (Gibco) and modified nucleotides.

Biotin 11 - dUTP and *digoxigenin 11 - dUTP* are two examples for base analogues which are incorporated into the DNA by nick translation instead of dTTP.

Biotin (vitamin H) has the advantage of a high binding affinity (k = 10 - 15 M) to the proteins *avidin* and *streptavidin*. Almost irreversible bonds are the results. Streptavidin is the preferable protein because avidin also binds less specifically with DNA and lipids.

Digoxigenin is a derivative of digitoxin from Digitalis purpurea and D. lanata. It is species - specific and therefore very efficient for selective antibody detection in animal cells.

Materials

- Nick translation kit (Gibco BRL, Nr. 8160 SB)
- Biotin 11 - dUTP (Fermentas, Heidelberg, Nr.: R 002)
- 10 x DIG DNA labelling mixture (Boehringer, Nr.: 1277065)
- β-mercaptoethanol (Sigma, Nr.: M-6250) (0,1 ml β - mercaptoethanol + 14,4 ml aqua bidest)

- Tris: see Preparation of slides
- 10 x NT buffer
  - 0,5 M Tris/HCL, pH 8,0
  - 50 mM $MgCl_2$
  - 0,5 mg/ml BSA (bovine serum albumin)
  - for 5ml: 2,5 ml 1M Tris/HCl
  - +250µl 1M $MgCl_2$ (10 mg/ml)
  - +2 ml aqua dest
- Sephadex G – 50 (Pharmacia, Nr.: 17 – 0043 – 01)
- Column buffer
  - 5 ml 1 M Tris/HCL, pH 7,5 (10 mM)
  - 1 ml 0,5 M EDTA (1 mM)
  - 5 ml 10% SDS (0,1%)
  - fill up to 500 ml with H_2O
- Sephadex solution: Steep 1 volume of sephadex in 3 volumes of column buffer over night, discard supernatant and refill with buffer
- Stop mix
  - 0,1 g bromphenolblue (0,1%)
  - 0,5 g dextranblue (0,5%)
  - 2 ml 5 M NaCl (0,1 M)
  - 4 ml 0,5 M EDTA (20 mM)
  - 2 ml 1 M Tris/HCL (20 mM), pH 7,7
  - fill up to 100 ml with H_2O, aliquots of 1 ml are stored at – 20°C
- Salmon sperm DNA (Sigma, D 7656)
- Cot-1-DNA (Gibco BRL 5279 SA (1 mg/ml))

Procedure

The method presented here is in accordance with the method of Langer et al., 1981. Labelled probes can be stored at -20°C for several months.

1. Pipette reagents in the presented sequence in eppendorf tubes on ice:
2. Incubate at 15°C for 90 min. During this time prepare sephadex columns as described below.
3. Add 50 µl stop mix (biotinylation) /100 µl stop mix (digoxigenin labelling).

Biotin labelling:	
sterile aqua dest (kit solution E)	x µl
Nucleotid mix (kit solution A4)	5 µl
Biotin 11 – dUTP (0,4 mM)	2,5 µl
Probe DNA	y µl (corresponding to 1µg DNA)
DNA polymerase I/DNase I (kit solution C)	5 µl
Total volume	50 µl

Digoxigenin labelling:	
sterile aqua dest (kit solution E)	x µl
10 x DIG DNA labelling mixture	5 µl
0,1 M β – mercaptoethanol	10 µl
10 x nick translation (NT) buffer	5 µl
Probe DNA	y µl (corresponding to 2µg DNA)
DNA polymerase/DNase I (kit solution C)	10 µl
Total volume	100 µl

4. Remove unbound nucleotides in a column (sephadex G – 50) by centrifugation for 3 min at 3000 rpm (Heraeus Christ Labofuge) as described in step "5" below.
5. Add 1 µl of salmon sperm DNA (10 µg/ml) and 2 µl cot-1-DNA (1µg/ml) as competitor DNA for the reassociation of repetitive sequences using whole chromosome paint probes.
6. For precipitation add 2,5 volumes of absolute ethanol.
7. Store at – 20°C.

Preparation of sephadex columns

1. Place a small piece of surgical gauze (1x1 cm) into the bottom of a 1ml syringe.
2. Fill the syringe carefully with sephadex G – 50 solution up to the 1ml mark. Avoid air bubbles!

3. Spin the syringe in a tube (15 ml) at 3000 rpm for 3 min.
4. Refill with sephadex solution and spin again until the 1 ml mark is reached.
5. Place an eppendorf cap (1,5 ml) in a clear 15 ml tube under the syringe and load the column with the entire volume of the labelled probe and spin at 3000 rpm for 3 min. The cap now contains the labelled probe, the unbound nucleotides remain in the column.

Subprotocol 3
Denaturation and Hybridization

Before DNA hybridization, double stranded sequences of the target – and the probe DNA must be denatured to single strands by heat. In the next step the single stranded labelled probes hybridize with the complementary target DNA sequences.

The *melting temperature* T_m of double stranded DNA molecules in solution is defined as the temperature, at which half of the basepairs are dissociated. T_m depends on various parameters: the concentration of monovalent kations, the amount of guanine - cytosine basepairs, the duplex length, the formamide concentration in the reaction and the amount of mismatches.

The *optimal temperature* for reassociation of DNA single strands T_{opt} in an aqueous solution is lower than T_m and is approximately: $T_{opt}=T_m$ -25°C. Usually a temperature between 60 – 75°C is chosen. An admixture of formamide reduces the melting temperature to a temperature which preserves the chromosome structure.

An additional influencing parameter is the *stringency* meaning the accuracy of basepairing. The lower the amount of mismatches, the higher the stringency. High stringency is the result of high temperature, high formamide and low salt concentrations. Washing steps with high stringency result in very accurate pairing of the heteroduplices.

Final components that increase the rate of hybridization are dextran sulfate (master mix) and the probe concentration.

Materials

- 70% Formamide/2 x SSC
  - 42 ml deionized formamide
  - 6 ml 20 x SSC
  - 12 ml aqua dest
- Formamide (Merck, Darmstadt, Nr.: 1.09684.250)
- Serdolit MB-3 (Serva, Nr.: 40721)
- Formamide deionized
  - 10 g Serdolit
  - fill up to 100 ml with formamide
  - stir for 30 min, filter and freeze aliquots of 50 ml at -20°C
- Master mix
  (20% dextran sulfate/2 x SSC)
  - 2 ml 20 x SSC
  - 8 ml 50% dextran sulfate
  - autoclave solution, store aliquots of 1 ml at 4°C
- Dextran sulfate (Sigma D6001)

Procedure

Denaturation of target DNA

1. Air dry the slides for 20 min.
2. Denature target DNA in 70% formamide/2 x SSC at 75°C for exactly 3 min in a coplin jar in the water bath.
3. Dehydrate target DNA in a series of ice – cold ethanol (70%, 90%, 100%) for 3 min each.
4. Air dry (meanwhile perform denaturation of probes).

Denaturation of probes

For commercially available probes follow the instructions of the providers. Perform denaturation of self labelled probes as follows:

1. Spin at 14000 rpm for 30 min.
2. Discard supernatant and dry the pellet for 5 min in a speed Vac (alternatively 1 h with tipped cap)

3. Resuspend in 5 µl deionized formamide by vortexing.
4. Incubate for 20 min at 37°C in the water bath.
5. Add 5 µl master mix, vortex.
6. Denature at 75°C in the water bath for 5 min.
7. Place on ice immediately.
8. Within the next minutes apply probe to denatured target DNA (slides).

Hybridization

1. Pour whole volume (10 µl) of the denatured probe mixture on the optimal area of the slide and cover with a coverslip (18 x 18 mm). Avoid air bubbles!
2. Seal with rubber cement.
3. Incubate immediately in a moist chamber at 37°C for 16 – 20 h (over night).

Subprotocol 4
Detection of Biotin Labelled Probes

Before detection the slides are washed with increasing stringency to remove base mismatches. Unspecific protein binding sites are blocked with bovine serum albumin (BSA). Then biotin labelled DNA probes are detected via avidin, a glycoprotein, which is linked to the fluorochrome *Fluorescein-isothiocyanate (FITC)*. Avidin has 4 binding sites for biotin, which results in a very stable complex. To enhance the fluorescent signal a biotinylated bridge antibody (from goat) is used. Subsequently a solution of FITC labelled avidin molecules is applied, leading to an amplification of the signal.

Materials

- 50% Formamide/2 x SSC
  - 125 ml formamide (no need for deionisation)
  - 25 ml 20 x SSC
  - 100 ml H_2O
  - adjust to pH 7,0 with HCl, store at room temperature

- 0,1 x SSC: 1,25 ml 20 x SSC fill up to 250 ml with H_2O, store at room temperature
- 10 x PBD (Appligene Oncor, S 1370 – 7)
- Tween (Serva 37470)
- 4 x SSC/0,2% Tween
  - 400 ml H_2O
  - 100 ml 20 x SSC
  - 1 ml Tween, stir at room temperature
- BSA (Sigma, Nr.: A-7906)
- 1% BSA/4 x SSC/0,2% Tween
  - 0,1g BSA
  - 10 ml 4 x SSC/0,2% Tween
- Avidin-FITC (Vector laboratories, Nr.: A-3100, via Serva)
- stock solution: store aliquots at -20°C
- working solution: final concentration 5 µg/ml, spin avidin FITC stock solution at 13000 rpm for 3 min, 1:200 dilution of supernatant in 4 x SSC/Tween/1% BSA
- Goat anti – avidin (Vector laboratories, Nr.: A-3101, via Serva)
- stock solution: store aliquots at -20°C
- working solution: final concentration 5µg/ml, spin stock solution at 13000 rpm for 3 min, 1:200 dilution of supernatant in 4 x SSC/Tween/ 1% BSA
- DAPI (Sigma D-1388)
- DAPI stock solution: 2 mg/10 ml Mc Ilvaine's buffer
- Mc Ilvaine's buffer
  - A: 0,1 M citric acid
  - B: 0,2 M Na_2HPO_4
  - mix 18 ml A + 82 ml B
- Propidiumiodide (Sigma P 4170)
- Propidiumiodide stock solution: 1 mg/ml aqua dest
- Propidiumiodide/DAPI solution
  - 5 µl propidiumiodide stock solution

- 17 μl DAPI stock solution
- in 10 ml 4 x SSC/0,2% Tween, store in the dark at room temperature

- Glycerol (Merck, Nr.: 4095)
- Tris (Boehringer Mannheim, Nr.: 708976)
  - 30 g Tris
  - 2,5 l aqua dest., adjust with HCl to pH 8,0
- Antifading
  - 0,233 g DABCO
  - 200 μl Tris/HCl, pH 8
  - 800 μl aqua bidest
  - 9 ml glycerol (86%)
  - mix on a shaker over night and store at 4°C
- DABCO (Sigma 2522)

Procedure

All washing steps are performed in a coplin jar on a shaker with prewarmed solutions. It is important that the slides never get dry during these procedures. Between each blocking and detection step place the slides vertically on a paper towel, so that the excess fluid can be absorbed.

1. After hybridization time carefully remove rubber cement and coverslip with a needle or forceps.
2. Wash 3 x 5 min in 50% formamide/ 2 x SSC (45°C).
3. Wash 3 x 5 min in 0,1 x SSC (60°C).
4. Overlay slides with 500 μl 5% BSA (diluted in 4 x SSC without Tween) cover with a coverslip (24 x 60 mm) and incubate in a moist chamber for 30 min at 37°C (aluminium-box in the water bath).
5. Remove coverslip and wash briefly in 4 x SSC/0,2% Tween at room temperature.
6. For the detection cover each slide with 100 μl avidin-FITC working solution and a coverslip and incubate at 37°C for 45 min.
7. Remove coverslip and wash 3 x 5 min in 4 x SSC/0,2% Tween (45°C).
8. For amplification add 100 μl goat-anti-avidin antibody working solution to each slide and cover with a coverslip.

9. Incubate at 37°C for 45 min in a moist chamber.
10. Remove coverslip and wash 3 x 5 min in 4 x SSC/0,2% Tween (45°C).
11. Place 100 µl avidin-FITC solution on each slide, cover with a coverslip and incubate at 37°C for 30 min in a moist chamber.
12. Remove coverslip and wash slides 3 x 5 min in 4 x SSC/0,2%Tween (45°C).
13. Place 100 µl DAPI-propidiumiodide working solution on each slide, cover with a coverslip and stain for 20 min in the dark.
14. Rinse with water, air dry and mount with 3 drops of antifading. Cover with a coverslip (24 x 60 mm).
15. Store slides at 4°C in the dark.

Subprotocol 5
Amplification

In case of insufficient signals an amplification of the signals is possible. Increased background fluorescence may then however become a problem.

Procedure

1. Remove coverslip carefully.
2. Wash 3 x 5 min in 4 x SSC/0,2%Tween at room temperature.
3. Place 100 µl goat-anti-avidin on each slide and incubate with a coverslip at 37°C for 45 min in a moist chamber (alu – box in a moist chamber).
4. Remove coverslip and wash 3 x 5 min in 4 x SSC/0,2% Tween (45°C).
5. Place 100 µl avidin-FITC on the slide and cover with a coverslip.
6. Incubate at 37°C for 30 min in a moist chamber.
7. Remove coverslip and wash 3 x 5 min in 4 x SSC/0,2% Tween (45°C).
8. Stain as described above.

Subprotocol 6
Simultaneous Two Colour Detection

For a simultaneous two colour signal analysis biotin and digoxigenin detection systems can be used in combination. Digoxigenin is detected by specific antibody conjugates which are linked to a red fluorochrome, e.g. *Tetramethylrhodaminisothiocyanate (TRITC)*. First TRITC binds to an anti – digoxigenin-antibody originating from the mouse. Detection than is accomplished with a second TRITC conjugated rabbit-anti-avidin. Finally, with a TRITC conjugated anti-rabbit-antibody (originated from the goat) an amplification of the signal is achieved.

Materials

- Mouse-anti-dig-antibody: 1:200 dilution in 4 x SSC/0,2% Tween/1% BSA, final concentration 1μg/ml (Sigma, Nr.: D-8156)
- Rabbit-anti-mouse-IgG-TRITC: 1:200 dilution in 4 x SSC/0,2% Tween/ 1% BSA, final concentration 1μg/ml (Sigma, Nr.: T-2402)
- Goat-anti-rabbit-IgG-TRITC: 1:200 dilution in 4 x SSC/0,2% Tween/1% BSA, final concentration 1μg/ml (Sigma, Nr.: T-5268)
- DAPI staining solution: 1 μl DAPI stock solution in 1 ml Mc Ilvaine's buffer
- Mc Ilvaine's buffer: see Detection of biotinylated probes

Procedure

1. Wash and block with BSA as described for the detection of biotinylated probes. The steps for washing and times for incubation are also identical.
2. Dilute the biotinylated goat-anti-avidin-antibody and the mouse-anti-dig-antibody in a total volume of 100 μl on each slide and incubate for 45 min at 37°C.
3. After washing (3 x 5 min, 45°C in 4 x SSC/0,2% Tween) place avidin-FITC and TRITC-conjugated rabbit-anti-mouse-antibodies together on each slide (45 min, 37°C).
4. Wash the slides as described above and incubate with TRITC-conjugated goat-anti-rabbit-antibodies for 45 min at 37°C.

5. Wash three times and stain with DAPI-working solution at room temperature for 15 min in the dark. Rinse with water, air dry and mount in antifading.

Subprotocol 7
In Situ Hybridization with Unique Sequence Probes

(e.g. Appligene Oncor probes: D5S23; D15S11; D22S75)

The hybridization of unique sequence probes has to be performed in a modified manner in order to receive optimal results.

Materials

See Subprotocol 4.

Procedure

Pretreatment of slides

1. Incubate slides in 2 x SSC, pH 7,0 at 37°C for 30 min in the water bath.
2. Dehydrate in a series of ethanol (70%, 80%, 100%) at room temperature for 2 min each.
3. Air dry.

Denaturation of target DNA

1. Incubate slides for 2 min in 70% formamide/2 x SSC, pH 7,0, prewarmed in a water bath to 70°C.
2. Dehydrate in 70%, 80%, 100% ethanol at room temperature for 2 min each.
3. Air dry.

Pretreatment of probe – DNA

Prewarm the probe at 37°C for 5 min in the water bath. It is not required to denature the probe! However, see the provider's instructions.

Hybridization

1. Pour 8 – 10 µl of the probe onto hybridization area, cover with a coverslip (18 x 18 mm).
2. Seal with rubber cement.
3. Incubate at 37°C for 16 h.

Detection

1. Remove rubber cement and coverslip carefully.
2. Wash 3 x 5 min in 50% formamide/2 x SSC at 43°C.
3. Wash in 2 x SSC at 37°C for 8 min.
4. Wash for 5-10 sec in 1 x PBD.
5. For detection incubate slides each covered with 100 µl avidin-FITC or mouse-anti-dig (coverslip: 24 x 60 mm) at 37°C for 45 min.
6. Remove coverslip and wash 3 x 2 min in 1 x PBD at room temperature.
7. For amplification add 100 µl goat-anti-avidin respectively goat-anti-mouse-IgG-FITC, cover with a coverslip and incubate at 37°C for 45 min.
8. Remove coverslip and wash 3 x 2 min in 1 x PBD at room temperature.
9. Add 100 µl avidin-FITC resp. rabbit-anti-goat-IgG-FITC on each slide, cover with a coverslip and incubate at 37°C for 30 min.
10. Remove coverslip and wash 3 x 2 min in 1 x PBD at room temperature.
11. Counterstain with 100 µl PJ/DAPI solution for 20 min in the dark.
12. Rinse with water, air dry and mount with antifading.

Subprotocol 8
Preparation of Interphase Nuclei

Numerical chromosome aberrations can be analysed in interphase cells with centromere probes of the target chromosome when additional tissues should be investigated for instance in mosaic cases, or if no or insufficient amounts of mitoses are available.

For extended postnatal interphase diagnostics of mosaicism, cells from buccal mucosa and/or from the hair root are recommended.

Both cell types are easy to prepare but each of them requires a specific pretreatment before FISH.

Materials

- Fixative: methanol:acetic acid = 3:1 (mix fresh every time)
- Proteinase K (Sigma, P-0390) (10-20 units per mg), working solution: 0,3 mg/ml aqua dest
- 4% Formaldehyde: 8 ml formaldehyde (37%) + 66 ml aqua dest
- Pepsin: Sigma, P-6887 (3200-4500 units per mg)
- Pepsin working solution: 0,5% in physiological saline
- 1 x PBS: see Subprotocol 1
- 1 x PBS + 50 mM $MgCl_2$: see Subprotocol 1
- 1% Formaldehyde: see Subprotocol 1
- Acetic acid 50%

Procedure

Preparation of cells from buccal mucosa

Smears are taken from the inner surface of the cheeks. It is important to obtain fresh and undamaged cells, therefore cells from the deeper epithelial layers should be preferred.

1. Cells are smeared in a thick layer on clean and grease-free slides.
2. The cells are immediately fixed for 30 min in ethanol (100%) or in fixative (methanol:acetic acid = 3:1) for 3 min.
3. Air dry.

Pretreatment of slides by proteinase K digestion

1. Bake slides for 20 min at 80°C.
2. For digestion add 150 µl proteinase K solution, cover with a coverslip and incubate for 20 min at 37°C.
3. Fixation in 4% formaldehyde for 10 min at room temperature.
4. Air dry slides.

Pretreatment of slides by pepsin digestion (alternatively)

1. Slides are incubated for 10 min at 37°C in 0,5% pepsin solution.
2. Wash 2 x 5 min in 1 x PBS.
3. Wash 5 min in 1 x PBS + 50 mM $MgCl_2$.
4. Air dry slides.

Post fixation

1. Incubate slides for 10 min at room temperature in 1% formaldehyde solution.
2. Wash 5 min in 1 x PBS.
3. Dehydrate in a series of ethanol (70%, 90%, 100 %).
4. Air dry.

FISH

1. Store slides over night in 70 % ethanol.
2. Conventional hybridization and signal detection are performed as described in subprotocols 3 and 4.

Evaluation

Only well preserved cells with an evenly granulated nuclear structure and without signs of pyknosis or damage should be analysed.

Preparation of cells hair follicles

Cells from hair roots are easily accessed by plucking the hair with a forceps. Check for the hair bulbus!

1. Place hair roots (3 – 6, according to the size of their bulbs) in an eppendorf cap, add 60 µl of 50% acetic acid, incubate at room temperature for 30 min.
2. Vortex for 2 – 3 min.
3. Spin at 4000 rpm for 4 – 5 min.
4. Remove the hairs.
5. Add 30 µl 100% methanol to the cell suspension.
6. After fixation for 20 – 30 min resuspend (vortex) the pellet, drop it on cold, grease free slides and air dry.

Pretreatment of slides

1. Bake slides for 20 min at 80°C.
2. Add 150 µl proteinase K solution, cover with a coverslip and incubate for 20 min at 37°C.
3. Fixation in formaldehyde (4%) for 10 min.
4. Air dry.
5. Store in 70 % ethanol at 4°C.

FISH

Perform FISH as described in Subprotocols 3 and 4.

Results

Analysis

Slides are analysed under a microscope equipped with DAPI, FITC and TRITC epifluorescence optics.

Different computer systems for imaging processing are helpful particularly for analysis but also for documentation.

Example

Karyotype: 47,XY,+mar[14]/46,XY[20]
Prenatal diagnosis was performed because of advanced maternal age.

The karyotype in both CVS direct preparation and long term culture was: 47,XY,+mar/46,XY. The marker chromosome was dicentric with satellites on both sides, the size was comparable to a chromosome of the G – group (banding technics: GTG, QFQ, CBG, NOR, DA/DAPI). G-banding revealed a light G-band between the 2 centromeres suggestive of euchromatin. The marker was DA/DAPI negative thus excluding a derivative chromosome 15, the most frequent marker chromosome.

Both parents were analysed cytogenetically since marker chromosome mosaicism can be hereditary. A *de novo* origin of the marker was proven since the parents showed normal karyotypes.

The initial ultrasonographic examination in the 17th week of pregnancy revealed normal growth for gestational age and no malformations.

For genetic counseling three questions had to be answered:

1. Does the marker chromosome contain genetic material (euchromatin) relevant for the child?
2. Is the marker chromosome present in the fetus or confined to extrafetal tissues?
3. Provided that true mosaicism exists in the fetus, what sonografic abnormalities can be expected in relation to the marker's origin?

Following topics had to be discussed with the consultants:

1. The investigation of additional tissues, e.g. amniotic fluid cells or fetal lymphocytes, to exclude confined placental mosaicism (CPM).
2. The clinical consequences depending on the chromosomal origin of the marker chromosome if true mosaicism was to be proven.

Subsequent studies showed the presence of the marker in amniotic fluid cells and in fetal lymphocytes proving true fetal mosaicism.

Hybridization with the centromere specific probe for chromosomes 14 and 22 gave two positive hybridization signals on the marker. Hybridization with a whole chromosome paint (wcp) for chromosome 14 gave no signals on the marker chromosome. In contrast, wcp for chromosome 22 revealed one positive signal between the two centromeres. To determine the extent of euchromatin in the marker chromosome 22, FISH was performed with probe D22S75, localized in the proximal region of 22q. This study showed three signals in all metaphases with the marker; one located on each of the normal homologues 22 and one positioned centrally on the marker chromosome.

Thus, the fetal karyotype could be described as: 47,XY,+ mar. ish idic(22) (pter-q11.2:q11.2-pter) (D14Z1/D22Z1+, wcp 22+, D22S75+)/46, XY.

Consequently, it could be assumed that the fetus was affected by the Cat Eye syndrome but sonographic investigation revealed no fetal abnormalities.

After counselling, the couple decided to terminate the pregnancy. The fetus was slightly growth retarded and had facial dysmorphisms.

Investigations of different placental areas showed varying ratios of the cell lines involved in the mosaic.

Troubleshooting

Cross hybridization

When using *centromere probes* and low stringency washes cross hybridization is observed in those chromosomes containing alpha satellite DNA in their centromeric regions belonging to the same supra family. This effect can be used for the identification of marker chromosomes of unknown origin. In these cases a number of chromosomes can be selected for or excluded from further investigations by one single hybridization.

After hybridization with *whole chromosome paint probes* cross hybridization is observed between X and Y chromosomes, as they possess common sequences such as the pseudoautosomal region.

But a fair amount of whole chromosome paint probes also show cross hybridization with centromere regions of single heterologous chromosomes or with the constitutive heterochromatin. For example the whole chromosome paint 15 probe (AGS) often gives positive hybridization signals on the short arm regions of all acrocentric chromosomes.

Specificity of probes

For diagnostic purposes it is necessary to test each probe on each cell system and preparation used. The position and extent of the signals is to be determined, as well as the efficiency of hybridization. Each probe has a detection limit below which a detection is not possible.

For diagnostic use it should be mentioned that the producers of most of the commercially available probes do not take legal liability for diagnostic purposes.

DNA probes selected for hybridization are expected to give positive signals on defined euchromatic or heterochromatic chromosome regions. The use of certain *whole chromosome paints* lead to the observation that not only the euchromatic regions of the specific chromosomes hybridize but also the pericentromeric constitutive heterochromatin. Therefore it is not possible to determine the genetic relevance of *de novo* developed small marker chromosomes by using these whole chromosome paint probes. Additional investigations with single copy probes of centromere associated regions are neccessary.

Another problem with the use of whole chromosome paints is caused by the different intensity of hybridization signals in AT- and GC-rich regions. When using some commercially available whole chromosome paints GC-rich bands often give less intensive hybridization signals and thus lead to diagnostic difficulties.

Even when applying *unique sequence probes* the evaluation of positive hybridization signals is still sometimes difficult. Therefore, at least one internal control of a successful hybridization should always be performed. Most commercially available single copy probes are equipped with a specific control cosmid.

Problems of interphase diagnostics

When analysing interphase nuclei all cells with degenerative alterations have to be excluded, as well as nuclei that overlap and nuclei of tissue sections that are damaged.

Hybridization is frequently performed with centromere probes. The number of signals corresponds with the number of target chromosomes if these are in G_0-phase or in G_1 of the cell cycle. Cells in S-phase show dispersed, not clearly defined signals which easily lead to a false interpretation of their number. This problem of signal interpretation is even greater in G_2-phase where each chromosome shows signal duplets after termination of replication, which are situated apart from one another in some cases. These double signals can overlap one another due to their 3-dimensional arrangement.

These peculiarities make highlight the limitations of interphase diagnostics.

Even in cases where patients show complete trisomy, positive hybridization may reveal less than 3 signals in up to 20% or more of interphase nuclei and thus can lead to the false conclusion of a chromosome mosaic with one normal cell line. Cells that have been prepared directly before investigation lead to better hybridization results than those taken from paraffin embedded tissues. Efficiency of hybridization also depends on the type of the investigated cell system and the DNA probe being applied. The highest frequency of signals is observed by hybridizing the probe YpH34 on male lymphocytes (over 95%).

Applications

FISH

- Analyses of structural chromosome aberrations may be complemented by FISH investigations if there is a *de novo* aberration or a complex chromosome rearrangement (CCR).
- In chromosomal mosaics a combination of metaphase and interphase FISH may be essential to reveal the true chromosomal constitution of the proband.
- There are a number of syndromes caused by tiny or submicroscopic deletions. These disorders called microdeletion syndromes might be diagnosed by FISH with small cosmid probes. Presently, probes for about 10 autosomal and 3 gonosomal disorders are commercially available to test for microdeletions.
- Rapid analysis on uncultivated amniocytes is feasible in prenatal diagnostics. Here kits containing different FISH probes (e.g. Vysis) are used to exclude the most frequent trisomies. Usually a combination of probes from chromosomes 13, 18, 21, X, Y is applied. As in all prenatal investigations a maternal cell contamination must be excluded. Additional conventional cytogenetics is recommended.

Advanced techniques

- Application of quantitative digital fluorescence microscopy
  For multicolour detection of simultaneously hybridized probes new automated cytogenetic imaging is required. When performing multicolour FISH, different signals can no longer be differentiated by the observer with desirable precision. With a confocal microscope the signals are digitalized in their quantitative differences and are then documented by pseudocolouring.

Signals can also be enhanced and cross hybridization effects reduced. By optical sections signals in a 3 dimensional nucleus can be registered and precisely localized.
Image digitalization can be done by a number of camera systems. Presently the most sensitive system is the Charge Coupled Device Camera (CCD). With a cooled CCD camera, fluorescence emission can be collected over a period of time visualising signals that are no longer analyzable by direct observation.
Thus, recognizing DNA probes smaller than 2 kb has become possible, and high resolution gene mapping on prometaphase chromosomes or fiber extended chromosomes is being performed.

- Spectral karyotyping (SKY), multicolour FISH (M-FISH)
  New techniques have been developed by combining FISH with spectral analysis. In contrast to the conventional methods the whole absorption and emission spectrum of light is registered. These methods improve the diagnostic potential in analyses of marker chromosomes, complex chromosome rearrangements (CCR), subtelomeric mutations and tumor karyotype alterations (Schröck et al., 1996; Speicher et al., 1996).

References

Bauman JGJ, Wiegant J, Borst P, van Duijn P, (1980) A new method for fluoroscence microscopical localization of specific DNA sequences by in situ hybridization of fluorochrome – labelled RNA. Exp Cell Res 128: 485 – 490

Collins C, Kuo WL, Segraves R, Fuscoe J, Pinkel D, Gray JW (1991) Construction and characterization of plasmid libraries enriched in sequences from single human chromosomes. Genomics 11: 997-1006

Cowles TR, Elder FFB, Taylor S (1995) Identification of abnormal chromosomal complement in formalin – fixed, paraffin embedded placental tissue. Prenat Diagn 15: 21 – 26

Cremer T, Lichter P, Borden J, Ward DC, Manuelidis L (1988) Detection of chromosome aberrations in metaphase and interphase tumor cells by in situ hybridization using chromosome – specific whole chromosome paint probes. Hum Genet 80: 235 – 246

Cremer T, Popp S, Emmerich P, Lichter P, Cremer C (1990) Rapid metaphase and interphase detection of radiation – induced chromosome aberrations in human lymphocytes by chromosomal supression in situ hybridization. Cytometry 11: 110 – 118

Gall G, Pardue ML (1969) Formation and detection of RNA – DNA hybrid molecules in cytological preparations. Proc Natl Acad Sci 63: 378 – 381

Höfler H (1990) Principles of in situ hybridization. In: Polak JM, McGee JOD (eds) In situ hybridization, Principles and practice. Oxford University Press, Oxford, New York, Toronto, pp 15 – 30

Jauch A, Daumer C, Lichter P, Murken J, Schroeder – Kurth T, Cremer T (1990) Chromosomal in situ suppression hybridization of human gonosomes and autosomes and its use in clinical cytogenetics. Hum Genet 85: 145 – 150

John HL, Birnstiel ML, Jones KW (1969) RNA – DNA hybrids at the cytological level. Nature 223: 912 – 913

Langer PR, Waldrop AA, Ward DA (1981) Enzymatic synthesis of biotin labeled polynucleotides: novel nucleic acid affinity probes. Proc Natl Acad Sci USA 78: 6633 – 6637

Lichter P, Cremer T, Borden J, Manuelidis L, Ward DC (1988) Delineation of individual human chromosomes in metaphase and interphase cells by in situ suppression hybridization using recombinant DNA libraries. Hum Genet 80: 224 – 234

Lichter P, Cremer T (1992) Chromosome analysis by non – isotopic in situ hybridization. In: Rooney DE, Czepulkowski BH (eds) Human Cytogenetics. A practical approach, Volume I. IRL Press, Oxford, New York, Tokyo, pp 157 – 190

Rigby PWJ, Dieckmann M, Rhodes C, Berg P (1977) Labeling deoxyribonucleic acid to high specific activity in vitro by nick – translation with DNA polymerase I. J Mol Biol 113: 237 – 241

Schröck E, du Manoir S, Veldman T, Schoell B; Wienberg J, Ferguson-Smith MA, Ning Y, Ledbetter DH, Bar-Am I, Soenksen D, Garini Y, Ried T (1996) Multicolor spectral karyotyping of human chromosomes. Science 273: 494 – 497

Speicher MR, Ballard SG, Ward DC (1996) Karyotyping human chromosomes by combinatorial multi-fluor FISH. Nature Genet 12: 368 – 375

Chapter 19

Fluorescence *in situ* Hybridization (FISH) Analysis in Human Sperm Cells

EVELYN KO AND RENÉE MARTIN

Introduction

Chromosomal abnormalities are extraordinarily common in humans and lead to pregnancy loss and the birth of children with physical and mental handicaps. Studies on the causes of chromosomal abnormalities have traditionally used information from newborns or spontaneous abortions. However, this leads to a major ascertainment bias since the great majority of chromosomal abnormalities are lost early in pregnancy, often before the pregnancy is recognized or implantation has even occurred. Clearly, the best information can be derived from the analysis of gametes since the origin of the vast number of constitutional chromosomal errors lies in meiosis. The ability to karyotype human sperm made a tremendous impact in the field of reproductive genetics. The first report by Rudak et al. (1978) which used hamster ova to reactivate human sperm was followed by other publications by a few laboratories (eg Martin et al.,1982,1983; Brandriff et al.,1985; Mikamo et al.,1990). The great advantage of sperm karyotyping is that detailed information about each individual chromosome is available, allowing analysis of both numerical and structural anomalies. However, there are a number of disadvantages - sperm must be capable of fertilizing a hamster egg and this is often an obstacle for sperm treated by mutagens such as chemotherapeutic agents. In addition, the technique is difficult, time-consuming, expensive, and yields a small number of karyotypes. This is exemplified by the fact that fewer than 12 laboratories worldwide have had success with this technique (despite the efforts of many) and fewer than 25,000 human sperm have been karyotyped in two decades of research.

Evelyn Ko, University of Calgary, Department of Medical Genetics, Faculty of Medicine, CalgaryCanada, *Correspondence to* Renée Martin, Alberta Children's Hospital, Medical Genetics, 1820 Richmond Road, S.W., CalgaryAlberta, T2T 5C7, Canada (*phone* +403-229-7369; *fax* +403-543-9100; *e-mail* rhmartin@ucalgary.ca)

Fluorescence *in situ* hybridization (FISH) with chromosome-specific DNA probes provides a rapid, easy alternative technique for detecting aneuploidy in human sperm. As a further bonus, the sperm need no fertilizing ability nor even motility for successful FISH analysis.

Aneuploidy frequencies detected in sperm by FISH analysis have been reported for almost all chromosomes, and it is expected that data from all chromosomes will be available in the near future (Spriggs et al.1996,Bischoff et al.,1994). Our own FISH studies (Holmes and Martin,1993; Martin et al.,1993) and those of Wyrobek's laboratory (Robbins et al.,1993) have demonstrated good correlation between disomy frequencies from sperm karyotype studies and those from FISH analyses. However, there has been considerable divergence in the frequency of disomy in FISH studies: for example, Pieters et al. (1990) reported a chromosome 1 disomy frequency of 0.8%, while Holmes and Martin (1993) found 0.06%. Studies from sperm karyotypes have not suggested a wide variation in the frequency of aneuploidy among normal men (Martin et al.,1991), thus casting doubt on the accuracy of the indirect assessment of sperm chromosome aneuploidy by the FISH technique. Tight control of the many technical aspects of the FISH assay which can lead to variability, therefore, is essential in the production of meaningful results. First, sperm nuclei must be decondensed to allow the DNA probes access to the sperm chromatin. However, if the nuclei are swollen to more than twice their original size, the signal from one chromosome may split and appear as two or more signals, causing the sperm to be falsely scored as disomic (Wyrobek et al, 1993). Scoring criteria are a second important source of variation: stringent criteria must be adopted for the scoring of two same-coloured signals within a single sperm as an indication of disomy. We have demonstrated that the frequency of disomy decreased significantly when one domain, rather than one-half domain, was considered the minimal distance between two like-coloured signals in a disomic sperm (Martin and Rademaker, 1995). Third, adequate sample sizes must be scored. Many researchers have discovered that most autosomes have a disomy frequency of approximately one per thousand. It is clear that scoring 1000 to 2000 sperm per chromosome probe will not provide an accurate estimate of aneuploidy with such a low frequency of disomy per chromosome. We routinely score 10,000 sperm per chromosome per donor. Finally, multicolour FISH analysis is much more accurate than single-colour FISH because it provides internal controls to detect a lack of hybridization and to distinguish disomy from diploidy (Martin et al.,1996). For autosomes, two probes are necessary to discern a disomic sperm (2 signals of one colour, one signal of the other colour) from a diploid sperm (2 signals of each colour). For sex chromosome studies, three colours are required to distinguish a disomic 24,XY sperm (one signal of each colour from two sex chromosomes and one autosome) from a diploid 46,XY sperm (one signal of each of two colours from two sex chromosomes and two signals from the autosome). A disomic 24,XY sperm with three coloured signals from chromosomes X (green), Y (red) and 1 (blue) is shown in Figure 1. Since the frequencies of diploidy in our donors have varied from 0.06% to 0.42%, it is clear that if these diploid sperm were counted as disomic sperm (as would be inevitable with single-colour FISH), the estimate of disomy would be greatly inflated.

FISH analysis of sperm, if carefully executed and scored, has great potential for aneuploidy analysis in basic studies and in men exposed to mutagens. As well as using FISH analysis in basic studies (Martin et al.,1995, Martin et al.,1996, Spriggs et al., 1996, Rademaker et al., 1997), we have also studied chromosomal abnormalities in infertile patients (Moosani et al.,1995, Martin,1996) and men exposed to potential mutagens such as methotrexate (Martin and Barclay,1996) and colchicine (Martin,1997). Recent reports have also suggested the use of telomeric probes combined with centromeric probes to ascertain at least some structural abnormalities in human sperm (Van Hummelen et al.,1996). Applications of this new technology will likely multiply in the future.

Fig. 1. XY disomic sperm with signals from X chromosome (green), Y chromosome (red), and chromosome 1 (blue)

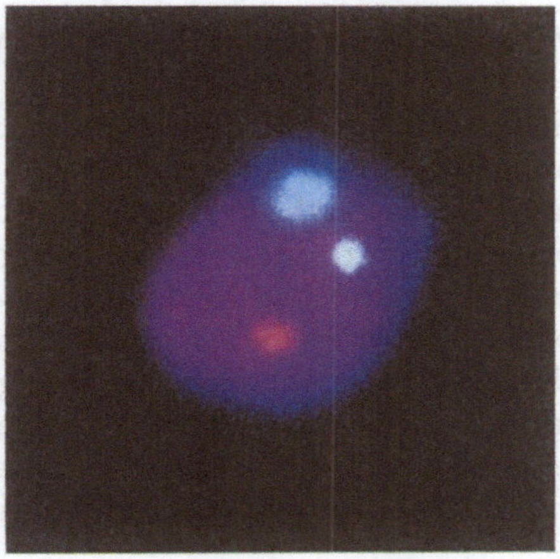

Materials

Safety considerations

Safe handling and disposal

Since there are jurisdictional differences in regulations for the safe handling and disposal of hazardous and potentially hazardous materials, a comprehensive review of safety precautions for each of the reagents utilized below is not included in this protocol. Instead, the reader is urged to treat all substances with respect, to be attentive to safe handling precautions available from the manufacturer, and to dispose of all substances in accordance with federal and local regulations. Where the possibility exists that a substance is toxic, the reader is encouraged to err on the side of caution: keep the substance off the skin and out of the eyes, nose and mouth – wear gloves, a lab coat, safety glasses and a mask when measuring, weighing, or dissolving such substances, and wipe down counter tops and balances when finished. (In the section below, where a substance is known to be toxic or potentially toxic, the reader is referred to this safe handling section: please exercise the precautions outlined here!) Finally, if your institution has a safety office, take advantage of this source of expertise to assist you in avoiding harm to yourself or the environment.

Protecting hybridizations from DNA contamination

Avoiding extraneous DNA

Just as it is of paramount importance to protect personnel and the environment from the reagents used for hybridizations, it is of critical importance to protect the hybridization from external sources of DNA. Since hybridizations are intended to detect infinitesimal quantities of DNA, results may be confounded by the introduction of DNA from the fingers of the person preparing the solutions and reagents, or carrying out the hybridization. Take care, therefore, to wear gloves whenever fingerprints could be introduced into solutions or reagents, or onto slides, microcentrifuge tube lids, and so on. Beware of sneezes! Use only autoclaved pipette tips and microcentrifuge tubes, and autoclave solutions as indicated below.

Solution preparation, reagents, and sources

Standard solutions

- 20 x standard saline citrate (SSC)
  - 175.2 g NaCl
  - 88.2 g Na citrate

- distilled water

  Dissolve sodium chloride and sodium citrate in distilled water, adjust to pH 7.4 with concentrated NaOH solution before bringing total volume to 1 litre with distilled water; aliquot, autoclave, and store at room temperature.

- 1 M Tris
  - 121.1 g Tris base (Trizma base, Sigma #T 1503)
  - 800 ml distilled water
  - 42 ml (approximately) concentrated HCl

  Dissolve Tris in water, add concentrated HCl, and allow to cool to room temperature before making final adjustments to bring the pH to 8.0. Make certain that the probe which you use to measure the pH will accurately measure the pH of Tris solutions (some probes do not). Bring the total volume to 1 litre with distilled water; aliquot and autoclave. Store at room temperature.

- 0.5 M Tris
  - 1 M Tris, pH 8.0
  - distilled water

  Mix equal volumes of Tris and water, check pH to ensure it is between 7.8 and 8.0; aliquot and autoclave. Store at room temperature.

- 0.1 M Tris
  - 100 ml 1 M Tris, pH 8.0
  - 900 ml distilled water

  Mix Tris with distilled water, ensure that pH is at 8.0, aliquot and autoclave. Store at room temperature.

Sperm washing solution

- 0.01 M Tris – 0.9% NaCl
  - 9 g NaCl
  - 100 ml 0.1 M Tris
  - 900 ml (approximately) distilled water

  Dissolve NaCl in 0.1 M Tris, add distilled water to a total volume of 1 litre, autoclave and store at room temperature.

Solutions for sperm head decondensation

- 1 M Dithiothreitol (DTT)

Note: see safe handling and disposal section above

- 3.08 g DTT (Sigma #D 9779) (caution: DTT is toxic)

- 20 ml sterile distilled water
  Dissolve DTT in water; store 500 µl aliquots in microcentrifuge tubes at -20°C
- 10 mM DTT in Tris
- 400 µl 1 M DTT
- 40 ml 0.1 M Tris
  Prepare fresh solution immediately before decondensing sperm heads; discard at end of day.
- 20 mM lithium diiodosalicylate (LIS)

Note: see safe handling and disposal section above

- 0.792 g LIS (Sigma #D 3635)
- 100 ml sterile distilled water
  Dissolve LIS in sterile distilled water; store at room temperature.
- 10 mM LIS, 1 mM DTT in Tris
- 20 ml 20 mM LIS
- 20 ml 0.1 M Tris
- 40 µl 1 M DTT
  Prepare fresh solution immediately before decondensing sperm heads; discard at end of day.
- 2 x SSC
- 4 ml 20 x SSC
- 36 ml sterile distilled water
  Prepare fresh solution on day of use; discard at end of day.

Nick translation solutions

- 2.5 M $MgCl_2$
  - 50.82 g $MgCl_2$ $6H_2O$
  - distilled water
    Dissolve $MgCl_2$ in 80 ml distilled water; adjust total volume to 100 ml, aliquot, autoclave, and store at room temperature.
- 10 x nick translation buffer (NTB)
  - 20 µl 2.5 M $MgCl_2$
  - 50 µl bovine serum albumin (BSA), DNase-free (10 mg/ml, Pharmacia, #27-8915-01)

- 930 µl 0.5 M Tris, pH 7.8 to 8.0

Combine and aliquot into 0.25 ml - 0.5 ml portions; store at -20°C. Thaw on ice to use.

- 10 x dNTP mix for direct incorporation of fluorochromes
  Use Ultrapure dNTP set: 2'-deoxynucleoside 5'-triphosphate (Pharmacia Biotech #27-2035-01)
  first, dilute dTTP: 1 µl dTTP + 4 µl sterile distilled water
  - 1 µl dGTP
  - 1 µl dCTP
  - 1 µl dATP
  - 1 µl dTTP dilution (see above)
  - 329 µl sterile distilled water

  Store at -20°C. Thaw on ice to use.

- 10 x dNTP mix for biotinylation
  Use Ultrapure dNTP set: 2'-deoxynucleoside 5'-triphosphate (Pharmacia Biotech #27-2035-01)
  - 1 µl dGTP
  - 1 µl dCTP
  - 1 µl dTTP
  - 197 µl sterile distilled water

  Store at -20°C. Thaw on ice to use.

- 10 x dNTP mix for digoxigeninylation
  Use Ultrapure dNTP set: 2'-deoxynucleoside 5'-triphosphate (Pharmacia Biotech #27-2035-01)
  - 1 µl dGTP
  - 1 µl dCTP
  - 1 µl dATP
  - 197 µl sterile distilled water

  Store at -20°C. Thaw on ice to use.

- 0.01 M DTT
  - 5 µl 1 M DTT
  - 495 µl sterile distilled water

  Store at -20°C. Thaw on ice to use.

- fluorochrome–conjugated dUTP
  fluorescein-11-dUTP (Fluorogreen™, Amersham #RPN 2121), rhodamine-4-dUTP (Fluorored™, Amersham #RPN 2122), FluoroLink™Cy3-dUTP (Amersham #PA 53022) and coumarin-4-dUTP (Fluoroblue™, Amersham #RPN 2123) are used as they are supplied,

at 1 mM concentration. Since fluorochromes are light sensitive, store at -20°C in the dark, and thaw on ice in the dark to use.

- Hapten-conjugated nucleotides
  - Biotin-14-dATP (0.4 mM, Gibco BRL #9524SA)
  - digoxigenin-11-dUTP (1 mM, Boehringer Mannheim #1093 088)
    Store at -20°C, thaw on ice to use.
- Nick translation enzymes
  - DNase I/DNA polymerase I, (0.4 U/µl, Gibco BRL #18162-016)
    Use as supplied; store at -20°C, thaw on ice to use.
- 0.5 M EDTA
  - 186.1 g disodium EDTA (Sigma #E 5134)
  - distilled water
  - 20 g (approximately) NaOH pellets
    Add EDTA to 800 ml distilled water and adjust to pH 8.0 with NaOH pellets, stirring vigorously; when the EDTA is in solution, adjust volume to 1 litre with distilled water. Aliquot and autoclave; store at room temperature.
- 20% sodium dodecyl sulfate (SDS)

Note: see safe handling and disposal section above

- 20 g SDS (lauryl sulfate, sodium salt, Sigma #L 6026) (Caution: very fine powder)
- 100 ml sterile distilled water
  Dissolve SDS in approximately 90 ml water, heating to 68°C to assist dissolution. Adjust the pH to 7.2 by adding a few drops of concentrated HCl, allow bubbles to subside, bring volume up to 100 ml with water, mix well, aliquot, and store at room temperature.

Hybridization solutions

- Formamide purification

Note: see safe handling and disposal section above
Before using ultrapure formamide (for example, ICN Biomedicals Inc. #71937) in any solutions, remove any degradation products: add 1 g Amberlite™ MB-1 resin (ICN Biomedicals Inc. #150330) for each 100 ml formamide, mix the beads well with the formamide, and filter through a paper filter (Whatman #1004 150). Store ultrapure formamide at 4°C.

- MM2.1 High stringency hybridization buffer

Note: see safe handling and disposal section above

- 5.5 ml ultrapure formamide
- 0.5 ml 20 x SSC
- 1 g dextran sulfate
  Ensure that the formamide and SSC are in the tube before the dextran sulfate is added, as dry dextran sulfate on the bottom of the test tube is virtually impossible to dissolve. Heat at 70°C to dissolve dextran sulfate, cool, adjust to pH 7, and dilute to 7 ml with sterile distilled water. Store 1 ml aliquots at -20°C.
- Salmon Sperm DNA
- 10 µl DNA from salmon testes (10 mg/ml, Sigma #D 7656)
- 190 µl sterile distilled water
  Dilute salmon sperm DNA in distilled water; store at 4°C.

Pre- and post-hybridization solutions

- 70% formamide / 2 x SSC (pre-hybridization solution)

Note: see safe handling and disposal section above

- 35 ml ultrapure formamide (see note, above)
- 5 ml 20 x SSC
- sterile distilled water
  Mix the formamide and SSC, and bring the pH to 7.5 with 1 M HCl. Adjust volume to 50 ml with water. Store in a Coplin jar at 4°C; use for a maximum of one month or 20 slides.
- 50% formamide / 2 x SSC (post-hybridization solution)

Note: see safe handling and disposal section above

- 75 ml ultrapure formamide (see note, above)
- 15 ml 20 x SSC
- sterile distilled water
  Mix the formamide and SSC, and bring the pH to 7 with 1 M HCl. Adjust volume to 150 ml with water. Divide evenly between 3 Coplin jars and store at 4°C; use for a maximum of one month or 20 slides.

Washing solutions

- Sodium phosphate, dibasic
  - 107.2 g $Na_2HPO_4 \cdot 7H_2O$
  - distilled waterDissolve sodium phosphate, dibasic in distilled water to a total volume of 4 L.

- Sodium phosphate, monobasic
  - 3.9 g $NaH_2PO_4 \cdot 2H_2O$
  - distilled water

  Dissolve sodium phosphate, monobasic in distilled water to a total volume of 250 ml.
- PN Buffer
  - 3800 ml sodium phosphate, dibasic solution (see above)
  - 200 ml (approximately) sodium phosphate, monobasic solution (see above)
  - 4 ml (approximately) Nonidet™ P-40 (NP-40) (Sigma #N 6507)

  Stir the sodium phosphate, dibasic solution while monitoring the pH, and add sodium phosphate, monobasic solution until the combined solution reaches a pH of 8.0. Measure the resultant solution volume and add 1 ml NP-40 per litre of solution.
- 2 x SSC / 0.1% NP-40
  - 500 ml 2 x SSC
  - 0.5 ml NP-40

  Mix NP-40 and 2 x SSC and allow bubbles to subside. Store at room temperature.

Detection, amplification, staining, and antifade solutions

- PNM Buffer

Note: see safe handling and disposal section above

- 100 ml PN buffer
- 5 g powdered non-fat milk
- 0.02 g Na azide (caution: sodium azide is poisonous)

  Dissolve the milk powder and sodium azide in PN buffer, then centrifuge. Remove the supernatant and store at 4°C. Avoid shaking the solution, as milk continues to precipitate out of solution over time, and re-centrifuge from time to time to remove precipitated particles.
- Streptavidin-Cy3 stock solution (1.667 mg/ml)
- 1 mg Cy3-conjugated streptavidin (Jackson ImmunoResearch Laboratories, Inc #016-160-084)
- 0.6 ml sterile distilled water

  Reconstitute streptavidin-Cy3 with water, let stand at room temperature for 1 to 2 hours, then draw solution into pasteur pipette to ensure it is completely clear. If it is not, centrifuge and keep the supernatant. Aliquot into 1 μl portions, and store at -20°C; avoid repeated freezing and thawing.

- Streptavidin-Cy3 working solution (3.3 μg/ml)
- 1 μl streptavidin-Cy3 stock solution (see above)
- 499 μl PNM buffer
  Dilute streptavidin-Cy3 aliquot with PNM buffer; store at 4°C for up to one month.
- Anti-digoxigenin-fluorescein isothiocyanate (FITC) stock solution (200 μg/ml)
- 200 μg anti-digoxigenin-fluorescein, Fab fragments (Boehringer-Mannheim, #1207741)
- 1 ml sterile distilled water
  Reconstitute anti-digoxigenin-FITC in water; store 15 μl aliquots at -20°C.
- Anti-digoxigenin-FITC working solution (13 μg/ml)
- 15 μl anti-digoxigenin-FITC stock solution (see above)
- 150 μl PNM buffer
- 50 μl normal sheep serum (Sigma #S 4265)
- 15 μl BSA (10 mg/ml, Pharmacia #27-8915-01)
  Mix reagents; store at 4°C for up to one month (solution may be unreliable if kept longer).
- Biotinylated anti-avidin stock solution (0.5 mg/ml)
- 0.5 mg biotinylated anti-avidin D from goat (Vector Laboratories #BA0300)
- 1.0 ml distilled water
  Reconstitute biotinylated anti-avidin in water; store 20 μl aliquots at -20°C.
- Biotinylated anti-avidin working solution (10 μg/ml)
- 20 μl biotinylated anti-avidin stock solution (see above)
- 980 μl PNM buffer
  Dilute biotinylated anti-avidin stock solution with PNM buffer; store at 4°C for 1 month.
- rabbit anti-sheep antibody: Use as supplied. (Oncor #S1331-4)
- FITC/anti-rabbit antibody: Use as supplied. (Oncor #S1331-5)
- 4',6-diamidino-2-phenylindole (DAPI) stock and working solutions

Note: see safe handling and disposal section above

- 1 mg DAPI (Sigma #D 9542)
- 1 ml sterile distilled water
  Dissolve DAPI in water to make stock solution (1 mg/ml). For long-term storage, keep at -20°C. Prepare serial dilutions to arrive at working solution dilutions: dilute 10μl of stock with 990μl of PN buffer to make a 10 μg/ml dilution, then dilute 10 μg of this 10 μg/ml solution with 990 μl of PN buffer to make the 100 ng/ml dilution, and finally dilute 5 μl of the 100 ng/ml dilution with 95 μl of PN buffer to make the 5 ng/ml dilution. Fresh 5 ng/ml solutions often stain more brightly than solutions that have been diluted for longer than a few weeks, so it may be necessary to prepare 2 ng/ml or 1 ng/ml solutions, in a similar fashion, if blue signals are being swamped by the brightness of the DAPI-stained sperm.
- Antifade

Note: see safe handling and disposal section above

- 5 mg p-phenylenediamine
- 10 ml PN buffer
- 10 ml glycerol
  Dissolve p-phenylenediamine in PN buffer, mix resultant solution with an equal volume of glycerol, and store 1-ml aliquots at −20°C.

Fig. 2. Outline: Fluorescence *in situ* hybridization of sperm

fresh ejaculate
⇙ ⇘
freeze → wash sperm and prepare slides
↓
decondense sperm heads
↓
hybridize sperm to indirect or direct probes
⇙ ⇘
probe detection → score

Procedure

Sperm washing and slide preparation

Sperm washing

1. If frozen sperm is to be used, thaw one 0.5-ml sperm straw.

2. Suspend 0.5 ml of fresh or thawed semen in 5 ml of 0.01 M Tris - 0.9% NaCl, centrifuge at 600 x g for 6 minutes, and discard supernatant. Resuspend in 5 ml of Tris-NaCl, and repeat wash twice.
3. Bring sperm to a concentration of approximately 60 x 10^6 sperm/ml with approximately 0.15 to 0.25 ml of Tris-NaCl.
4. Place 1-2 µl of the sperm suspension on a clean glass slide, and smear over a 1 cm^2 area using the side of the pipette tip. Circle the area on the underside of the slide with a diamond marking pencil. **Slide preparation**
5. Air dry on a flat surface.

In situ sperm nucleus decondensation

(modified from protocol described by Robbins et al.,1993, and Williams et al.,1993)

1. Place sperm slides in a Coplin jar containing 10 mM of DTT in 0.1 M Tris for 30 minutes at room temperature. If sperm heads are unusually susceptible to the decondensation chemicals and are over-decondensing or rupturing, DTT time may be decreased to 25 or 20 minutes, with a correspondingly shorter time in LIS (step 2, below). **Sperm head decondensation**
2. Transfer slides into a Coplin jar of 10 mM of LIS, 1 mM of DTT in Tris, for 3 hours at room temperature. If DTT time has been decreased to 25 or 20 minutes (see above), decrease the LIS time to 2.5 hours or 2 hours, respectively.
3. Rinse slides briefly in 2 x SSC and air dry vertically before use. Slides may be stored in a dark slide box at room temperature for several months. (We have very low relative humidity here, which may facilitate the storage of slides at room temperature. If older slides tend not to hybridize when stored at room temperature in your climate, store them at -20°C and thaw just before use.)

Probe preparation

DNA culture

DNA is transformed and grown in a plasmid culture, and insert fragment is isolated and purified using standard techniques (Maniatis et al.,1982; Sambrook et al.,1989).

Commercial probes

Ready-to-use probes are now available from a variety of suppliers. We have had success with a number of probes now available from Vysis, Inc. (in the US: 3100 Woodcreek Drive, Downers Grove, IL 60515; in Europe: Vysis GmbH, Vor dem Lauch 25, 70567 Stuttgart-Fasanenhof, Germany).

Nick translation

The nick translation reaction is used to create indirect and direct probes, respectively, by introducing into the DNA a hapten-conjugated nucleotide (biotin-14-dATP or digoxigenin-11-dUTP), or a fluorochrome-conjugated nucleotide (fluorescein-11-dUTP, rhodamine-4-dUTP, or coumarin-4-dUTP). Protocols for the two types of nick translation are similar, as are the reagents, but there are some notable differences. As the fluorochrome-labelled nucleotides are incorporated more slowly into the DNA strand, the direct-labelling reaction is allowed to proceed for a longer time than the procedure involving hapten-conjugated nucleotides, and the direct-labelling reaction's deoxynucleotide triphosphate (dNTP) mix includes a small amount of dTTP to fill gaps in the DNA strand where the fluorochrome-conjugated dUTP fails to bind. Indirectly labelled probes are stored at 4°C; directly labelled probes are light sensitive, and have been observed to lose their ability to label sperm DNA after storage at 4°C for approximately 6 months, so small aliquots are protected from light at -20°C when long-term storage is required. The first of the two following procedures is for nick translation with fluorochrome-conjugated nucleotides to create direct probes; the second is for nick translation with hapten-conjugated nucleotides to create indirect probes.

Direct probe preparation

1. To make direct probes, with all reagents on ice, the reaction mixture is assembled in a sterile 0.5-ml microcentrifuge tube as follows:
   Nick translation mix: direct probes:
   - __ μl DNA (2 μg)
   - 10 μl 10 x nick translation buffer
   - 20 μl 10 x dNTP mix
   - 10 μl 0.01 M DTT
   - 2 μl fluorochrome-conjugated dUTP
   - __ μl sterile distilled water (to adjust total volume to 100 μl)
   - 10 μl DNase I/DNA polymerase I
   - 100 μl Total volume
2. The reaction mixture is mixed gently, centrifuged briefly to move it to the bottom of the tube, and incubated at 15°C for 4 hours.
3. The reaction is terminated by the addition of 2.5 μl of 0.5 M EDTA and 0.5 μl of 20% SDS, and subsequent incubation at 65°C for 15 minutes.

Indirect probe preparation

1a. To prepare indirect probes, the protocol is similar, but the reaction time and some of the reagents differ. With all reagents on ice, the reaction mixture is assembled in a sterile 0.5-ml microcentrifuge tube as follows:
Nick translation mix: biotinylated probes:
- __ μl DNA (2 μg)
- 10 μl 10 x nick translation buffer
- 10 μl 10 x dNTP mix
- 10 μl 0.01 M DTT
- 12.5 μl biotin-14-dATP
- __ μl sterile distilled water (to adjust total volume to 100 μl)
- 10 μl DNase I/DNA polymerase I
- 100 μl Total volume

or

Nick translation mix: digoxigeninylated probes:
- __ μl DNA (2 μg)
- 10 μl 10 x nick translation buffer
- 10 μl 10 x dNTP mix
- 10 μl 0.01 M DTT
- 5 μl digoxigenin-11-dUTP
- __ μl sterile distilled water (to adjust total volume to 100 μl)
- 10 μl DNase I/DNA polymerase I
- 100 μl Total volume

2a. The reaction mixture is mixed gently, centrifuged briefly to move it to the bottom of the tube, and incubated at 15°C for 2 hours.

3a. The reaction is terminated by the addition of 2.5 μl of 0.5 M EDTA and 0.5 μl of 20% SDS, and subsequent incubation at 65°C for 15 minutes.

Hybridization, posthybridization, and detection

1. The following reagents are assembled in a sterile 0.5-ml microcentrifuge tube:
Probe mix for 2-colour FISH:
- 7 μl MM2.1 hybridization buffer
- 1 μl Salmon sperm DNA (500 μg/ml)
- 1 μl direct-labelled green probe **or** digoxigeninylated probe
- 1 μl direct-labelled red probe **or** biotinylated probe
- 10 μl Total volume

Note: in some instances we have omitted the salmon sperm DNA and increased the MM2.1 volume to 8 µl, with comparable or better signals after hybridization. As more probes are commercially available, we have needed to modify our hybridization mixes to accommodate probes which are sold pre-mixed in hybridization buffer, or which need to use hybridization buffers of the manufacturer's specific formulation. Therefore, when probes from a number of sources are hybridized simultaneously, it is necessary to compromise between the conditions which are recommended for the various probes, and to use the recommended hybridization buffer when possible. When using Spectrum Orange™ or Spectrum Green™ probes from Vysis, for example, we also use the hybridization buffer supplied with the probe (even when we are using probes which we have prepared in this lab in the same hybridization), and we find that the hybridizations are usually successful. In general, we use MM2.1 hybridization buffer for most other hybridizations, and we try to use the following pre- and post-hybridization protocols as the starting point from which we deviate only when hybridizations are not successful.

Denaturation of sperm nuclei and probe mix

2. The microcentrifuge tube containing the hybridization probe mix is heated at 70°C to 75°C for 5 minutes, then is snap-cooled in an ice/water bath. The probe mixture is kept on ice for at least 5 minutes.
3. The Coplin jar containing 70% formamide and 2 x SSC at pH 7.5 must be pre-heated by 1°C higher than the desired denaturation temperature for each slide which is to be denatured. For example, if 2 slides are to be denatured at 70°C, the Coplin jar containing the pre-hybridization buffer needs to be pre-heated to 72°C. It is not advisable to denature more than 4 slides at once. Sperm DNA is denatured by a 5-minute immersion of the slide in a Coplin jar containing 70% formamide and 2 x SSC, pH 7.5, at 70 to 75°C. The DNA is snap cooled and dehydrated on the slide by 2-minute washes in Coplin jars containing 70%, 85%, and 95% ethanol at −20°C, then is air dried and prewarmed to 37°C on a slide warmer.

Note: in cases where hybridization is not proceeding adequately, the slide may be immersed for up to 7 minutes at 75°C. The optimum temperature and time combination needs to be determined empirically in these more difficult hybridizations, but we have found that 5 minutes at 72 to 73 °C results in bright clear signals with most probes.

Hybridization

4. The hybridization probe mix is removed from the ice bath and transferred to the slide area over the sperm, and is allowed to warm briefly. If there are bubbles in the hybridization probe mix, they are removed at this time by popping them with the corner of the cover glass. A 22-mm^2,

0.1-mm-thick cover glass is placed over the area to be probed, and after the hybridization mix has spread to the edges of the coverslip, rubber cement, applied from a syringe, is used to seal the cover glass onto the slide.

5. The slide is incubated in a dark, humidified container at 37°C for 16 to 24 hours.

Duration of hybridization

Note: hybridizations have been incubated very successfully for up to 72 hours in instances where signals had been faint in previous hybridizations, or for convenience (for example, over a weekend).

Posthybridization

6. At the conclusion of the desired hybridization time, (a) a Coplin jar containing 2 x SSC at pH 7 is preheated to 70° to 72°C or (b) three Coplin jars containing 50% formamide and 2 x SSC at pH 7 are preheated to 45°C. The slide is removed from the hybridization chamber and the rubber cement is carefully peeled from around the cover glass. The slide is dipped in the posthybridization bath and is held horizontally for a few seconds to allow the liquid to penetrate under the cover glass and render it more mobile, before gently sliding it off the slide.

7. The slide is washed for 2 minutes (a) in the 2 x SSC bath, or (b) in each of the three posthybridization baths, then is rinsed briefly in a Coplin jar containing (a) 2 x SSC / 0.1% NP-40 or (b) PN buffer.

Note: 1) We have found that the 2 x SSC posthybridization wash at 70 °C gives bright, clear signals more consistently, and now use this method almost exclusively. 2) Vysis recommends air-drying slides hybridized with their direct probes. As long as no indirect probe has been simultaneously hybridized, the air-drying procedure significantly enhances the brightness of all direct signals.

Counterstaining: direct hybridizations

8a. Hybridizations using direct-labelled probes need no detection step, and nuclear DNA may be counterstained at this time: apply 10 µl of DAPI, 5 ng/ml (for hybridizations including blue signals) or 100 ng/ml to the hybridized area for 10 seconds, rinse in PN buffer, apply 8.5 µl antifade and a 0.1 mm thick cover glass, and store flat in a dark slide box until viewing. Note: air-dried slides (see note above) which have no blue signals may be coverslipped using 10 µl of DAPI-II counterstain (Vysis #30-804931) instead of staining with DAPI and mounting with antifade.

Hints: antifade and DAPI

It has been found that in most cases, the antifade itself stains the sperm nuclei red, eliminating the need for propidium iodide staining. In cases where the antifade causes the background and/or the sperm nuclei to be stained too red for the sperm nuclei or the red signals to be clearly seen, it is possible to use a diluted solution of antifade in PN buffer; usually a 75% antifade solution will allow the sperm nuclei to be distinguished from the background, and the red signals to be seen within the sperm nuclei, but will still prevent the fading of signals. If, upon viewing, it is observed that the DAPI is too bright and has obliterated the blue signals, it is sometimes possible to salvage the situation by removing the cover glass and incubating the slide in a Coplin jar of PN buffer at 37°C for 20 minutes, rinsing in fresh PN buffer at room temperature, and re-coverslipping in antifade. Check the brightness of the DAPI, and if blue signals are still not visible, this DAPI-removing step may often be repeated without damaging the sperm or signals.

Detection and counterstaining: indirect hybridizations

8b. Hybridizations using indirectly-labelled probes must undergo a detection step in which fluorochromes are bound to the haptens. It is important that slides with hapten-bound fluorochromes never be allowed to dry after hybridization is complete. Apply 20 µl of PNM buffer, cover with a flexible plastic cover slip (for example, a 2.5 cm square of Parafilm™ laboratory film, American Can Company, Dixie/Marathon Greenwich, CT, 06830), and incubate in the dark at room temperature for 30 seconds to 5 minutes to block non-specific binding sites. Rinse well in PN buffer, and then apply 10 µl of streptavidin-Cy3 (binds to biotin) mixed with 10 µl anti-digoxigenin-FITC (binds to digoxigenin), cover with a flexible plastic cover slip, and incubate at 37°C in a humidified chamber for 30 minutes. Rinse well in PN buffer, then counterstain with DAPI, and coverslip in antifade as outlined above.

Amplification of signals

9. Faint or absent signals may be amplified in hybridizations using hapten-labelled probes: wet the slide in PN buffer, gently remove the cover glass, and rinse slide well in PN buffer to remove antifade. If both digoxigenin and biotin signals must be amplified, it is important to perform the digoxigenin amplification before the biotin amplification procedure, or the digoxigenin amplification will fail. To amplify the digoxigenin signal, apply 20 µl of rabbit anti-sheep antibody to the slide, cover with a flexible plastic cover slip, and incubate in a humidified chamber at 37°C for 15 minutes. Remove cover slip, rinse slide well in PN buffer at room temperature, apply 20 µl of FITC/anti-rabbit antibody to the slide, cover with a flexible plastic cover slip, and incubate in a humidified chamber at 37°C for 15 minutes. Remove cover slip, rinse slide well in PN buffer, and either proceed to the biotin amplification or apply 8.5µl of antifade, cover with a 0.1-mm-thick cover glass, and store flat in a dark slide box until viewing. If amplifying the biotin signal, apply 15 µl of biotinylated anti-avidin, cover with a flexible plastic cover slip, and incubate in a humidified chamber at 37°C for 30 to 60 minutes. Remove cover slip and rinse well in PN buffer, then apply 10 µl of streptavidin-Cy3, cover with a flexible plastic cover slip, incubate in a humidified chamber at 37°C for 30 to 60 minutes, remove the cover slip, and rinse well in PN buffer. Apply 8.5µl of antifade, cover with a 0.1-mm-thick cover glass, and store flat in a dark slide box until viewing.

Slide viewing

Slides are viewed under 100 x magnification on an epifluorescence microscope (for example, the Zeiss Axiophot microscope) fitted with both an FITC / rhodamine dual bandpass filter set (to allow simultaneous viewing of red and green signals in pale-red antifade-stained sperm) and an AMCA single bandpass filter set (to view blue signals, and DAPI-stained sperm). DAPI is useful in determining the edges of some sperm. Additional filter sets which are not crucial, but which improve the ability to visualize faint signals or different combinations of signals, are a triple bandpass filter set (FITC / rhodamine / DAPI) and an FITC single bandpass filter set.

Scoring and signal assessment

The presence of red and green signals (domains) are identified in each sperm nucleus, and nullisomic, unisomic, and disomic nuclei are scored accordingly. A single red domain and a single green domain are expected within each unisomic sperm which is being probed for autosomal chromosomes; an extra red or green signal indicates a disomic sperm, and lack of a signal indicates a sperm which is nullisomic for the chromosome corresponding to the absent domain. Simultaneous scoring for two chromosomes allows the differentiation between diploid sperm (containing two red domains and two green domains) and disomic sperm, and permits the elimination of false nullisomies arising from non-hybridized sperm (which contain neither red nor green domains). Interobserver variability is minimized by following strict scoring criteria. Sperm are considered disomic only if the duplicated domains are of similar size, shape, and intensity, are separated by a distance of at least one domain (one-half domain in the case of enormous signals such as the Yq signal), and are not found on the sperm's periphery (to avoid counting flecks of background on the surface of the sperm). Sperm nuclei are scored only if they are intact, non-overlapped, have a clearly-defined border, and have not decondensed to more than twice the size of a non-decondensed sperm head. Experience in determining the degree of decondensation may be obtained by staining a non-decondensed sperm slide with DAPI, and comparing decondensed sperm to this non-decondensed sperm slide.

Troubleshooting

In a single word, the most useful piece of advice for those working with FISH hybridizations is "PERSEVERE". Frustrations often occur when there is no discernible reason for hybridization failure. The recommendations outlined above (especially regarding solution pH, ultraformamide purification, and the length of time to keep solutions), will help you to avoid many of the "invisible" problems which resulted in failed hybridizations in our lab over the years. It is important NOT to immediately change hybridization parameters if a hybridization fails: there have been many instances in our laboratory where we have repeated all parameters of a failed hybridization exactly, and the repeated hybridization has been successful (although it must be noted that the successful hybridization sometimes took more than one repeat to achieve!). Therefore, save a lot of time and difficulties by first repeating the hybridization exactly (if there is no apparent reason for the failure) and fine-tune the protocol as suggested in the section above only if the repeated hybridization fails.

References

Bischoff FZ, Nguyen DD, Burt KJ and Shaffer LG (1994) Estimates of aneuploidy using multicolor fluorescence in situ hybridization on human sperm. Cytogenet Cell Genet 66:237-243.

Brandriff B, Gordon L, Ashworth G, Watchmaker G, Moore D, Wyrobek A, Carrono A (1985) Chromosomes of human sperm: Variability among normal individuals. Hum Genet 70:18-24.

Holmes JM and Martin RH (1993) Aneuploidy detection in human sperm nuclei using fluorescence in situ hybridization. Hum Genet 91:20-24.

Maniatis T, Fritsch EF and Sambrook J (1982) Molecular Cloning: A Laboratory Manual. Cold Spring Harbor Laboratory, New York.

Martin RH (1996) The risk of chromosomal abnormalities following ICSI. Hum Reprod 11:924-925.

Martin RH (1997) Sperm chromosomal abnormalities in a Familial Mediterranean Fever patient treated with colchicine. Middle Eastern Fertility Society Journal 2:82-83.

Martin RH, Barclay L (1996) Sperm aneuploidy and diploidy frequencies after treatment of psoriatic arthritis with methotrexate. Ind J Hum Genet 2:65-67.

Martin RH, Rademaker AW (1995) Reliability of aneuploidy estimates in human sperm: results of fluorescence in situ hybridization studies using two different scoring criteria. Molecular Reproduction and Development 42:89-93.

Martin RH, Balkan W, Burns K, Lin CC (1982) Direct chromosomal analysis of human spermatozoa. Am J Hum Genet 34(3):459-468.

Martin RH, Balkan W, Burns K, Rademaker AW, Lin CC, Rudd NL (1983) The chromosome constitution of 1000 human spermatozoa. Hum Genet 63:305-309.

Martin RH, Ko E, Rademaker A (1991) The distribution of aneuploidy in human gametes: comparison between human sperm and oocytes. Am J Med Genet 39:321-331.

Martin RH, Spriggs E, Ko E, Rademaker AW (1995) The relationship between paternal age, sex ratios and aneuploidy frequencies in human sperm as assessed by multicolour fluorescence in situ hybridization. Am J Hum Genet 57:1395-1399.

Martin RH, Spriggs E, Rademaker AW (1996) Multicolour fluorescence in situ hybridization analysis of aneuploidy and diploidy frequencies in 225,846 sperm from ten normal men. Biol Reproduction 54:394-398.

Martin RH, Ernst S, Rademaker A, Barclay L, Ko E, Summers N (1997) Chromosomal abnormalities in sperm from testicular cancer patients before and after chemotherapy. Hum Genet 99:214-218.

Mikamo K, Kamiguchi Y, Tateno H (1990) Spontaneous and in vitro radiation-induced chromosome aberrations in human spermatozoa: application of a new method. In: Mendelsohn ML, Albertini RJ (eds), Mutation and the Environment Part B: Metabolism Testing Methods and Chromosomes. John Wiley, New York, pp 447-456.

Moosani N, Pattinson HA, Carter MD, Cox DM, Rademaker AW, Martin RH (1995) Chromosomal analysis of sperm from men with idiopathic infertility using sperm karyotyping and fluorescence in situ hybridization. Fertility and Sterility 64(4):811-817.

Rademaker A, Spriggs E, Ko E, Martin RH (1997) Reliability estimates of diploid human sperm using multicolour fluorescence in situ hybridization. Human Reproduction 12:77-79.

Robbins WA, Segraves R, Pinkel D, and Wyrobek AJ (1993) Detection of aneuploid human sperm by fluorescence in situ hybridization: evidence for a donor difference in frequency of sperm disomic for chromosomes 1 and Y. Am J Hum Genet 52:799-807.

Rudak E, Jacobs PA, Yanagimachi R (1978) Direct analysis of the chromosome constitution of human spermatozoa. Nature 274:911-913.

Sambrook J, Fritsch EF, and Maniatis T (1989) Molecular Cloning: A Laboratory Manual. Cold Spring Harbor Laboratory, New York.

Spriggs EL, Rademaker AW, Martin RH (1996) Aneuploidy in human sperm: the use of multicolor fluorescence in situ hybridization (FISH) to test various theories of nondisjunction. Am J Hum Genetics 58:356-362.

Williams B, Ballenger C, Malter H, Bishop F, Tucker M, Zwingman T, Hassold T (1993) Nondisjunction in human sperm: Results of fluorescence in situ hybridization studies using two and three probes. Hum Mol Genet 2:1929-1936

Wyrobek A, Robbins W, Tang C, et al. (1993) Hierarchical organization of human sperm chromatin is a critical factor in the detection of chromosomal aneuploidy by fluorescence in situ hybridization. Am J Hum Genet 53(Suppl):130

Chapter 20

Microdissection and Reverse Chromosome Painting

GABRIELE SENGER, JÖRG WEIMER, UWE CLAUSSEN and ILSE CHUDOBA

Introduction

The technique of chromosome microdissection was initially developed for generating DNA libraries from specific parts of polytene chromosomes of Drosophila (Scalenghe 1981) in order to isolate certain genes known to be localized in those regions. First improvements of this original technique such as the use of GTG-banded chromosomes and the invention of the first sequence independent DNA amplification system (Lüdecke 1989, Senger 1990, Lüdecke 1990) allowed the generation of microdissection libraries from precisely defined regions of human chromosomes and a stepwise reduction of the number of chromosomes that had to be dissected for one library. By Trautmann and co-worker (1991) it was shown for the first time that microdissection libraries can be used as painting probes for *in situ* hybridization.

Due to a further simplification of the technique of sequence independent DNA amplification with a degenerate oligonucleotide primer (Telenius 1992) microdissection and reverse painting soon proved to be an extremely valuable procedure in prenatal, postnatal and tumor cytogenetic diagnosis for the analysis of chromosome rearrangements that are difficult to be identified (Meltzer 1992, Viersbach 1994, Müller-Navia 1995, Rubtsov 1996, Chudoba 1996). The complete analysis of a rearranged chromosome can be finished in less than three days. In contrast to forward chromosome painting with commercially available probes this technique reveals not only the chromosomal origin of a marker chromosome or a translocated

---

Correspondence to Gabriele Senger, Praxis für Medizinische Genetik, Roritzerstr. 2, Regensburg, 93047, Germany (*phone* +49-941-53710; *fax* +49-941-53708; *e-mail* GSenger@t-online.de), Jörg Weimer, Universitätsfrauenklinik Kiel, Onkologisches Labor, Michaelisstr. 16, Kiel, 24105, Germany, Uwe Claussen, Institut für Humangenetik und Anthropologie, Kollegiengasse 10, Jena, 07740, Germany, Ilse Chudoba, MetaSystems GmbH, Robert-Bosch-Str. 6, Altlussheim, 68804, Germany band, it also immediately identifies the chromosomal region involved. Thus even deletions can be defined precisely (Rubtsov 1996). Several groups transfer the chromosome fragment directly to a microcentrifuge tube by breaking off the needle with the adhering fragment inside the tube (Meltzer 1992, Müller-Navia 1995, Yokoyama and Sakuragawa 1995). This seems to be much easier compared to the method of collecting the fragments in a small droplet of several nanoliter placed in a small moist chamber to prevent its evaporation. But it has the disadvantage that the successful transfer of the fragments to the PCR reaction mixture can not be visually controlled. Fragments might get lost due to an electrostatic charge of the plastic tube. The procedure of microdissection described in this manual includes a new method of collecting the microdissected fragments. It still enables one to visually control the successful transfer of a microdissected chromosome fragment to a small volume of buffer solution. In contrast to an earlier protocol (Senger 1990), however, this collection drop is not placed inside a moist chamber but in the tip of a micropipette that is attached to the microscope. Evaporation of the microdrop is prevented due to a high concentration of glycerol. Siliconization of the pipette can be omitted as the tip of the pipette is broken off inside the PCR tube.

The flow diagram in Figure 1 summarizes the individual steps that are performed in a typical protocol of microdissection and reverse painting.

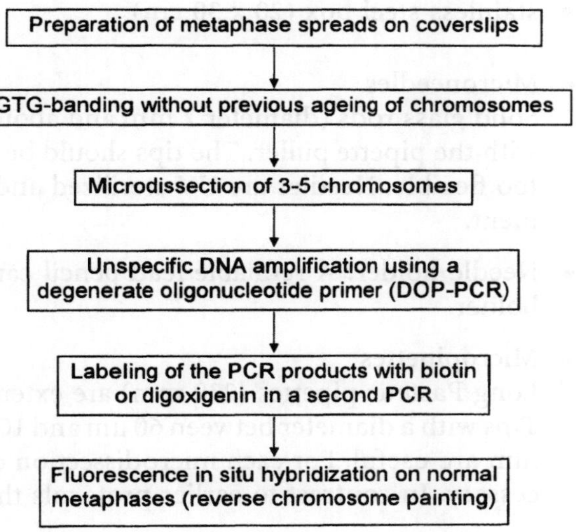

Fig. 1. Schematic flow sheet of the individual steps during microdissection and reverse painting.

Materials

Equipment

- inverted microscope (e.g. Zeiss Axiovert 135)
  Microdissection is performed using an inverted microscope equipped with a rotating table, two objectives with a magnification of 10x and 100x (oil-immersion, Plan-Neofluar) and an electronically controlled micromanipulator (eg MR Mot; Zeiss) attached to the right hand side of the microscope. A second mechanical micromanipulator is attached to the left hand side of the microscope. Additionally, an Optovar magnification lens (2.5x) is useful but not essential. The microscope should be placed on a vibration-free location (best in a room at the basement of the building).
- pipette puller (eg Narishige, PC-10)
  For the preparation of glass needles and extended micropipettes a pipette puller is necessary. A vertical 2-step pipette puller is better than a horizontal one.
- thermocycler with heated lid (eg PTC-200 from MJ Research)
- laminar flow-hood equipped with a short wave UV-lamp
- waterbath (exclusively used in microdissection experiments)
- one set of microliter-pipettes (0.01 – 2 µl, 0.1 – 20 µl, 10 – 200 µl) (exclusively used in microdissection experiments **before** DNA amplification)
- stainless steel box (20 x 30 cm)

Microinstruments

- Microneedles
  Solid glass rods (diameter 2 mm and about 100 mm long) are extended with the pipette puller. The tips should be very short otherwise they are too flexible. Needles are UV sterilized and discarded after each experiment.
- Needle holder: A refillable lead pencil can be used as a simple needle holder.
- Micropipettes
  Long Pasteur pipettes (230 mm) are extended using the pipette puller. Tips with a diameter between 60 µm and 100 µm and a length of about 15 mm are useful. For each microdissection experiment one pipette is necessary. In contrast to earlier protocols these micropipettes are not si-

liconized. Sterilize by incubation at 100°C for 1 h and by UV-illumination for at least 30 min.

- Petri dish with excised bottom
  A large glass petri dish (150 mm diameter) serves as a table for the coverslips carrying the metaphase spreads. Cut a rectangular hole (55 x 55 mm) in the middle of the petri dish to enable direct contact of the oil immersion objective with the coverslip from underneath (see Figure 2). Alternatively a metal plate (approximately 10 x 10 cm) may be used whereas a plastic petri dish is not recommended as the immersion oil is rather aggressive and will corrode plastic.

Reagents

- Bacto trypsin (Difco, #0153-60-2)
- Phosphate-buffer, pH 6.88 (Merck, #7294)
- Giemsa-solution (Merck, #9204)
- Glycerol (free of nucleases) (Merck, #12011)
- Proteinase K (Boehringer, #1413 783)
- DOP-primer (Telenius 1992) 5' ccg act cga gn_6 atg tgg 3'
- Sequenase, Version 2.0 (Amersham, #US70775)
- AmpliTaq DNA polymerase, Stoffel fragment with 10x Stoffel buffer and 25 mM $MgCl_2$ (Perkin Elmer, #N808-0038)
- AmpliTaq DNA polymerase (Perkin Elmer, #N808-0160)

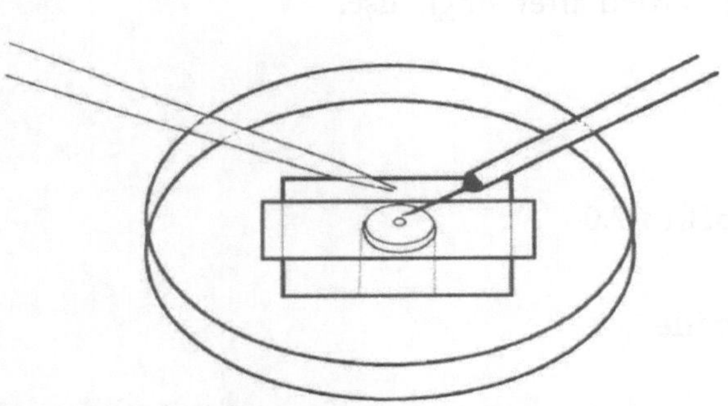

Fig. 2. Illustration of the microdissection set-up on the inversed microscope. The coverslip with metaphase spreads facing upwards is placed in a large petri dish. A rectangular hole in the bottom allows direct contact of the oil immersion objective with the coverslip. Above the coverslip the microdissection needle (right) and the micropipette containing collection buffer (left) are visible.

- GeneAmp dNTPs (10 mM each)(Perkin Elmer, #N808-0007)
- Replitherm buffer, 20x (Biozym/EpiCentre, #100001)
- Biotin-16-dUTP (50 nmol)(Boehringer, #1093 070)
- E. coli tRNA (Boehringer, #109 541)
- Formamide (for hybridization mixture)(International Biotechnologies INC, IBI, #72024)
- Dextran sulfate (Pharmacia, #17-0340-02)
- Tween 20 (Sigma, #P-1379)
- Human Cot-1 DNA (BRL, #15279-011)
- FITC-conjugated Avidin (Vector laboratories, #A-2011

Solutions

- Collection buffer
  - 10 mM Tris-HCl, pH 7.5
  - 10 mM NaCl
  - autoclave
  - add 0.1% SDS

  prepare aliquots of 50 µl, illuminate with short wave UV-light for 30 min and store at room temperature.

Note: It is essential to avoid any contamination of solutions with foreign DNA or nucleases. Therefore, all chemicals necessary for microdissection and DOP-PCR should not be in common use within the lab and only handled with gloves. All solutions must be prepared sterile and stored in small aliquots that are discarded after single use.

- 20x SSC
  - 3 M NaCl
  - 0.3 M Na_3-citrate

  autoclave and adjust pH to 7.0
- Hybridization mixture
  - 50% deionized formamide
  - 2xSSC
  - 10% dextran sulfate
  - 1% Tween 20

  adjust pH to 7.0 and store in aliquots of 100 µl at − 20°C

Procedure

Preparation and GTG-banding of metaphase spreads

For chromosome preparation routine cell harvesting protocols can be used (see Chapters Chapter 1, Chapter 5 and Chapter 12). Any treatment of the cell culture with reagents that modify DNA (eg bromodeoxyuridine) or intercalate into the DNA (eg ethidiumbromide or Hoechst 33258, sometimes used to obtain high resolution chromosomes) should be avoided as they reduce the PCR efficiency substantially. Although fixative and especially acetic acid is known to cause DNA depurination, fixed cell suspensions can be used for more than one year if stored at – 20°C. The oldest cell suspension that has been used successfully for microdissection was stored for 3 years in fixative. Metaphase spreads, however, should always be prepared immediately before microdissection and subsequently GTG-banded without previous ageing. It should go without saying that during all these steps gloves must be worn.

Preparation of metaphase spreads

Soak coverslips (24x60x0.17mm) for several days in 10% SDS. Rinse the coverslips thoroughly in sterile water and use wet for the preparation of metaphase spreads. After evaporation of fixative continue directly with GTG-banding. Previous storage of metaphase spreads in ethanol might help to improve the banding quality, however, dissection of chromosomes that have been stored in ethanol is more difficult as their consistency will be rather hard and fragile. For a satisfactory banding quality it is important to prepare metaphases free of cytoplasm.

GTG-banding

Dissolve one ampule of lyophilized Bacto trypsin in sterile distilled water to a stock solution of 5% (w/v). Store frozen in aliquots of 100 µl each.

Prepare four sterile 50 ml plastic tubes (A-D) with the following solutions, each at room temperature:

- tube A: 35 ml phosphate-buffer, pH 6.88
- tube B: 35 ml phosphate-buffer, pH 6.88 mixed with 100 µl trypsin stock solution
- tube C: 35 ml phosphate-buffer, pH 6.88 mixed with 3 ml Giemsa and filter-sterilized
- tube D: 50 ml sterile distilled water

Incubate the air dried coverslips successively in phosphate-buffer (A) for 1 min, Trypsin (B) for 20-40 sec and Giemsa (C) for 3 min. After a short wash in water (D) air dry the coverslips.

Note: The short incubation in phosphate buffer prior to trypsination can be skipped, however, it was found that this pre-treatment slightly improves the banding quality.

Microdissection

Chromosome microdissection should be performed in a room that is separated from all laboratories where large amounts of DNA are generated and used (eg molecular or FISH laboratories).

Microdissection set-up

1. Place the petri dish with the excised bottom on the rotating table of the inverted microscope and place one coverslip with the metaphase spreads facing upwards over the rectangular hole of the petri dish as illustrated in Figure 2.
2. Mix one aliquot (50 µl) of collection solution with:
   - 0.5 mg/ml proteinase K and
   - 50 µl glycerol
3. Soak up a large volume of collection solution into the pipette. The extended tip must be filled entirely and at least 3 – 4 mm above the first widening. Flick several times against the pipette to produce foam. Remove some collection solution from the pipette in order to place a small volume into the front tip of the pipette isolated by air from the remaining solution inside of the pipette. Attach this pipette to the mechanical micromanipulator at the left hand side of the microscope.
4. Attach the microdissection needle to the electronically controlled micromanipulator at the right hand side of the microscope.
5. At a magnification of 100x (10x objective) position both the micropipette and the tip of the needle in the centre of the visual field. The distance between the tip of the pipette and the coverslip should be at least 5 mm to avoid any contact of the pipette with the coverslip but one still has to be able to focus the tip of the pipette with the 10x objective without moving the pipette.

Dissection of chromosomes

During the following steps **do not use gloves** as this will cause loss of dissected chromosome fragments due to electrostatic charge.

1. Locate a suitable metaphase and turn the microscope table in order to position the chromosome of interest perpendicularly to the needle.
2. Place the needle at 100x magnification above the metaphase.
3. At a magnification of 1000x (100x oil immersion objective) lower the needle carefully until the tip is visible. Place the tip directly in front of the chromosome band to be dissected. As soon as the tip touches the coverslip further lowering of the needle results in a forward movement of the tip and leads to the excision of the chromosome region in front of the needle.

Note: If a larger part of a chromosome (eg one arm) or the entire chromosome has to be dissected (see Figure 3) it is not recommended to excise this region in several small fragments. By causing **minimal** vibration of the needle it is possible to scratch off one chromosome(arm) in one single piece as shown in Figure 3b.

4. Touch the excised fragment with the needle until it sticks to the tip and can thus be taken up (Figure 3c). As the bottom side of the fragment is more sticky than its top it might be easier to take it up by turning the fragment upside down.

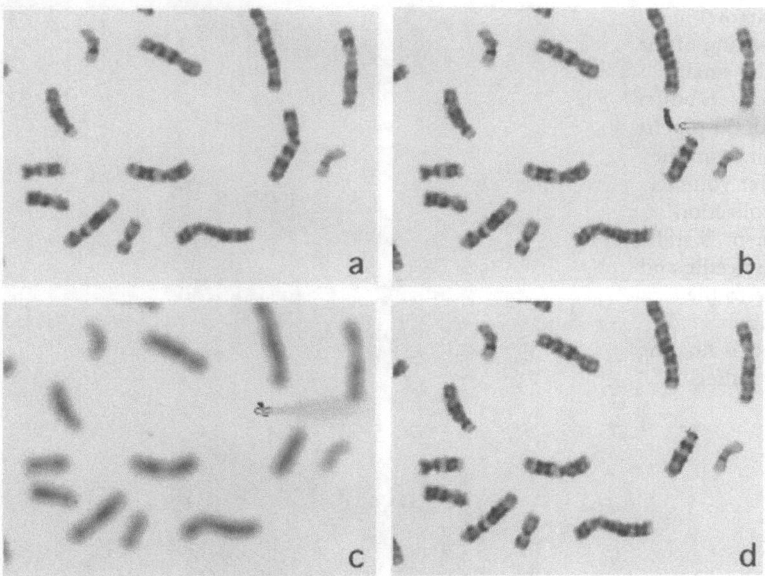

Fig. 3. Microdissection of the short arm of chromosome 7. a) Chromosomes before microdissection. b) The whole chromosome arm 7p is isolated in one single piece and c) taken up with the needle. d) Chromosomes after excision of 7p.

5. Switch the microscope to low magnification and focus at the tip of the micropipette. Move the tip of the needle with the adhering chromosome fragment in front of the opening of the pipette (Figure 4a) and wash the tip in the frontal portion of collection solution until the chromosome fragment is detached from the needle (Figure 4b – c).

Note: Poor visualization of the needle and pipette may be caused by immersion oil present at the coverslip. Move the coverslip until the pipette and needle are clearly visible.

6. Repeat steps 1 – 5 until the desired number of fragments have been collected.
7. Place the micropipette in a stainless steel box with lid and incubate at 60°C for at least 1 hour in a water bath.

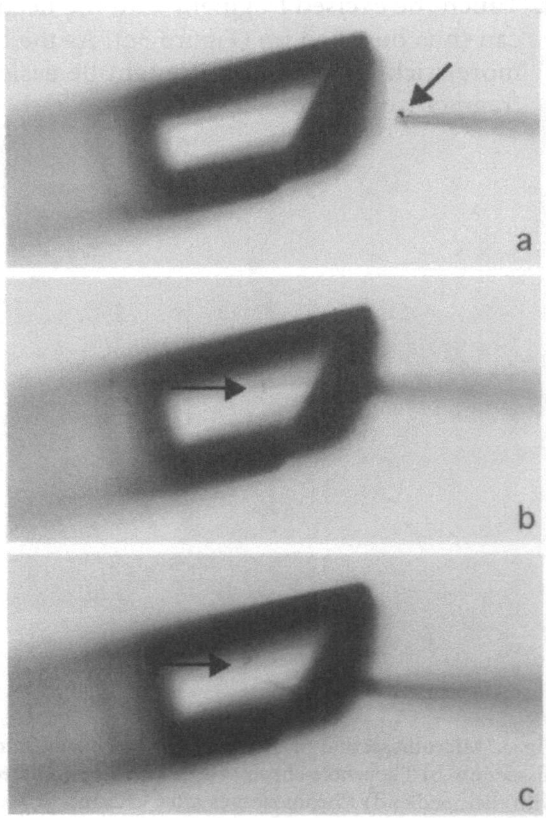

Fig. 4. Transfer of the chromosome fragment into the collection buffer. a) Positioning of the needle with the adhering chromatin fragment (arrow) in front of the opening of the pipette. Note the small volume of collection buffer at the front tip of the pipette separated by air from the remaining buffer volume. b) Inside the collection drop the fragment is still attached to the needle and well visible due to the Giemsa staining. c) The fragment is floating free inside the buffer.

Note: Even one single fragment is sufficient to obtain a visible signal after reverse chromosome painting. As these signals, however, will be rather weak it is recommended to isolate a minimum of 3 – 5 copies.

DNA Amplification using DOP-PCR

Initial amplification

All following steps should be performed under a laminar flow-hood that has been UV illuminated for a minimum of 30 min before the PCR is started.

Use a separate set of adjustable microliter pipettes that should never be used for pipetting DNA. As a further precaution against DNA contamination use plugged, sterile microliter tips that are classified as RNAse-free. Gloves must be worn again during all following steps.

PCR mixture 1:
For each microdissection library prepare 5 µl of PCR mixture 1 containing:

- 0.6x Sequenase buffer
- 5 µM primer (DOP) and
- 200 µM each dATP, dCTP, dGTP, dTTP

Note: A low buffer concentration of 0.6x is chosen due to the fact that in spite of the heated lid of the thermocycler a small part of the reaction mixture will evaporate. For a volume as small as 5 µl this results in a substantial increase of buffer concentration which is thus compensated. In case a thermocycler without heated lid is used and the reaction mix has to be overlaid with mineral oil in order to avoid evaporation, the buffer concentration should be increased to 1x.

Sequenase mixture:
Dilute Sequenase 1:7 in Sequenase dilution buffer. For each microdissection library 2 µl of diluted enzyme solution is required.

Note: Keep the concentrated Sequenase on ice and re-freeze immediately. Also the diluted enzyme has to be kept on ice.

PCR mixture 2:
For each microdissection library prepare 45 µl of PCR mixture 2 containing:

- 1x AmpliTaq Stoffel fragment buffer
- 220 µM each dNTPs

- 1.1 μM primer (DOP)
- 2.5 mM $MgCl_2$

PCR mixture 3:
For each microdissection library prepare 5 μl of PCR mixture 3 containing:

- 1x AmpliTaq Stoffel fragment buffer
- 2.5 mM $MgCl_2$
- 1Unit/μl AmpliTaq Stoffel fragment DNA polymerase

Transfer of DNA Remove the micropipette from the waterbath and break off the tip inside of the PCR tube containing 5 μl of PCR mixture 1. Centrifuge briefly to collect all solution at the bottom.

PCR cycles Insert the tube in the thermocycler and cycle with the parameters shown below.

Note: If a thermocycler with heated lid is used the actual temperature inside of the reaction mixture is higher than the programmed block-temperature. This is especially true during steps 1 to 4 due to the small reaction volume of 5 μl. (For a 50 μl volume this effect is negligible). In brackets the actual temperatures are shown that were measured with an in-sample probe during these four steps.

step 1: 94°C (101°C) for 5 min

step 2: 25°C (29°C) for 2 min 20 sec, here add 0.25 μl (0.3 – 0.4 U) of Sequenase **in each cycle**

step 3: 34°C (37°C) for 2 min

step 4: 94°C (98°C) for 1 min

step 5: repeat from step 2 for 7 more times

step 6: 30°C for 2 min, here add 45 μl of PCR mixture 2

step 7: 94°C for 1 min

step 8: 56°C for 1 min, here add 5μl of PCR mixture 3 during the first high temperature cycle

step 9: 72°C for 2 min

step 10: repeat from step 7 for 31 more times

step 11: 72°C for 8 min

step 12: hold at 4°C

Add EDTA to a final concentration of 5 mM and freeze at – 20°C for long term storage.

Labelling of the PCR products with biotin

The DOP-PCR products are labelled in a second PCR reaction. For each microdissection library 2 µl of the PCR product are added to 20 µl of the labelling mixture. For this step **never use the microliter pipette for microdissection.**

Labelling mixture:

- 1x Replitherm buffer
- 200 µM dATP, dCTP, dGTP
- 100 µM dTTP
- 100 µM biotin-16 dUTP
- 2.0 µM primer (DOP)
- 2.5 mM $MgCl_2$

Note: Biotin-16 dUTP may be replaced either by digoxigenin 11-dUTP or other directly labelled dUTPs for multicolor FISH with different microdissection libraries.

PCR cycles

Perform 20 cycles of

- 92°C for 1 min
- 56°C for 1 min
- 72°C for 2 min

Purification of labelled probe

1. Precipitate the PCR products by adding 100 µg E. coli tRNA, 1/10 volume of 3 M sodium acetate pH 5.6 and 2 volumes of 100% ethanol.
2. Mix well and incubate at – 20°C for 2 h (or at – 70°C for 30 min).
3. Centrifuge at 12,000 rpm for 15 min at 4°C.
4. Discard the supernatant and lyophilize the pellet.

5. Add 30 μl hybridization mixture and resuspend the DNA thoroughly with the pipette. The DNA amount obtained after PCR labelling is used for five FISH experiments (hybridization area: 24 x 24 mm). Store at – 20°C.

Reverse chromosome painting

Standard protocols for *in situ* hybridization and probe detection may be used as described in the previous Chapters Chapter 18 and Chapter 19. The detection of a biotinylated probe with FITC conjugated Avidin is described below. Usually one layer of Avidin-FITC should be sufficient for obtaining good painting signals.

Slide preparation

For reverse painting use slides with normal, well spread metaphases that have been prepared at least one week before.

1. Bake slides at 65°C for 2 h.
2. Denature the chromosomal DNA in 70% formamide, 2x SSC, pH 7.0 for 3 min.
3. Dehydrate the slides by passing through 70% ethanol (ice-cold), 90% and 100% ethanol for 3 min each.
4. Air dry the slides.

Probe denaturation and hybridization

1. Lyophilize 25 μg Cot-1 DNA
2. Add 6 μl hybridization mixture and 6 μl of the labelled probe DNA to the dried Cot-1 DNA and mix well.
3. Denature the probe at 75°C for 5 min.
4. Chill the denatured probe/Cot-1 mixture on ice.
5. Incubate at 37°C for 30 min to pre-anneal repetitive sequences within the probe.
6. Place the probe (12 μl) on one half of the slide cover with a 24 x 24 coverslip and seal with rubber cement.
7. Hybridize overnight in a moist chamber at 37°C.

Probe detection

1. Wash the slide:
   - 3 x 5 min at 42°C with 2x SSC, 50% formamide, pH 7.0
   - 3 x 5 min at 42°C with 2x SSC, pH 7.0

- short with 4xSSC, 0.05% Tween 20, pH 7.0 (SSCT) at room temperature.

2. Incubate for 15 min at 37°C with SSCT containing 5% non fat dried milk (Marvel) (SSCTM).
3. Wash slides briefly with SSCT.
4. Incubate with FITC-conjugated Avidin diluted 1:500 in SSCTM for 30 min at 37°C.
5. Wash the slides 1 x 3 min with SSCT and 2 x 3 min with PBS.
6. Dehydrate the slides in an ethanol series (70%, 90%, 100%).
7. Mount with antifade solution (e. g. Vectashield) containing DAPI or Propidiumiodide as a counterstain.
8. Analyze the slides with an epifluorescence microscope equipped with appropriate filters.

Results

Two examples of reverse chromosome painting with microdissection probes are given in Figure 5. The first example (Figure 5a) shows the result obtained with 5 copies of a small marker chromosome that was detected during prenatal diagnosis. Although two separated painting signals on chromosome 9 in bands p12 and q21.1 are visible the interpretation of this result is not, that both bands are fused together and part of the marker-chromosome. As we know from chromomosome 9 arm-specific microdissection libraries these two bands are more or less identical. Probably the identity of sequences within these two bands is the reason for the relatively frequent finding of pericentric inversions of chromosome 9 where the breakpoints are in most cases at these two bands.

In diagnostic cases (especially during prenatal diagnosis) where such marker chromosomes or translocated bands of unknown origin have to be identified it is highly recommended to prepare always two independent microdissection libraries. Only if both libraries show the same result one can be sure that all signals are truly derived from the dissected chromosome and not from another piece of chromatin that has been mistakenly collected.

The second example (Figure 5b) shows the result of a small interstitial deletion within chromosome 15. During routine cytogenetic analysis of

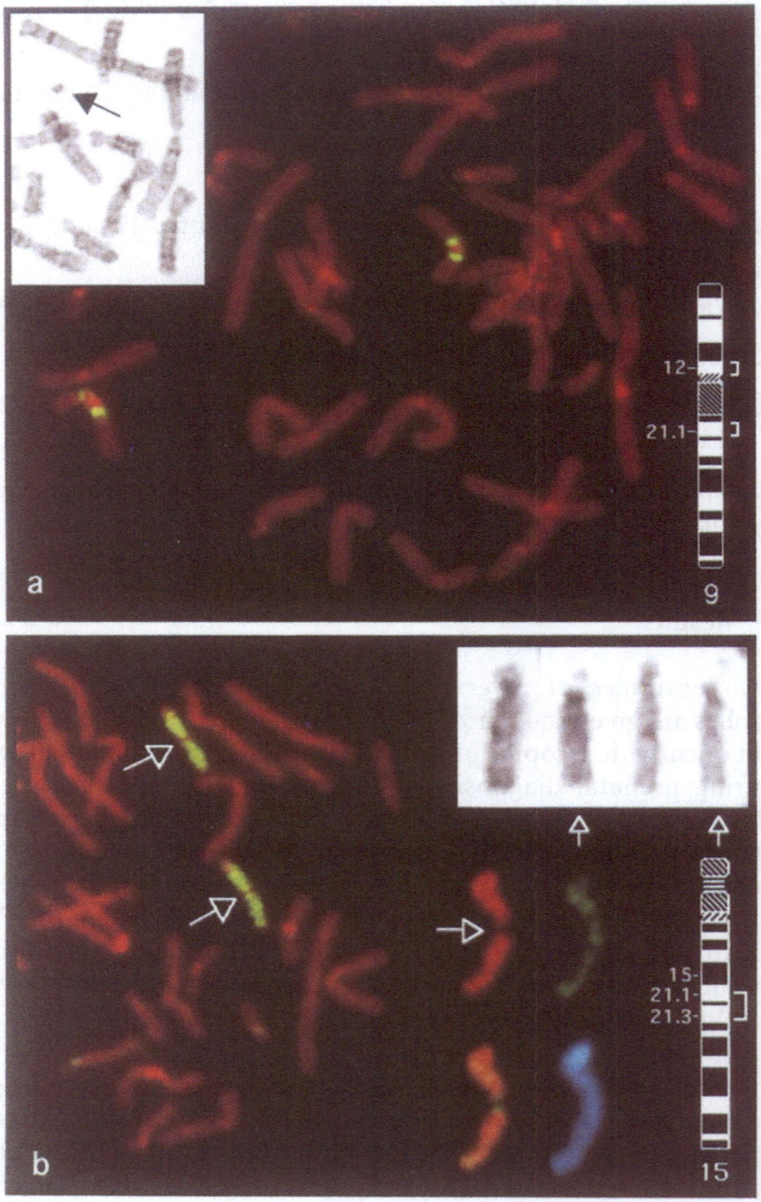

Fig. 5. Examples of reverse chromosome painting. a) Identification of a small marker chromosome (see inset with a GTG-banded partial metaphase) in a prenatal case. Reverse painting with 5 microdissected copies of the marker on normal metaphases shows signals covering bands 9p12 and 9q21.1. b) A suspected deletion in chromosome 15 was proved and precisely defined by reverse painting. Two pairs of GTG-banded chromosomes 15 are shown with the deleted chromosome indicated by arrows. In the painting signal from 5 dissected chromosomes a small gap at the position of the deletion is visible (arrowed) on normal chromosomes 15. Four images of one enlarged chromosome 15 show the painting signal in red, a replication G-banding pattern in green, the merged red and green images and the DAPI counterstain in blue. The deleted region is indicated at the idiogram of chromosome 15.

banded chromosomes this deletion was suspected but is was not possible to define the deleted region from the GTG-banding pattern. After reverse painting the precise position of the deletion del(15)(q21.1q21.3) was clearly visible. In most cases the banding pattern obtained with DAPI is sufficient for the identification of breakpoints. As shown in this figure a very good replication G- banding pattern can also be achieved by incorporating bromodeoxyuridine in the late replicating regions of the chromosomal DNA. After probe detection (here in red) this banding pattern (in green) is acquired with an anti-BrdU antibody coupled to FITC (Senger 1993). It is evident that a small deletion would not be visible if the painting signal is weak or the background high. For the identification of deletions it is therefore better to isolate, if possible, at least 5 chromosomes in order to obtain signals that are strong enough to see the gap clearly for defining the breakpoints.

Comments to technical details

- Proteinase K treatment
  The treatment of the microdissected fragments with proteinase K is recommended as it improves the efficiency of DNA amplification substantially, which is apparent in a more evenly stained and more intense painting signal. This is especially important if small deletions or translocation breakpoints have to be analyzed precisely. Here it is necessary to have a clear-cut border between a bright signal and the unstained region instead of a gradual transition between signal and background. The addition of fresh proteinase K solution after all fragments have been collected (as performed earlier when the collection drop was placed in a moist chamber) can, however, be omitted.

- The use of sequenase
  The DNA amount obtained after dissection of 3-5 chromosomes is usually less than 1 picogram (eg one average sized chromosome contains about 260 femtogram DNA). Therefore a highly efficient PCR protocol must be used. In our hands the use of T7 DNA polymerase (Sequenase) during the first eight low-temperature cycles and Stoffel fragment Taq polymerase during the high temperature cycles has proved to give far better and more reliable results compared to other protocols where various kinds of Taq polymerases are used in all cycles (eg Viersbach 1994, Müller-Navia 1995, Yokoyama and Sakuragawa 1995). Although it is fairly laborious to add Sequenase eight times to the reaction mixture the more or less 100% success rate with high quality background-free

signals justifies this procedure. Regarding contamination with foreign DNA it might seem to be rather unsafe to open the tube several times but if this is done in a sterile, DNA-free area (under a laminar flow-hood that previously has been illuminated with UV-light) this risk is neglectably small.

- Quality control by running an agarose gel
  After initial DNA amplification of the microdissected fragments the yield and average size of DNA fragments may be checked by loading 5 μl on a standard 1 – 2% agarose gel along with a suitable size marker. The PCR products should be visible as a smear usually with a peak intensity at 200-400 nucleotides. One should, however, not be too much alarmed if some PCR products are present even in the negative control without any microdissected DNA. Many enzymes used for DNA amplification are contaminated with more or less significant amounts of bacterial DNA. This is the reason why during the first cycles with low annealing temperature, the reaction volume should be as small as possible (5 μl in our protocol). Thus the amount of enzyme necessary for this unspecific DNA amplification can be reduced to a minimum.
  The only information acquired by agarose gel electrophoresis is whether the PCR has worked successfully. In this respect an agarose gel is only useful as long as the PCR conditions are still not sufficiently optimized. As soon as amplification of any DNA can be detected the quality of a library should be further analyzed by FISH.

Troubleshooting

- The quality of chromosome banding is not sufficient to identify chromosomes. One should not expect the same high quality GTG-banding as obtained after ageing of metaphases. Nevertheless, a reasonably good banding can still be obtained with fresh metaphase spreads if these are well spread and free of cytoplasm (see Figure 3). The main reason for poor banding is a large amount of cytoplasm in which the chromosomes are embedded.
  - Try to reduce cytoplasm by adding two or three drops of fresh fixative (3 parts methanol and 1 part acetic acid) on the still moist metaphase spreads before complete evaporation of the first fixative.
  - Vary the time of treatment with trypsin to find optimal conditions.
- As soon as the needle touches the excised fragment it jumps away. This is due to electrostatic charge.

- If you are still wearing gloves, remove them.
- Use another needle and, if possible another coverslip with metaphase spreads.

- It is not possible to reach the collection drop inside of the pipette without touching the outer surface of the pipette where the fragment gets lost.
  - The margin of the opening must not be visible as a closed circle (remember that the view at an inverted microscope is from below).
  - Turn the pipette so that the opening is visible as shown in Figure 4.
  - Focus in the middle between the front and the rear margin. Thus it should be possible to enter the inner side of the pipette with the needle. Always try this first before you start with microdissection to make sure that the orientation of the pipette is correct.

- The agarose gel does not show any PCR products.
  - Check the PCR conditions first with a small amount (1 – 10 pg) of genomic DNA.
  - If DNA amplification is successful in this control experiment start with the dissection of large chromosome fragments. Make sure that you can see the transfer of each chromosome fragment to the collection buffer. Start with a minimum of 5 identical fragments.

- There is DNA visible on the agarose gel but after FISH only background is obtained but no specific signals.
  - Exclude any possible contamination of solutions and microinstruments with DNA or nucleases, eg autoclave and use only microreaction tubes from an originally sealed bag to make sure that they have not been handled without gloves. For PCR and collection buffer use only re-distilled sterile water that is commercially available in small ampules. Once again, to reduce the risk of contamination **aliquot every solution** (except enzymes) used for microdissection and PCR. Do this under the laminar flow-hood that has been UV-illuminated.
  - Check all solutions for *in situ* hybridization (especially the hybridization mixture) with a commercially available probe (preferably a painting or single copy probe).

References

Chudoba I, Rubtsov N, Senger G, Junker K, Bleck C, Claussen U (1996) Improved detection of chromosome 16 rearrangements in acute myeloid leukemias using 16p and 16q specific microdissection libraries. Oncol Rep 3:829-832

Lüdecke H-J, Senger G, Claussen U, Horsthemke B (1989) Cloning defined regions of the human genome by microdissection of banded chromosomes and enzymatic amplification. Nature 338:348-350
Lüdecke H-J, Senger G, Claussen U, Horsthemke B (1990) Construction and characterization of band-specific DNA libraries. Hum Genet 84:512-516
Meltzer PS, Guan X-Y, Burgess A, Trent JM (1992) Rapid generation of region specific probes by chromosome microdissection and their application. Nature Genet 1:24-28
Müller-Navia J, Nebel A, Schleiermacher E (1995) Complete and precise characterization of marker chromosomes by application of microdissection in prenatal diagnosis. Hum Genet 96:661-667
Rubtsov N, Senger G, Kuzcera H, Neumann A, Kelbova C, Junker K, Beensen V, Claussen U (1996) Interstitial deletion of chromosome 6q: precise definition of the breakpoints by microdissection, DNA amplification, and reverse painting. Hum Genet 97:705-709
Scalenghe F, Turco E, Edström JE, Pirrotta V, Melli M (1981) Microdissection and cloning of DNA from a specific region of Drosophila melanogaster polytene chromosomes. Chromosoma 82:205-216
Senger G, Lüdecke H-J, Horsthemke B, Claussen U (1990) Microdissection of banded human chromosomes. Hum Genet 84:507-511
Senger G, Ragoussis J, Trowsdale J, Sheer D (1993) Fine mapping of the MHC class II region within 6p21 and evaluation of probe ordering using interphase fluorescence in situ hybridization. Cytogenet Cell Genet 64:49-53
Telenius H, Carter NP, Bebb CE, Nordenskjöld M, Ponder BAJ, Tunnacliff A (1992) Degenerate oligonucleotide-primed PCR: general amplification of target DNA by a single degenerate primer. Genomics 13:718-725
Trautmann U, Leuteritz G, Senger G, Claussen U, Ballhausen WG (1991) Detection of APC region-specific signals by nonisotopic chromosomal in situ suppression (CISS)-hybridization using a microdissection library as a probe. Hum Genet 87: 495-497
Viersbach R, Schwanitz G, Nöthen MM (1994) Delineation of marker chromosomes by reverse chromosome painting using only a small number of DOP-PCR amplified microdissected chromosomes. Hum Genet 93:663-667
Yokoyama Y, Sakuragawa N (1995) Improved simple generation of GTG-band specific painting probes. Cytogenet Cell Genet 71:32-36

Suppliers

AMERSHAM BUCHLER GMBH & CO. KG, Gieselweg 1, Braunschweig, 38110, Germany

BIOZYM DIAGNOSTIK GMBH, Postfach, Hess. Oldendorf, 31833, Germany (*phone* +49(0)5152-2075; *fax* +49(0)5152-2070

BOEHRINGER MANNHEIM GMBH, Sandhofer Straße 116, Mannheim, 68305, Germany (*phone* +49 (0)621-7590; *fax* FAX: +49 (0)621-7598509

GIBCO BRL, LIFE TECHNOLOGIES GMBH, Postfach 1212, Eggenstein, 76339, Germany (*phone* +49(0)721-780444; *fax* +49(0)721-780499

DIFCO LABORATORIES GMBH, Ulmer Straße 160a, Augsburg, 86156, Germany (*phone* +49 (0)821-443391; *fax* +49 (0)821-443891

INTERNATIONAL BIOTECHNOLOGIES INC, IBI, P.O. Box 9558, 25 Science Park, New HavenCT, 06535, USA (*phone* +1-800-243-2555

MERCK, Postfach 4119, Frankfurter Straße 250, Darmstadt 1, 64271, Germany (*phone* +49 (0)6151-720; *fax* +49 (0)6151-722000

MJ RESEARCH INC., 149 Grove Street, WatertownMassachusetts, 02172, USA (*phone* +1-617-923-8000; *fax* +1-617-923-8080; *e-mail* tech@mjr.com

NARISHIGE, UNIT 7, Willow Business Park, Willow Way, London, SE26 4QP, UK (*phone* +44(0)181-699-9696; *fax* +44 (0)181-291-9678

PERKIN-ELMER GMBH, Alte Nußdorfer Straße, Überlingen, 88662, Germany (*phone* +49 (0)7551-810; *fax* +49 (0)7551-1612

PHARMACIA BIOSYSTEMS GMBH, Munzinger Straße 9, Freiburg, 79111, Germany (*phone* +49 (0)761-49030; *fax* +49 (0)761-4903246

SIGMA CHEMIE GMBH, Grünwalder Weg 30, Deisenhofen, 82039, Germany (*phone* +49 (0)89-613010; *fax* +49 (0)89-6135135; *e-mail*

CARL ZEISS JENA GMBH, Department of Microscopy, Jena, 07740, Germany (*phone* +49 (0)3641-642936; *fax* +49 (0)3641-643144; *e-mail* mikro@zeiss.de

VECTOR LABORATORIES, BOEHRINGER INGELHEIM PRODUCTS, Carl-Benz Str. 7, Heidelberg, 69042, Germany (*phone* +49(0)6221-5020; *fax* +49(0)6221-502113

Chapter 21

Comparative Genomic Hybridisation (CGH)

TRAUDL HENN and OSKAR A. HAAS

Introduction

CGH is a fluorescence in situ hybridisation (FISH) technique that allows the genome-wide assessment of changes in the relative copy number of chromosome regions in tissues that are not accessible to conventional cytogenetic analysis (Kallioniemi et al., 1994). Test-DNA and reference DNA from a normal individual are differentially labelled, mixed together in equal amounts and co-hybridised under conditions of in situ suppression hybridisation. With this new technique chromosomal gains and losses as well as amplifications are mapped along normal metaphase chromosomes in a single hybridisation step (Figure 1, Figure 2). It has proven to be particularly useful for the evaluation of genetic changes associated with solid tumours and haematological neoplasms as well as to clarify unbalanced abnormalities in clinical cytogenetics. The disadvantages of this method are that only those clonal changes show up that are present in a substantial proportion of tumour cells and that balanced structural rearrangements are not detectable.

To perform CGH one basically needs:

- metaphase preparations from a normal individual
- test DNA extracted from the tissue of interest
- reference DNA from a normal individual
- reagents and laboratory equipment for FISH analysis

Traudl Henn, St. Anna Children's Hospital, Children's Cancer Research Institute (CCRI), Kinderspitalgasse 6, Vienna, 1090, Austria, *Correspondence to* Oskar A. Haas, St. Anna Children's Hospital, Children's Cancer Research Institute (CCRI), Kinderspitalgasse 6, Vienna, 1090, Austria (*phone* +43-1-40170-480; *fax* +43-1-40170-481 or -430; *e-mail* o.a.haas@magnet.at)

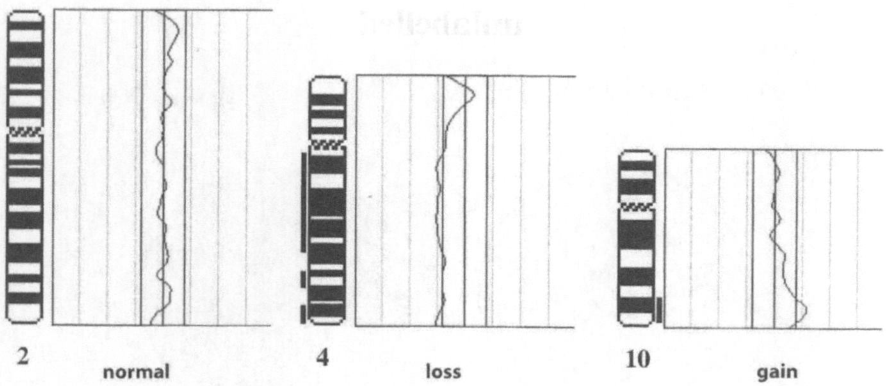

Fig. 1. Copy number changes in the tissue of interest are elicited and mapped by measuring the tumor/normal fluorescence intensity ratios along the target metaphase chromosomes with an image analysis system and a dedicated software. The center line indicates the ratio value for a balanced state of chromosomal material (chromosome 2 on the left). The right and left black lines depict upper and lower 50% tresholds. A low green to red ratio indicates a DNA loss (loss of the long arm of chromosome 4 in the middle) and a high green to red ratio indicates a DNA gain (chromosomal region 10(q25-26) on the right). Dark bars to the left and right of the respective chromosomes denote the lost and gained regions, respectively.

- a dedicated image analysis system for evaluation of the preparations (Du Manoir et al. 1995a, Du Manoir et al. 1995b, Piper et al. 1995)

Insufficient amounts of test DNA can be amplified with the degenerated oligonucleotide primed polymerase-chain-reaction (DOP-PCR) technique (Telenius et al. 1992, Joos et al. 1996).

Fig. 2. Schematic representation of the CGH method. Whole genomic DNA is isolated from the tissue of interest (e.g. tumor) and from normal reference tissue and differentially labelled. This is achieved either directly by incorporation of fluorochrome-conjugated nucleotides (tetramethyl-rhodamine-6-dUTP and fluorescein-12-dUTP) or indirectly by incorporation of modified nucleotides (biotin-21-dUTP and Digoxigenin-11-dUTP) that are detected with fluorochrome-conjugated secondary reagents (fluorescein-avidin-D and anti-dixogen-in-rhodamine). Identical amounts of tumor DNA (labelled with a green fluorochrome) and reference DNA (labelled with a red fluorochrome) are mixed and, together with unlabelled Cot-1 DNA (to prevent binding of labelled DNA to repetitive elements) hybridised onto normal metaphase chromosomes. All regions that contain equal amounts of tumor and reference DNA (A) are indicated by an equal red and green staining intensity (yellow-orange colour). Regions with gains in the tumor DNA (B & D) show a shift towards green, whereas those with losses show a shift towards red (C & D).

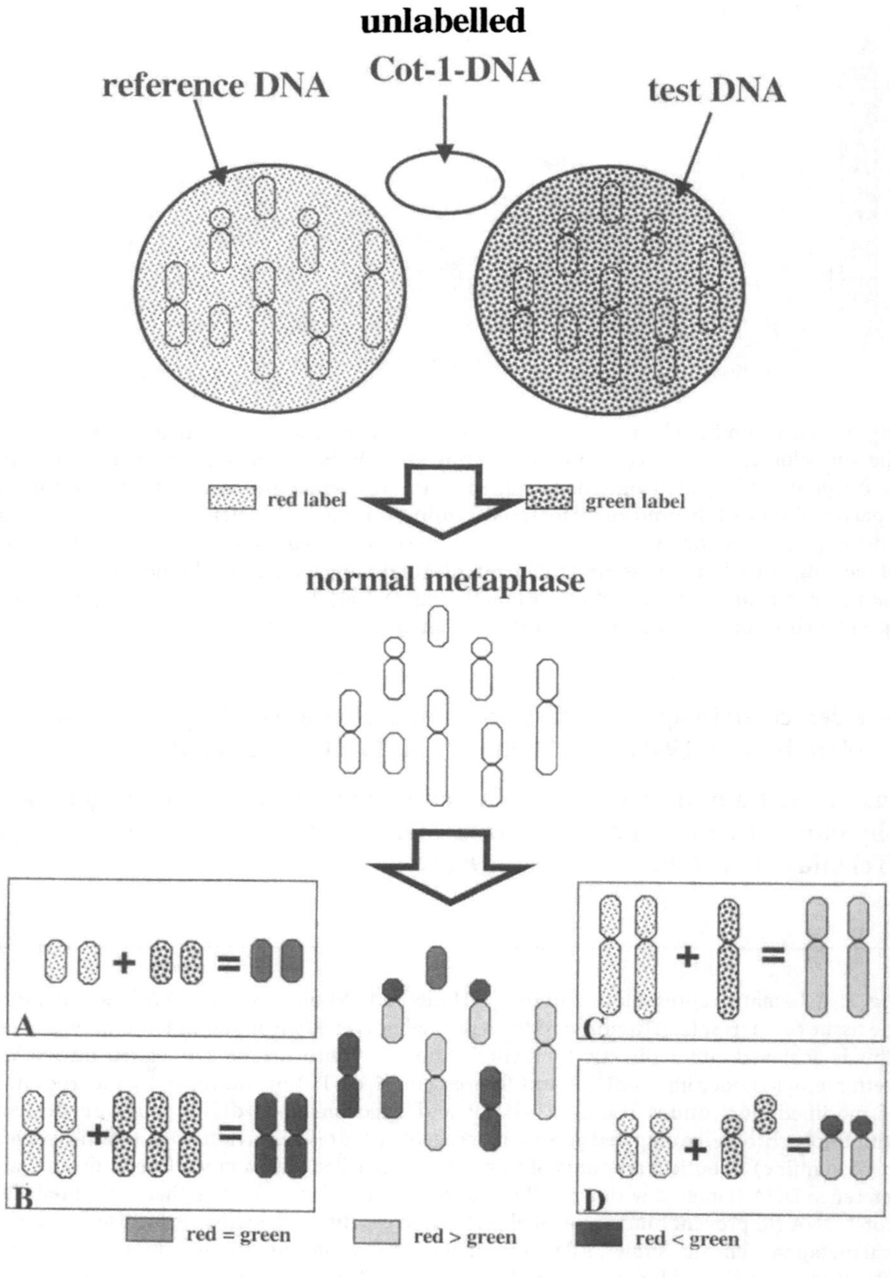
unlabelled
Cot-1-DNA
reference DNA
test DNA
red label
green label
normal metaphase
A
B
C
D
red = green
red > green
red < green

Materials

Equipment

- centrifuge
- microcentrifuge
- CO_2 incubator
- waterbath up to 80°C
- heating plate
- thermal cycler
- system for electrophoresis
- light microscope
- fluorescence microscope with adequate filters
- image analysis system (Applied Imaging, Metasystems, Perceptive Systems International, Vysis)

Utensils

- slides
- coverslips 24 mm x 24 mm, 24 mm x 50 mm and 24 mm x 60 mm
- Parafilm-M (Merck, Germany #3-1012)
- rubber cement (Marabu, Germany #2901 17000)

Cell culture and preparation of normal metaphases

Media and solutions

- Heparin (preservative-free) (Seromed, Germany, 5000 U/ml #L6510)
- Ficoll (Nycomed, Norway, #1001969)
- PBS: (137 mM NaCl/2,7 mM KCl/8,2 mM Na_2HPO_4/1,5 mM KH_2PO_4)
  - 40 g NaCl
  - 1 g KCl
  - 5,8 g Na_2HPO_4
  - 1 g KH_2PO_4
  - dissolve in 4 l demineralised H_2O, adjust to pH 7,2
  - demineralised H_2O up to 5 l
  - autoclave, store at room temperature

- culture medium: (RPMI 1640 with 10% FCS, 2 mM L-glutamine and 4 µg/ ml PHA-M)
  - 90 ml RPMI 1640 (BRL, USA, #31870-074)
  - 10 ml FCS (BRL, USA, #10106-151)
  - 1 ml 200 mM L-glutamine (BRL, USA, #25030-032)
  - 0,2 ml 2 mg/ml PHA-M (BRL, USA, #105760-510)
  - adjust to pH 7,4 with 1N HCl (salmon pink colour)

Note: warm up to 37°C before use, at 4°C storable up to two weeks

- ethidiumbromide (etbr):(1 mg/ml etbr in RPMI 1640)
  - 10 mg etbr (Sigma, USA, #E-8751)
  - 10 ml RPMI 1640
- Colcemide (10 µg/ml) (BRL, USA, #15210-016)
- hypotonic solution: (60 mM KCl/0,04% NaCit)
  - 38 ml 4,73 g/l KCl-stock solution (Merck, Germany, #4936)
  - 2 ml 0,8% NaCit (Merck, Germany, #1.06448)
  - adjust to pH 7,4 and prewarm to 37°C before use

Note: produces straight chromosomes and excellent spreading, but requires gentle centrifugation of cells.

- fixative 1: (3 parts methanol + 1 part acetic acid)
  - 75 ml methanol (-20°C) (Merck, Germany, #1.06009)
  - 25 ml acetic acid (Merck, Germany, #1.00063)

Note: Prepare fresh immediately before use!

- fixative 2: (5 parts methanol + 2 parts acetic acid)
  - 75 ml methanol (-20°C)
  - 30 ml acetic acid

Note: Prepare fresh immediately before use!

- Giemsa (5%)
  - 3,5 ml Giemsa (Merck, Germany, #1.09204)
  - 66,5 ml fresh water

Note: Prepare fresh before use!

- ethanol (70%) (Merck, Germany, #818761)
- silica gel (Merck, Germany, #1.1925)

Pre-treatment of slides

Stock solutions

- HCl-ethanol (3%)
  - 15 ml HCl (Merck, Germany #1.00317)
  - 485 ml ethanol
- HCl-ethanol (20%)
  - 100 ml HCl
  - 400 ml ethanol
- 20xSSC: (3M NaCl/0,3M NaCit)
  - 175,3 g NaCl
  - 88,2 g NaCit
  - adjust to pH 7
  - demineralised H_2O up to 1000 ml
  - autoclave
- RNase-stock solution (10 mg/ml)
  - 10 mg RNase A (Roche, Germany #109 169)
  - 1 ml demineralised H_2O
  - incubate for 10 min at 95°C to eliminate DNases, store aliquots at -20°C
- Pepsin-stock solution (10%)
  - 100 mg Pepsin (Sigma, USA, #P-6887)
  - 1 ml demineralised H_2O
  - store aliquots at -20°C

Working solutions

- RNase A per slide: (0,1 mg/ml RNase/2xSSC)
  - 100 µl 2xSSC
  - 1 µl RNase-stock solution
- Pepsin-working solution (PEPS): (10 pg/ml pepsin/0,01N HCl)
  - 75 ml demineralised H_2O
  - 0,75 ml 1N HCl
  - 7,5 µl pepsin-stock solution

Note: Warm up H_2O in coplin jar without pepsin to 37°C, add the enzyme immediately before use, because its activity deteriorates within minutes!

- denaturing solution per slide: (70% formamide/1xSSC, pH 7)
  - 70 µl deionised formamide (Merck, Germany, #1.09684)
  - 5 µl 20xSSC
  - 23 µl demineralised H_2O
  - 2 µl 1N HCl

- denaturing solution for coplin jar: (70% formamide/1xSSC, pH 7)
  - 49 ml formamide
  - 3,5 ml 20xSSC
  - 17,5 ml demineralised H_2O
  - adjust to pH 7 and heat up slowly to 72°C

DNA extraction

Solutions

- Sarcosyl (20%) (Sigma, Germany #L-9150)
- 2xDNA-EB: (40 mM Tris HCl/40 mM EDTA/2% Sarcosyl)
  - 4 ml 1M Tris HCl pH 7,6
  - 8 ml 0,5M EDTA pH 8
  - 10 ml 20% Sarcosyl
  - demineralised H_2O up to 100 ml
  - filter sterilise, store at RT
- Proteinase K (PK) (10 mg/ml)
  - 100 mg Proteinase K (Roche, Germany #754 723)
  - 10 ml demineralised H_2O
  - filter sterilise, store aliquots at -20°C

Note: Always prepare fresh

- buffer saturated phenol (BRL, USA #5513 UA/UB)
- Ph/$CHCl_3$/IA
  - 50 ml buffer saturated Phenol
  - 49 ml chloroform (Merck, Germany #1.02445)
  - 1 ml isoamylalcohol (Merck, Germany #1.00977)
- $CHCl_3$/IA
  - 49 ml chloroform
  - 1 ml isoamylalcohol

Nick translation

Solutions

- 10xNT: (500 mM Tris, 50 mM $MgCl_2$, 0,5 mg/ml BSA)
  - 5 ml 1 M Tris HCl, pH 7,8 (Amresco, USA #77-86-1)
  - 0,5 ml 1 M $MgCl_2$ (Merck, Germany #5833)
  - 5 mg bovine serum albumin (BSA) (Sigma, Germany #A-3350)

- demineralised H_2O up to 10 ml
- store aliquots at -20°C

- 10xMOH (0,1 M)
  - 69,5 µl 14,4 M ß-mercapto-ethanol (Sigma, Germany #M-3148)
  - demineralised H_2O up to 10 ml
  - store aliquots at -20°C
- 10x NT-bio-dNTP's: (0,5 mM dATP, 0,5 mM dCTP, 0,5 mM dGTP, 0,5 mM bio-21-dUTP)
  - 2,5 µl 100 mM dATP
  - 2,5 µl 100 mM dCTP
  - 2,5 µl 100 mM dGTP
  - 25 µl 10 mM biotin-21-dUTP (Clontech, USA #5021-3)
  - 467,5 µl demineralised H_2O
  - store aliquots at -20°C
- 10x NT-dig-, -FITC-, or -TRITC-dNTP's: (0,5 mM dATP, 0,5 mM dCTP, 0,5 mM dGTP, 0,375 mM dTTP, 0,125 mM labelled dUTP)
  - 1 µl 100 mM dATP
  - 1 µl 100 mM dCTP
  - 1 µl 100 mM dGTP
  - 0,75 µl 100 mM dTTP
  - 25 µl 1 mM dig-11-, or FITC-12-, or TRITC-6-dUTP
  - 171,25 µl demineralised H_2O
  - store aliquots at -20°C
- DNaseI-stock solution: (3 mg/ml DNaseI/150 mM NaCl/50% glycerol)
  - 6 mg DNase I (Roche, Germany #104 159)
  - 60 µl 5M NaCl
  - 1000 µl glycerol (Merck, Germany #4094)
  - 940 µl demineralised H_2O
  - store aliquots at -20°C
- DNase I-solution
  - 1 µl DNase-stock solution
  - 200 µl ice-cold water

Note: Prepare immediately before use!

- EDTA, (0,5M)
  - 18,6 g Titriplex III (Merck, Germany #1.8418)
  - 80 ml demineralised H_2O

- adjust to pH 8, otherwise Titriplex will not dissolve
- demineralised H_2O up to 100 ml

- SDS (10%)
  - 1 g sodium-dodecyl-sulfate (Sigma, Germany #L-4390)
  - 10 ml demineralised H_2O

Note: Very toxic when aspirated, wear mask!

- DNA-polymerase I (DNA-Pol I) (BRL, USA #18010-025): 10 U/µl

DOP-PCR

- 2 U/µl Taq-polymerase (Dynazyme) (Finnzymes, Finland #F501 L)
- 10xPCR-buffer (included to Dynazyme)
- 10xPCR-dNTP's (Promega, Germany #U1240): (2 mM dATP, 2 mM dCTP, 2 mM dGTP, 2 mM dTTP)
  - 20 µl 100 mM dATP
  - 20 µl 100 mM dCTP
  - 20 µl 100 mM dGTP
  - 20 µl 100 mM dTTP
  - 920 µl demineralised H_2O
- 10xPCR dNTP's with biotin label: (2 mM dATP, 2 mM dCTP, 2 mM dGTP, 2 mM bio-21-dUTP)
  - 10 µl 100 mM dATP
  - 10 µl 100 mM dCTP
  - 10 µl 100 mM dGTP
  - 100 µl 10 mM bio-21-dUTP (Clontech, USA #5021-3)
  - 370 µl demineralised H_2O
- 10xPCR dNTP's with dig, FITC, or TRITC label: (2 mM dATP, 2 mM dCTP, 2 mM dGTP, 1,4 mM dTTP, 0,6 mM labelled dUTP)
  - 2 µl 100 mM dATP
  - 2 µl 100 mM dCTP
  - 2 µl 100 mM dGTP
  - 1,4 µl 100 mM dTTP
  - 60 µl 1 mM dig-11-, or FITC-12-, or TRITC-6-dUTP
  - 32,6 µl demineralised H_2O

- labelled nucleotides
  - Digoxigenin-11-d-UTP (1 mM) (Roche, Germany #1558 706)
  - FITC-12-d-UTP (1 mM) (Roche, Germany #1373 242)
  - TRITC-6-d-UTP (1 mM) (Roche, Germany #1534 378)
- 10xDOP-primer
  - 20 μM 5'- ccg act cga gnn nnn nat gtg g -3'
- mineral oil (Sigma, Germany #M-3516)

Control incorporation-efficiency of labelled nucleotides with alkaline phosphatase

- AP-buffer-1: (100 mM Tris HCl (pH 7,5)/100 mM NaCl/2 mM $MgCl_2$/ 0,05% Triton X-100)
  - 25 ml 1M Tris
  - 5 ml 5M NaCl
  - 0,5 ml 1M $MgCl_2$
  - 0,125 ml Triton
  - demineralised H_2O up to 250 ml
- AP-block: (3% BSA/AP-buffer-1)
  - dissolve 150 mg BSA at 45°C in 5 ml AP-buffer-1
- AP-buffer-2: (100 mM Tris HCl (pH 9,5)/100 mM NaCl/50 mM $MgCl_2$)
  - 10 ml 1M Tris
  - 2 ml 5M NaCl
  - 5 ml 1M $MgCl_2$
  - demineralised H_2O up to 100 ml
- bio-labelled control DNA (Promega, USA #Z5261)
- dig-labelled control DNA (Roche, Germany #1585 738)
- AP-conjugate for 10 cm^2 filter:
  - 1 μl anti-dig-AP-conjugate (Roche, Germany #1093 274)
  - 5 μl streptavidin-AP-conjugate 0,4 mg/ml (Clontech, USA #1113-1)
  - 5 ml AP-buffer-1
- colour-reagent: (4,4 μl/ml NBT; 3,3 μl/ml BCIP)
  - 22 μl NBT (Roche, Germany #1383 213)
  - 16,6 μl BCIP (Roche, Germany #1383 221)
  - 5 ml AP-buffer-3

- Triton® X-100 (Merck, Germany #12298)
- nylon membrane (Schleicher & Schuell, Germany #278 8501)

Pre-treatment of probes

- cot-I DNA (1 mg/ml) (BRL, USA #15279-011)
- 3M NaAc
  - 40,8 g NaAc (Merck, Germany #1539)
  - 100 ml demineralised H_2O
  - adjust to pH 7
- SSC-Dex: (20% dextransulfate/4xSSC)
  - 2 g dextransulfate (Pharmacia, Belgium #17-0340-02)
  - 2 ml 20xSSC
  - demineralised H_2O up to 10 ml
  - filter sterilise, store at 4°C
- Hybridisation solution per probe: (50% deionised formamide/10% dextransulfate/2xSSC)
  - 6 µl formamide
  - 6 µl SSC-Dex

Solutions for washing and detection

- 4xSSC/Tween 20 (SSC-T): (4xSSC/1% Tween 20)
  - 100 ml 20xSSC
  - 400 ml demineralised H_2O
  - 0,5 ml Tween 20 (Pharmacia, Belgium #17-1316-01)
- fluorescent-solution per slide
  - 100 µl SSC-T
  - 1 µl fluorescein-avidin-D (Vector, USA #A-2011)
  - 0,5 µl anti-Dig-rhodamine (Roche, Germany #1207 750)
- DAPI-counter-stain: (80 ng/ml DAPI/2xSSC)
  - 18 µl of 0,2 mg/ml DAPI (Serva, Germany #18860)
  - 45 ml 2xSSC

Note: Usable for months, if kept at 4°C in a light-proof coplin jar!

- mounting medium (Vectashield, Vector, USA #H1000)

Procedure

Preparation of normal metaphases

Cell source

0,1 ml preservative-free heparin (Seromed, Germany, 5000 U/ml #L6510)

10 ml peripheral blood of a normal individual

Note: Immediately invert syringe several times to avoid coagulation!

Note: Work without interruption; the quicker you are, the better the results. Try to get your cultures into the incubator within one hour!

Separation of mononuclear cells

1. Overlay 3 ml Ficoll with maximum 7 ml heparinised peripheral blood and centrifuge for 15 min at 2500 rpm.
2. Pipette ring of mononuclear cells (MNC) (maximum 4 ml) into a fresh tube.
3. Immediately fill up the tube with PBS and centrifuge again for 8 min at 1200 rpm.
4. Repeat step 4, take off supernatant leaving 0,5-1 ml PBS, resuspend and count MNC.
5. Instead of the Ficoll-separation of MNC, you can centrifuge heparinised peripheral blood for 8 min at 1200 rpm, take off the buffy-coat, fill up the tube with PBS, centrifuge again for 8 min at 1200 rpm, take off the supernatant leaving 0,5-1 ml PBS, resuspend and count the MNC.

Culture of mononuclear cells

6. Use 3 ml culture medium per 12 ml flat-bottom centrifugation-tube. Add 1x 106 MNC/ml culture medium. Incubate at 37°C with 5% CO2 for 72 hours.

Note: Cell density at the beginning of culture is important. If cells are seeded too dense, mitotic cells are sparse and chromosomes small. If cells are proliferating well, medium turns orange.

Cleaning of slides

7. Soak glass slides over night in 3% HCl-ethanol (alternatively 30 min in 20% HCl-ethanol).
8. Rinse for 10 min in demineralised water and store the slides at 4°C in demineralised water.

Cell harvest

9. Optional: add 10µl etbr per ml culture 60 min prior to harvest and incubate at 37°C with 5% CO_2.

Note: Ethidiumbromide intercalates between basepairs of DNA and improves DAPI-banding

10. Add Colcemide (final concentration 60 ng/ml; corresponds to approximately 1 drop with a fine tip per 3 ml culture medium) 20 min prior to harvest and incubate at 37°C with 5% CO_2.
11. Spin down for 8 min at 1200 rpm.
12. Remove supernatant leaving at least 0,5 ml and dissolve pellet completely.
13. Add prewarmed hypotonic solution while shaking gently, the first ml slowly drop by drop, then fill up quickly. Incubate for 8 min at 37°C in an incubator or waterbath.

Fixation

14. Add 200 µl of concentrated acetic acid before centrifugation to stop reversible swelling of the cells and to lyse remaining red blood cells, centrifuge for 15 min at 800 rpm.

Note: Gentle centrifugation prevents bursting of the cells!

15. Remove supernatant carefully, leave approximately 0,5 ml and dissolve pellet gently.
16. Add first ml of ice-cold fixative drop-wise while shaking gently, then quickly add the rest and store for 10 min at 4°C. Centrifuge for 15 min at 800 rpm.
17. Fill up with fixative 2 and centrifuge again.
18. Take off supernatant and dissolve pellet in fixative 2 until the solution looks cloudy (approximately 1 to 1,5 ml).
19. Rinse off water from slide. With a 100 µl pipette place 1-2 small drops (each 10-15µl) of fixed cells from 1 cm distance onto cold, moist (not wet) slides. Hold slide horizontally and put it immediately on a rack into the steam over a waterbath preheated at 70°C.

20. Let cells spread and the fixative dissolve in the steam for about 30 seconds.

21. Check quantity and quality of metaphases on one slide by staining in 5% Giemsa for 5 min, rinse shortly in water and let dry.

Note: Metaphases should be well spread and without any residual cytoplasm; chromosomes should be straight and lack cross-overs.

22. Before the final preparation of slides fill up tube again with fixative 2, centrifuge and repeat steps 18 to 20.

23. Let the slides dry overnight at RT and store them up to one month in 70% ethanol at 4°C until further use. For a longer storage period, dehydrate the slides by passing them through a series of 70%, 80% and absolute ethanol (2 min each), seal dry slides together with silica gel in plastic bags and store them at -20°C. For further use put them out of the freezer directly into ice-cold absolute ethanol for 2 min and air-dry them.

Probe preparation

DNA extraction of native material

1. Wash cells (tumour, bone-marrow, MNC) twice in PBS.
2. Pick up about 106-5x10^6 cells in 500 µl PBS in a 2 ml centrifugation tube.
3. Add 500µl 2xDNA-EB and 20 µl PK quickly, immediately close the tube and invert several times very slowly until the viscous solution is mixed completely.
4. Incubate for 3 to 6 hs at 55°C or overnight at 37°C, invert occasionally.
5. For each sample prepare one tube with 700 µl Ph/$CHCl_3$/IA and one with 700 µl $CHCl_3$/IA.
6. Add 700 µl phenol to each DNA-solution, mix gently by inverting the tube several times and centrifuge for 5 min at 13 000 rpm.
7. Use a wide bore pipette or cut off the very tip of a blue Gilson pipette tip and pipette the upper DNA-containing phase into the Ph/$CHCl_3$/IA containing tube without transferring the interphase, rather leave some DNA.
8. Mix gently and centrifuge for 5 min at 13 000 rpm.
9. Pipette the upper DNA-containing phase into the tube with $CHCl_3$/IA, mix gently and centrifuge for 5 min at 13 000 rpm.

10. Pipette the upper DNA-0,5-containing phase into a new tube, eliminate remaining RNA by adding 5µl RNase-stock solution to the DNA-solution and incubate for 1 h at 37°C. To prevent precipitation of RNase dilute the enzyme 1:10 in demineralised H_2O before adding.
11. Add 700 µl Ph/$CHCl_3$/IA to each DNA-sample.
12. Repeat steps 8 and 9.
13. Pipette the upper DNA-containing phase into a new tube, measure the DNA-concentration in a photometer or check on a 0,8% agarose-gel.

Nick translation

1. Label normal control-DNA with tetramethyl-rhodamine or Digoxigenin and the probe-DNA with fluorescein or biotin.
2. For a 50µl reaction mix pipette the following into a tube, keep all reagents on ice!:
   - 1 µg DNA
   - 5 µl 10xNT
   - 5 µl 10xMOH
   - 5 µl 10x dNTP's (with labelled nucleotide)
   - 1 µl DNA-PolI
   - 1-2 µl DNase I-solution
   - demineralised H_2O up to 50 µl.
3. Incubate reaction mix for 90 min at 15°C.
4. Put reaction mix onto ice.
5. Check fragment size on a 1,2% agarose-gel. Optimal length is 300-1000 basepairs.

Note: Never use unchecked probes!

6. If fragments are too long, add DNase and DNA-Pol I and continue digestion for 30 min at 15°C. If fragments are too small, discard the sample. Start again with step 2 and add less DNase.
7. If fragments are in the correct size range, stop nick translation by adding 3 µl 0,5M EDTA and 1 µl 10% SDS and incubate 10 min at 65°C.
8. Store probes at -20°C.

DOP-PCR amplification (first round PCR)

Note: Keep all reagents on ice. To avoid contamination, preferably use Microman pipettes (Gilson, France #M10, M50 and M250, tips #CP10, CP50 and CP250).

1. Pipette the following into a PCR-reaction tube for a total volume of 25 µl:
   - 2,5 µl 10xPCR-buffer
   - 2,5 µl 10xPCR-dNTP's
   - 2,5 µl 10xDOP-primer
   - 0,5 µl Taq-polymerase (1U)
   - demineralised H_2O 12 µl.
2. Overlay the reaction-mixture with 30 µl mineral oil.
3. To avoid contamination pipette through the oil: 5 µl DNA (0,1-10 ng; preferably dissolved in water).
4. Prewarm the thermal cycler to 95°C and put the tubes from the ice into the thermal cycler.

Note: Always include a negative control containing all components but DNA.

5. PCR-conditions:
   - 1 cycle 8 min at 93°C
   - 5 cycles 1 min at 93°C, 1 min at 30°C, 5 min transition 30 to 72°C, 1 min at 72°C
   - 35 cycles 1 min at 93°C, 1 min at 56°C, 2 min at 72°C
   - 1 cycle 7 min at 72°C

DOP-PCR labelling

6. Pipette the following into a PCR-reaction tube for a total volume of 25 µl:
   - 2,5 µl 10xPCR-buffer
   - 2,5 µl 10xPCR labelled dNTP's
   - 2,5 µl 10xDOP-primer
   - 0,5 µl Taq-polymerase (1U)
   - demineralised H_2O 17 µl.
7. Overlay the reaction-mixture with 30 µl mineral oil.
8. pipette through the oil: 1 µl of the first round PCR.
9. Preheat the thermal cycler to 95°C and put the tubes from the ice into the thermal cycler.

Note: Include the negative control from the first round PCR.

10. PCR-conditions:
   - 1 cycle 8 min at 93°C
   - 25 cycles 1 min at 93°C, 1 min at 56°C, 2 min at 72°C
   - 1 cycle 7 min at 72°C

11. Check 5 µl of the reaction on 1,2% agarose-gel. Size range should be from 100 to 800 basepairs.

Note: Never use unchecked probes!

Check incorporation of labelled nucleotides

1. Label nylon membrane with pencil. Concentration of nick translated DNA should be 20 ng/µl. Dilute DNA 1/10 (2 ng/µl), 1/100 (200 pg/µl), 1/1000 (20 pg/µl) and put 1 µl drops onto a small piece of filter. Add the same dilutions of control DNA and one drop of 2 ng/µl unlabelled DNA as negative control.
2. Put the filter into a petridish and soak briefly in AP-buffer-1, incubate upside down in AP-block for 10 min at 37°C.
3. Seal the filter into a plastic bag together with 500 µl AP-conjugate per cm2 of the filter or fill a petridish up to 1 mm with AP-conjugate and put the filter upside down into the solution. Incubate 10 min at room temperature.
4. Wash in AP-buffer-1 2 x 3 min and in AP-buffer-2 3 x 3 min at RT with gentle agitation.
5. Fill 5 ml of the colour-reagent into a 10 cm petridish. Put the filter upside down into the petridish and keep it in the dark for 15-120 min at RT without shaking.
6. When the colour has developed appropriately, rinse the filter briefly in fresh water and air-dry it

Note: All dilutions of labelled DNA should give at least a faint signal. The negative control must be clear.

Probe pretreatment

1. Pipette the following into a microcentrifugation-tube and precipitate for 20 min at -20°C:
   - 20 µl labelled probe-DNA (400 ng)
   - 20 µl labelled control-DNA (400 ng)
   - 20 µl cot-I DNA (20 µg)
   - 6 µl 3M NaAc
   - 150 µl ethanol
2. Spin down for 25 min at 13 000 rpm.
3. Decant supernatant, wash once with 500 µl 70% ethanol.
4. Spin down again and take off the rest of the supernatant with a fine pipette-tip.

5. Dissolve each probe in 12 μl hybridisation solution, preferably using a thermomixer for 15 min at 45°C.
6. Denature for 5 min at 73°C.
7. Precompetition: Spin down briefly, incubate in the thermomixer for 20 min at 45°C to promote renaturing of highly repetitive sequences with cot-I DNA. Pre-treated probe can be stored at -20°C for months.

Hybridisation procedure

RNase and pepsin pre-treatment

Note: Always wash and incubate without agitation.

1. Equilibrate slides in 2xSSC for 1 min at RT
2. Apply 100μl RNase A solution, cover with a 24 mm x 60 mm piece of Parafilm and incubate for 30 min at 37°C.

Note: Parafilm instead of a glass-coverslips prevents scratching of the slide.

3. Put 2 min at RT in 2xSSC.
4. Incubate in PEPS for 3 min at 37°C.
5. Put 2 min at RT in 2xSSC.
6. Put slides into 70% ethanol.

Note: Treated slides can be stored in 70% ethanol for up to two weeks. For further use dehydrate in 80% and 100% ethanol for 2 min each and air-dry slides.

Denaturing

7. Apply 100 μl denaturing solution onto each slide and cover with a 24 mm x 60 mm coverslip. Put slide for 90 sec on a heating block at 72°C. Alternatively prewarm dry slides to 60°C and denature in a coplin jar 90 sec at 72°C.

Note: Verify the correct temperature by placing a clean thermometer directly into the coplin jar.

8. Dehydrate slides for 2 min each in precooled (-20°C) 70%, 80% and 100% ethanol and airdry the slides.

Note: Keep coplin jars with ethanol on ice!

Hybridisation

9. Prewarm slides and pre-treated probes to 45°C. Apply each probe onto a 24 mm x 24 mm coverslip and pick it up with the pre-treated slide.

10. Seal each coverslip generously with rubber cement.

11. Incubate for 3 days at 37°C in a moist chamber.

Wash

12. Remove the rubber cement carefully with forceps and put the slides into a coplin jar with 2xSSC at RT to rinse off the coverslip.

13. Wash in 1xSSC for 5 min at 72°C without agitation. For FITC and TRITC labelled probes continue with counterstaining (step 17).

Detection

14. Put bio- and dig- labelled probes in SSC-T for 2 min at RT.

15. Apply 100 μl fluorescent-solution, cover with a 24 mm x 60 mm piece of Parafilm and incubate for 10 min at 37°C.

16. Gently lift off the Parafilm of each slide and wash 3 x 2 min at RT in SSC-T.

Counterstaining

17. Use DAPI-counter-stain directly out of the refrigerator, incubate the slides 3 min.

18. Wash the slides by dipping each slide three times into fresh water, air-dry or blow-dry gently.

19. Immediately after slides are dried, pipette 25μl of mounting medium onto a 24 mm x 50 mm coverslip and pick up with the slide.

Note: Fix one short side of the coverslip with rubber cement or nail polish to prevent sliding and scratching.

20. Keeping the slides for at least 30 min at 4°C stabilises the staining for up to one week.

Results

For evaluation of CGH preparations use only an image analysis system with a dedicated software.

Demands on optical system

Take good care that the camera, the optical part, the slide table and the focusing systems are well fixed and aligned. The use of an automatic filter-wheel prevents shifts between the individual images taken with the different filters. Adjust the lamp, the collector lens and the collector mirror carefully to guarantee a uniform illumination and to avoid chromatic errors. Use apochromatic lenses with a high numerical aperture.

Check filters

Check the specificity of the filters individually with a control-slide containing a mixture of FITC and TRITC. Filters must be impermeable for the second fluorescence dye during exposure for at least 20 seconds. Inspect the UV excitation filter regularly. It commonly cracks and becomes light-permissible because of prolonged intensive heat exposure.

Check optical field

Control the homogeneity of the light distribution with the working magnification by comparing the top, bottom, left and right background areas surrounding a centrally located nucleus. Expose them for 20 seconds each. The individually measured values should not deviate by more than 10% from each other.

Capturing the images

The combination of an objective with a high (100x or 63x) and an ocular with a low magnification (10x without zoom) provides an optimal resolution. To increase the contrast of the image enclose the metaphase of interest with the diaphragm. Exposure times of both the FITC and TRITC image should not exceed 10 seconds and differ by less than 20%. The chromosomes should be at least twice as bright as the background. Restrict exposure time. To detect and locate faulty pixels in the CCD camera, take one image for 10 seconds with a closed light pathway. To exclude these faulty pixels, subtract this "empty" image from every image used for comparative evaluation. Determine the optimal exposure time once for each slide and use it for every metaphase. Do not use auto-exposure, because bright background artefacts might influence the measurement and thus, alter the ratios in your profiles.

Choose Metaphases

Use only well spread plasma-free metaphases with straight chromosomes of the same band level. Optimal band levels lie between 400 and 800. Although the resolution of detectable abnormalities increases with the band-level, the smoothness of the hybridization and with it the correctness of the CGH results decrease.

Analysis

Use 8 to 20 top quality metaphases for analysis. Check each individual chromosome for background and artefacts.

References

Du Manoir S, Schröck E, Bentz M, Speicher MR, Joos S, Ried T, Lichter P, Cremer T (1995a) Quantitative analysis of comparative genomic hybridization. Cytometry 19:27-41

Du Manoir S, Kallioniemi O-P, Lichter P, Piper J, Benedetti PA, Carothers AD, Fantes JA, García-Sagredo JM, Gerdes T, Giollant M, Hemery B, Isola J, Maahr J, Morrison H, Perry P, Stark M, Sudar D, van Vliet LJ, Verwoerd N, Vrolijk J (1995b) Hardware and software requirements for quantitative analysis of comparative genomic hybridization. Cytometry 19:4-9

Kallioniemi OP, Kallioniemi A, Piper J, Isola J, Waldman FM, Gray JW, Pinkel D (1994) Optimizing comparative genomic hybridization for analysis of DNA sequence copy number changes in solid tumors. Genes Chrom Cancer 10:231

Joos S, Schütz B, Bentz M, Lichter P (1996) Detection of chromosomal imbalances using DOP-PCR and comparative genomic hybridization (CGH). In: Nonradioactive in situ hybridization. Application manual, Boehinger Mannheim, pp 72-78

Piper J, Rutovitz D, Sudar D, Kallioniemi A, Kallioniemi O-P, Waldman FM, Gray JW, Pinkel D (1995) Computer image analysis of comparative genomic hybridization. Cytometry 19:10-26

Telenius-H, Pelmear-AH, Tunnacliffe-A, Carter-NP, Behmel-A, Ferguson-Smith-MA, Nordenskjold-M, Pfragner-R, Ponder-BA (1992) Cytogenetic analysis by chromosome painting using DOP-PCR amplified flow-sorted chromosomes. Genes-Chromosomes Cancer 4: 257-263

Zhang A, Lin SM (1994) KCl/Na-Cit 20:1 hypotonic solution for blood chromosome preparations. Appl Cytog 20:198-200

Suppliers

Suppliers of image analysis systems

APPLIED IMAGING INT. LTD., Hylton Park, Wessington Way, Sunderland, SR5 3HD, UK (*phone* +44-191-5160505; *fax* +44-191-5160512

METASYSTEMS GMBH, Robert-Bosch-Straße 6, Altlussheim, 68804, Germany (*phone* +49-620-539610; *fax* +49-620-532270

PERCEPTIVE SCIENTIFIC INT. LTD., Halladale, Lakeside, Wrexham Road, Chester, CH4 9QT, UK (*phone* +44-1244-682288; *fax* +44-1244-681555

VYSIS GMBH, Vor dem Lauch 25, Stuttgart Fasanenhof, 70567, Germany (*phone* +49-711-720250; *fax* +49-711-7202510

Abbreviations

10xNT	10 times buffer for nick translation
10xPCR	10 times buffer for PCR
AP	alkaline phosphatase
BCIP	5-bromo-4-chloro-3-indolyl phosphate
bio	biotin
BSA	bovine serum albumin
CGH	comparative genomic hybridization
$CHCl_3$	chloroform
d-ATP	deoxy adenosine triphosphate
d-CTP	deoxy cytosine triphosphate
d-GTP	deoxy guanosine triphosphate
d-NTPs	deoxyribonucleotide triphosphates
d-TTP	deoxy tymidine triphosphate
d-UTP	deoxy uridine triphosphate
DAPI	4'6-diamidino-2'-phenylindole-dihydrochloride
Dex	dextransulfate
dig	Digoxigenin
DNA-EB	DNA-extraction buffer
DNase	deoxyribonuclease
DOP-PCR	degenerated oligonucleotide primed polymerase-chain-reaction
EDTA	disodium ethylene diaminetetraacetate (Titriplex III)
etbr	ethidiumbromide
FCS	fetal calf serum
FISH	fluorescence in situ hybridization
FITC	fluorescein-iso-thio-cyanate
HSP	herring sperm DNA
IA	iso-amylalcohol
MNC	mononuclear cells
MOH	ß-mercapto-ethanol
NaAc	sodium acetate
NaCit	sodium citrate
NBT	4-nitroblue-tetrazolium-chloride
PBS	phosphate buffered saline
PCR	polymerase-chain-reaction
Ph	phenol
PHA-M	phytohaemagglutinine, M-form
PK	Proteinase K
RNase	ribonuclease
Roche	Roche Diagnostics
Sarkosyl	n-lauroyl-sarcosine
SDS	sodium dodecyl sulfate
SSC	saline sodium citrate
SSC-T	SSC/Tween20
TRITC	tetra-rhodamine-iso-thio-cyanate

Part VI

Techniques in Development

Chapter 22

Fetal Cells in Maternal Blood

DOROTHEE GÄNSHIRT, HENK S.P.GARRITSEN AND WOLFGANG HOLZGREVE

Introduction

It is well known that fetal cells circulate in the maternal peripheral blood during pregnancy, although it is believed that their absolute number is very small. However, recovery and analysis of those cells would permit noninvasive prenatal diagnosis, which would avoid the procedure related risks inherent to current invasive techniques. Expected fetal cells present in maternal circulation are nucleated and nonnucleated red blood cells, white blood cells, hematopoietic stem cells and trophoblast cells (Gänshirt et al. 1995 a,1995b). Our work has been focused on the isolation of nucleated red blood cells (NRBCs). One reason Fetal nucleated red blood cells have already been successfully isolated from maternal circulation and is that they are very abundant in early peripheral fetal blood, thus they seem to be a promising target for isolation in early gestation. However NRBCs are rare in adult circulation. This means fetal NRBCs can be enriched using tissue specific antibodies – whereas enrichment of fetal white cells requires individual specific antibodies in order to discriminate between the few fetal and the vast majority of maternal white cells. Also fetal red cells have a relatively short life span and thus they are likely to derive from a current pregnancy.

Fetal nucleated red blood cells have already been successfully isolated from maternal circulation and fetal aneuploidies were analysed thereafter using interphase FISH analysis (Gänshirt-Ahlert et al. 1993: Bianchi et al. 1994; Elias et al. 1994). The majority of current techniques for isolation of fetal cells are based on density gradient separation, antibody staining and

Correspondence to Dorothee Gänshirt, Haseldorfer Chaussee 36, Haselau, 25489, Germany (*phone* +49-4122-83188), Henk S.P. Garritsen, Kantonsspital, Universitäts-Frauenklinik, Schanzenstraße 46, Basel, 4031, Switzerland (*phone* +41-61-265-7088; *fax* +41-61-325-9399), Wolfgang Holzgreve, Kantonsspital, Universitäts-Frauenklinik, Schanzenstraße 46, Basel, 4031, Switzerland (*phone* +41-61-265-7088; *fax* +41-61-325-9399) subsequent separation of labelled cells from unlabelled cells by either fluorescence activated flow sorting (FACS) or magnetic separation. We introduced the technique of magnetic cell sorting (MACS) for separation of fetal NRBCs, which will be presented in detail here. There are some advantages of magnetic sorting as opposed to FACS, which caused us to choose this route. On one hand MACS is much less expensive than FACS, both in equipment and in running costs. FACS requires a specially trained operator, while the MACS technique is relatively easy to learn and may well be perfomed by technical persons. The cell condition after magnetic separation is excellent, which is a prerequisite for subsequent analysis by Fluorescence *In Situ* Hybridization (FISH). The good vitality of the enriched cell material may eventually be advantageous for cultivation. These qualities are important with respect to future application in routine prenatal diagnosis.

Outline

Briefly the technique consists of a triple density gradient for preenrichment of NRBCs, followed by labelling with an antibody specific for the transferrin receptor (CD71). This antigen is located on membranes of nucleated and nonnucleated red cells. The CD71 – antibody is linked to superparamagnetic beads. The separation procedure is performed in a syringe filled with steel wool, which is brought into the magnetic field. During the passage of the cells through the syringe the antibody – and beads – labelled cells are caught in the steel wool, while the negative fraction can be washed out. The postive fraction can be collected subsequently after the syringe has been removed from the magnetic field. (See schematic protocoll in Figure 1.)

Materials

- MACS-CD71 Microbeads (Miltenyi Biotec GmbH, Bergisch Gladbach,-Germany)
- Injection canulas (2.0 x 80 mm), Semadeni AG, Ostermundingen, Switzerland

Equipment

- Mini MACS Separation system (Miltenyi Biotec GmbH)
- Mini MACS separation columns (Miltenyi Biotec GmbH)
- Cytocentrifuge Cytospin 3 (Shandon, Frankfurt, Germany)

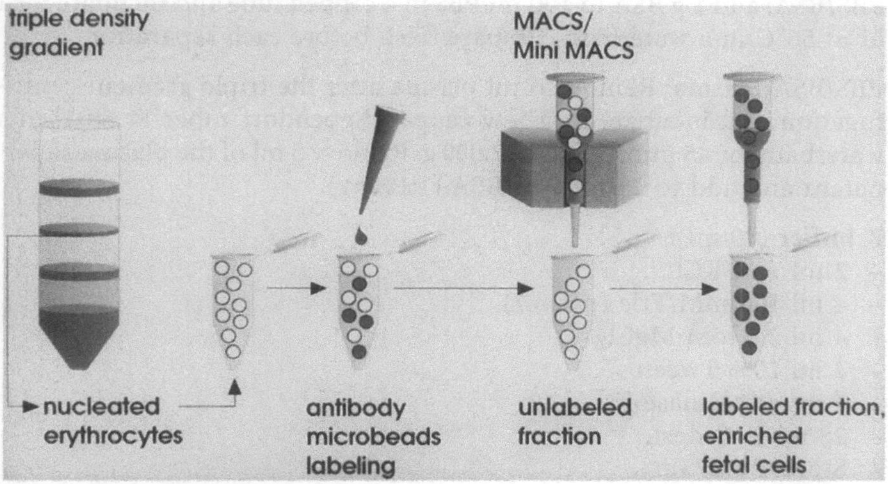

Fig. 1. Schematic protocol

Buffers and solutions

- Heparin: Vetren 200 (Promota GmbH, Hamburg, Germany)
- Histopaque: Histopaque 1077 and Histopaque 1119 (Sigma Chemie, München, Germany)
- Histopaque 1110: prepare 50 ml for separation of a 40 ml blood sample: mix 40 ml of Histopaque 1119 with 10 ml of Histopaque 1077 and mix thoroughly
- Differential staining solution: DiffQuick (Baxter Diagnostics AG, Düdingen, Germany)
- Mounting solution: Eukitt (Kindler GmbH, Freiburg, Germany)
- PBS (phosphate buffered saline):
  - 8 g NaCl
  - 0.2 g KCl
  - 1.44 g Na_2HPO_4
  - 0.24 g KH_2PO_4
  - add 800ml of distilled H_2O
  - adjust the pH to 7.4 with HCl.
  - add H_2O to 1 liter.
  - sterilize the solution by autoclaving for 20 min at 15 lb/sq.in. on liquid cycle.
  - Store at room temperature.

- PBS/BSA: add 1 g BSA to 100 ml PBS in a capped tube and incubate for 1 hr at 56°C in a waterbath. Prepare fresh before each separation.
- PBS/BSA/Plasma: Remove 6 ml plasma after the triple gradient centrifugation and incubate in a few capped Eppendorf tubes at 56°C in a waterbath for 45 min. Spin at 12.000 g. Remove 5 ml of the plasma supernatant and add to 45 ml PBS/BSA (1:10/v:v)
- K buffer (40 ml):
  - 2 ml 1 M KCl
  - 4 ml 100 mM Tris (pH 8.7)
  - 4 ml 2.5 mM $MgCl_2$
  - 2 ml 10% Tween
  - 4 mg Proteinase K
  - 28 ml H_2O dest.
  - Store at -20°C.
- TE buffer:
  - 10 mM Tris HCl (pH 7.6)
  - 1 mM EDTA (pH 8.0)

Procedure

CD71 enrichment of fetal cells

The whole enrichment procedure should be performed under a sterile hood with sterilized tubes, pipettes and solutions. Make sure that buffers and tubes are prepared before you start, because the whole procedure should be performed smoothly, in order to avoid unnecessary cell loss.

Blood sampling

1. Obtain 40 ml of heparinized veneous blood from a pregnant woman.
2. Attach the tubes to a test tube rotator, rotating at approximately 10 rpm at room temperature. Samples should be subjected to the enrichment procedure within 48 hours of blood sampling.

Note: Long storage of blood samples causes blurring in the triple gradient which makes it difficult to properly distinguish the bands of nucleated cells. Usually the nucleated red cells are more resistent to lysis than white cells and successful enrichment has even been achieved with blood samples that have been stored at room temperature for as many as 10 days. This led to extremely high cell purities in the enriched fractions of those samples. However, in old blood samples we also frequently observed cases with zero enrichment of NRBCs. Thus, extreme outliers – that is with very high and very low enrichment – are increased in old blood samples. For reproducible results samples should be processed as soon as possible. If samples are shipped within 48 hrs it is not necessary to cool them.

Triple density gradient

For the triple density gradient use sterile 12 ml capped polystyrol tubes.

Note: It is important not to use polypropylene tubes, because the cells will stick to the polypropylene wall after the density centrifugation step. Tubes should be lucid and should have volumes in ml marked on the side. In tall tubes with a small diameter the column of each Ficoll layer is higher which yields a better separation.

For setting up a gradient with 40 ml of blood 20 tubes are needed.

1. Add the 40 ml blood sample to 80 ml of PBS in a capped container and mix carefully by inverting.
2. Pipette 6ml of the blood/PBS mixture into each of the 20 tubes.
3. Subsequently underlayer the following three layers of Histopaque:
   - 2 ml Histopaque 1077
   - 2 ml Histopaque 1110
   - 2 ml Histopaque 1119

   For underlayering the histopaque fractions use a canula,that is 80 mm long with a 2.0 mm diameter attached to a 10 ml syringe. Place the tip of the needle at the bottom of the test tube and very carefully eject the histopaque solution. Try to avoid any turbulence between the adjoining layers, when underlayering the gradient.

   Figure 2 indicates the gradient with all three layers of histopaque and the blood/PBS mixture on top. Independent layers of Ficoll should be visible, when the tube is held against the light.

   You may use the same canula for setting up the gradient and for removing the bands of nucleated cells later. Before changing to a new densitiy of histopaque or a new cell layer, wash the syringe and the needle with PBS twice.
4. Centrifuge at 560 g for 30 min, but do not use the centrifuge brakes, as this will cause unnecessary turbulence when the centrifuge stops. It is advisable to use a large stable centrifuge instead of a small table-top machine. After the centrifugation the triple gradient should look as indicated in

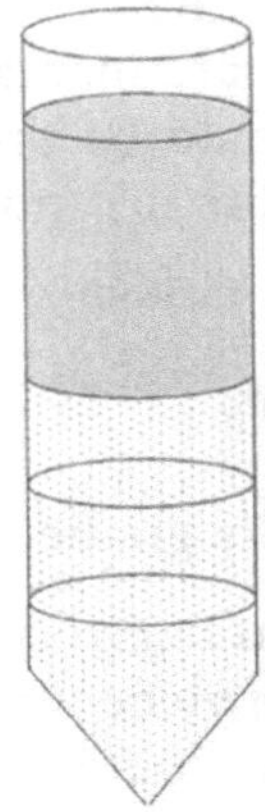

Fig. 2. Triple density gradient before centrifugation

Figure 3. Usually after separation of maternal blood samples the upper layer will contain lymphocytes and monocytes and the lower band will contain eosinophilic and basophilic granulocytes. The middle layer of nucleated cells after the triple gradient contains largely neutrophilic granulocytes and few NRBCs. Due to the scarcity of NRBCs in peripheral blood of pregnant women the middle layer is very faint in maternal blood samples as compared to the other two layers of nucleated cells.

5. First remove the plasma from each gradient with a 10 ml syringe attached to a canula for preparation of Plasma/PBS/BSA as indicated in the section Materials.

6. With a 10 ml syringe attached to the 2.0x80 mm canula, subsequently remove all three layers of nucleated cells from each tube.
   Remove the middle layer to the extent indicated in Figure 3 by slowly rotating the needle around the wall of the test tube. Before transferring the cell suspension into a fresh tube, remove the needle from the canula in order to prevent the cells from being disrupted! Wash the syringe and needle twice with PBS before changing to a new cell layer.
   The cells of the middle layers of all 20 test tubes – containing the NRBC's – are collected as follows.

7. Pipette the middle layer from each tube into separate new 12 ml test tubes. There should be 20 tubes at this stage, each containing approximately 1 – 1.5 ml of the cell suspension from the middle layer. After transferring the cell suspension into a tube, attach the needle to the syringe again, aspirate 2 ml of sterile PBS, remove the syringe and press out the buffer into the tube, in order to collect residual cells from the syringe or needle.

Fig. 3. Triple density gradient after centrifugation

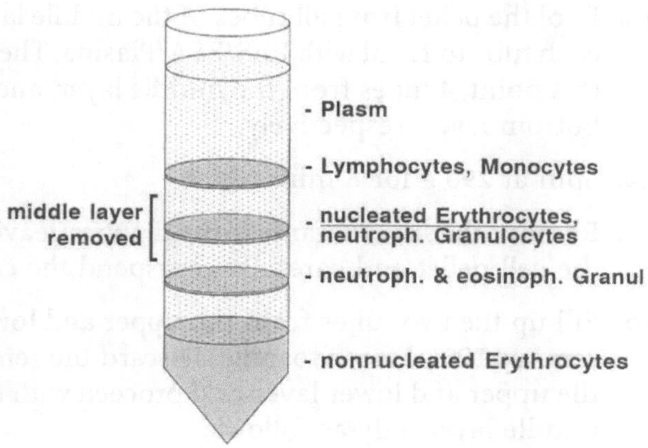

8. The cells from the upper and lower layer are only washed if you want to make differential cell counts for confirmation of proper separation in the triple gradient.
   For this purpose the cell suspension of only one test tube of each layer – the upper and the lower layer – have to be saved, the upper and lower layers of the remaining 19 tubes can be discarded. Proceed with the cells from the single tube according to the protocol concerning the middle layers: Pipette each layer into a separate 12 ml tube (approximately 1 – 1.5 ml), indicated "upper layer" or "lower layer", respectively.
   At this point there should be 22 tubes, 20 with cell suspensions from the middle layer and 2 with cell suspensions from the upper and lower layer, respectively.
9. Fill each tube to 12 ml with PBS.
10. Centrifuge all tubes at 390 g for 10 min.
11. Remove the supernatants with a sterile pasteur pipette, leaving 1ml of buffer on top of the cell pellet and resuspend the cells gently by scraping the bottom of the test tube several times carefully along a metal bar (eg a test tube rack).
    Make sure to resuspend the cells as soon as possible after centrifugation, because the viability of the cells decreases quicker in a pellet than in suspension.
12. Fill up each of the tubes from the upper and lower band with PBS to 12 ml.

13. Pool the pellet from all tubes of the middle layer into four tubes and fill each tube to 12 ml with PBS/BSA/Plasma. There should be 6 test tubes at this point, 4 tubes from the middle layer, and one tube from the top or bottom layer, respectively.
14. Spin at 290 g for 8 min.
15. Remove the supernatants from all tubes, leaving 1 ml of buffer on top of the cell pellet and carefully resuspend the cells.
16. Fill up the two tubes from the upper and lower layer to 5 ml. Mix and remove 100 µl for cytospins. Discard the remaining cell suspension of the upper and lower layer and proceed with the cell suspensions of the middle layer only as follows.
17. Collect the pellets of all 4 test tubes of the middle layer in one test tube. Fill the test tube up to exactly 10 ml with PBS/BSA/Plasma. Mix well and remove 100 µl for a cell count and 100 µl for a cytospin.
18. Spin the tube again for 8 min at 290g.
19. Carefully remove the supernatant until approximately 100 µl of the cell suspension remain at the bottom of the tube (estimate by comparison with a second test tube containing 100 µl of water).
20. Resuspend the cells carefully and transfer the test tube to an ice bath. Should there be any need to interrupt the procedure, this is the best time to do so. Leave the cells at 4°C after they have been resuspended. Do not interrupt after one of the preceeding steps because Ficoll is toxic to cells. From now on keep the cells in an ice bath after each centrifugation or incubation step. Work with cold buffers only. Keep the MACS columns and the MiniMACS magnet in the refrigerator at 4°C. They should be removed only immediately before the separation is begun. The cooling will prevent capping phenomenas of the antibody after staining.

Antibody staining

1. For the antibody incubation add 25 µl of the CD71 microbeads to the 100 µl cell suspension. Should your cell suspension for any reason have a different volume, make sure to add the CD71 microbeads to a final concentration of 1:5(v:v). This is the antibody concentration recommended by the manufacturer.
2. Incubate for 10 min at 4°C.
3. Fill up the test tube to 5 ml with cold PBS/BSA.
4. Spin for 8 min at 290 g at 4°C.

5. Remove supernatant until a cell suspension of 50 – 100μl remains and resuspend the cells. Leave test tube in an ice bath.

Magnetic separation

1. Attach the prepared and cooled Mini MACS column to the Mini MACS magnet and place a 12 ml test tube under the column. This test tube should be labelled "1. negative fraction".
2. Wash the column three times with 500 μl fractions of cold PBS/BSA and immediately afterwards apply the cell supension to the column. Make sure that you have carefully resuspended the cells before separating them on the column, as cell clumps will significantly reduce the purity of the sort. Pipette the cells up and down repeatedly, but try to avoid air bubbles.
3. With 500 μl PBS/BSA remove residual cells from the tube that contained the cell suspension and apply this fraction to the column as well.
4. As soon as the cell suspension has entirely entered the column add the first wash fraction of 500 μl of cold PBS/BSA.
5. Wash twice more with 500 μl fractions of PBS/BSA by adding each fraction to the column as soon as the previous fraction has entered the column. At the end of the washing procedure the column will stop dripping.
6. Collect the whole wash fraction as "1. negative fraction" in the 12 ml test tube under the column.
7. Remove the column from the magnet and place it in a 12 ml test tube labelled "positive fraction".
8. Add 1.5 ml PBS/BSA to the top of the column and with the appropriate plunger supplied by the manufacurer eject the fluid into the test tube labelled as "positive fraction". It is preferable to eject the cells gently rather than under pressure in order to prevent cell rupture.
9. For more effective positive selection the whole positive fraction may be applied to the column again and washed with 3 fractions of 500 μl PBS/BSA (see steps 4 and 5). This wash fraction is collected as "2. negative fraction".
10. The column is then again removed from the magnet, 1.5 ml PBS/BSA are added to the top of the column and the "positive fraction" is ejected. Collect the positive fraction in the same test tube you used after the first passage over the magnet. This fraction should contain the NRBCs.

11. Adjust the volumes of the "positive fraction", the "1. negative fraction" and "2. negative fraction", respectively to whole mls and record the total volumes of all three fractions. You will need the total volume of each fraction in order to calculate the total amount of cells in this fraction after cell counts. Remove 100 μl of each of the three tubes for cytospins and 100 μl of each tube for cell counts.

Analysis of the enriched cell fraction

The expected distribution of specific blood cell types after the triple gradient in all three nucleated cell layers after the gradient (Figure 3) indicates proper separation. In the same way the occurence of NRBCs in the positive fraction after MACS indicates successful enrichment. Therefore differential cell counts during and after the enrichment procedure may be performed while establishing the procedure. The yield and puritiy of NRBCs in enriched samples may be recorded. In order to record the number of NRBCs in the positive fraction from pregnancy samples one should evaluate all cells from at least one cytospin. The percentage of NRBCs in the positive fraction after MACS will reflect the **purity** of the enriched sample.

However, because of the extreme scarcity of fetal cells in the maternal circulation, the **yield** of NRBCs after the enrichment procedure is almost more important than purity of the enriched sample. The yield of NRBCs is calculated on the basis of the purity of NRBCs in the cytospin of the positive fraction as well as the total number of nucleated cells in the positive fraction. The yield of NRBCs is partly dependent on recovery nucleated cells in general. Unnecessary cell loss can be traced by total cell counts before and after individual steps of the protocol. In this separation protocol the largest cell loss occurs during the density gradient, where part of the nucleated cells is lost in the pellet of nonnucleated red blood cells and another part during centrifugation steps. During the MACS separation the cell recovery is much higher and more than 90% of all cells applied to the column should be recovered in the positive and negative fraction after MACS.

Nucleated red blood cells in peripheral maternal circulation are a mixture of maternal and fetal NRBCs. Therefore identification of fetal cells cannot be performed on the basis of cell morphology in differentially stained cytospins.

Identification of fetal cells so far has primarily been achieved by DNA analysis for the Y chromosome in male pregnancies. The most specific modern techniques are Y-specific Fluorescence *In Situ* Hybridization (FISH) of interphase nuclei and Y-specific PCR.

Should you intend to analyse the entire enriched cell fraction by FISH or PCR, cytospins should be omitted, in order to avoid unnecessary cell loss.

Differential cell counts

1. For cytocentrifugation remove 100µl of the cell suspension, where indicated in the Materials section, apply it to the cytocentrifuge and spin the cells onto a slide at 400 rpm for 5 min.

2. Let the slides air dry.

3. Stain the slides differentially with DiffQuick and mount them with a coverslip using Eukitt.
   The cell quality on the cytospin is one parameter to check successful enrichment. Ruptured or clumped cells on cytospins indicate significant cell loss due to nonviable or dying cells. In general the quality of cells is excellent after the separation procedure if blood is processed within 48 hrs after obtainment.

Total cell counts

Total cell counts may either be performed in a Coulter Counter or in a Zeiss Thoma Counter under the light microscpe. In a Zeiss Thoma Counter only a very small amount of cells is needed for a cell count , which is preferable. This system also has the advantage that the viability of the cells may be checked with an appropriate dye (f.e. 0.16% trypane blue in PBS). More than 90% of the cells should still be vital after the separation procedure.

Preparation of the enriched cell fraction for PCR

1. Spin the positive cell fraction 10 min at 900 g and remove the supernatant to 500 µl.

2. Resuspend the cells and transfer the suspension to a sterile Eppendorf tube.

3. Aspirate the remaining cells of the 12 ml tube with 1000 µl TE buffer and add this to the cell suspension in the Eppendorf tube.

4. Close the Eppendorf tube and shake vigorously. Spin at 13.000 rpm for 3 min and remove the supernatant.

5. Resuspend the pellet in 1000 µl TE and shake again in order to lyse the nonnucleated erythrocytes.

6. Spin again at 13.000 rpm.

7. Remove the supernatant and resuspend the pellet with 30 µl of K-buffer.

8. Incubate at 50°C for 1 hr and at 37°C overnight.
9. Stop the reaction at 96°C for 10 min.
10. Store the DNA at 4°C until used for PCR.

Preparation of the enriched cell fraction for FISH

Usually the cells are very well preserved after the enrichment procedure and may easily be processed for interphase-FISH analysis as follows.

1. Prepare the slides by incubating them overnight with ethanol abs. in a closed container.
2. Wash for 10 min with running tap water and FISH for 3 min with running aqua dest.
3. Dry the slides on a slide warmer or in an incubator at 80°C.
4. Incubate at room temperature 2 min with 2% 3-Aminopropyltriethoxysilan and wash with running aqua dest for 4 min.
5. Dry slide again as indicated above and store in a dust free container at room temperature.
6. Spin the positive fraction after MACS at 200 g for 10 min.
7. Remove the supernatant and resuspend the pellet in 5 ml 75 mM KCl.
8. Incubate at 37°C for 10 min.
9. Spin at 150g for 10 min.
10. Remove the supernatant and resuspend the pellet in 5 ml methanol/glacial acetic acid (3:1/v:v).
11. Spin at 200 g for 10 min.
12. Resuspend the pellet in the reflux and apply to a slide.
13. Prior to *In Situ* Hybridization the slides may be stored in a dessicated container at -20°C.

Results

In peripheral blood of pregnant women NRBCs are so rare that the enrichment of NRBCs cannot be tracked from the beginning to the end of the enrichment procedure. Occasionally, NRBCs are found on cytospins of the middle layer after the triple gradient, but in general they may only be detected in the positive fraction after MACS.

In cord blood obtained after delivery NRBCs are already detectable on blood smears. In the middle layer after the triple gradient the percentage of NRBCs should have increased about 2.5 fold in the mean and after MACS 90% of all nucleated cells of the positive fraction are NRBCs. Thus if individual enrichment procedures are to be evaluated for their effectiveness, spiking experiments with cord blood in adult blood may be traced NRBCs throughout the enrichment procedure by differential cell counts, Y-spcific FISH or PCR, respectively.

The number of enriched NRBCs was found to be highly variable in individual pregnancies. In a large series of pregnancies (n=400), which we investigated, the mean percentage of NRBCs in the enriched fraction (corresponding to the purity of the enriched fraction) was 0.1% in early pregnancy (6 weeks post L.M.P.) and raised to 0.5% at term. The yield of NRBCs enriched from 40 ml of maternal blood was 100 (median) in early gestation (6 weeks post L.M.P) and 1000 (median) at term (Gänshirt et al. 1994).

To date techniques have achieved enrichment, not purification of fetal cells from adult circulation, because all antibodies that have been applied so far showed cross-reactivity to adult cells. On the other hand the constraint to recover as many fetal cells as possible, while at the same time eliminating as many maternal cells as possible can often not be realized experimentally. Increase of cell purities requiring additional separation steps will cause additional overall cell loss. On the other hand, best recovery of cells is obtained when the least of all possible steps are performed within a separation procedure. However, the occurence of fetal cells in maternal circulation is extremely low, current results indicate that the feto/maternal cell ratio is about $1:10^6$ or less (Gänshirt et al. 1995a, 1995b), which means that not more than about 100 fetal cells may be expected in a 40 ml maternal blood sample. Therefore a separation technique for fetal cells should rather aim for high yields than for high purities. Moreso, the use of probe – based techniques as FISH do not require physical isolation of fetal cells, but only the knowledge of their location on the microscope slide. Techniques for staining of enriched cells with fetal specific antibodies (f.e. for fetal hemoglobin) are currently investigated for this purpose (Ferguson-Smith et al. 1994; Park et al., 1994). FISH analysis would then be performed on identified fetal cells only. With improved automatized image analysis, slide search capability might then be traded against enrichment purity.

The potential applicability of the method for a noninvasive prenatal test can only be judged after validation of the feasibility and accuracy of the method. For this purpose the National Instutute of Health and Human Development (NICHD) is currently funding a clinical investigation of this technology (De la Cruz et al. 1995). The study involves five programs. Initially different enrichment techniques are compared, while in a later phase of the study the accuracy of cytogenetic diagnosis from fetal cells from maternal blood will be compared with results obtained by amniocentesis or chorionic villus sampling.The evaluation of this study is expected to be completed in 1997.

Troubleshooting

- It may sound trivial, but in some syringes for blood collection the Heparin and the blood only mix, if you carefully invert the syringe a few times, immediately after the blood has been drawn. It is crucial for proper cell separation that the blood sample is not coagulated and the person obtaining the blood should be informed.
- If individual layers of the gradient are not visibly separated by sharp bands before the centrifugation, the ficoll layer may have been pressed out off the syringe too fast. The first parts of each layer should be pressed out step by step rather than uninterrupted in order to avoid turbulences.
- Occasionally the gradient looks blurry after centrifugation and bands of nucleated cells have not yet established. In these cases the gradient should be centrifuged again for another 15 min.
- In old blood samples the gradient might be contaminated with fibrin or erythrocyte plaques after centrifugation. Remove the cell fractions as usual but try not to aspirate fibrin or erythrocyte plaques.
- Due to the scacity of NRBCs in maternal circulation the middle layer of cells may be invisible after gradient centrifugation. Nevertheless, the layer does contain cells, as you will see after you spin the cells down. Remove the upper cell layer, which is visible and collect the layer in between the upper and lower layer as middle layer.
- Before the cells are applied to the MACS column they should be resuspended very carefully in order to avoid cell clumps, which may clog the

column. However, in some cases the cell flow through the column may stop. In these cases remove the column from the magnet, press out the cells with 2 fractions of 6ml PBS/BSA, spin down the collected cell fraction, resuspend the cell pellet in PBS/BSA and apply it to another column. Sometimes visible fibrin threads will cause clogging of the column. Fibrin should not be applied to the column with the cells. Instead try to attach the fibrin to the wall of the test tube with the pipette tip before you aspirate the cell suspension, that is applied to the column.

- You may also observe clotted cells in the positive fraction, when counting the cells with a Thoma Zeiss Counter. Pipette the cells carefully up and down again in order to further resupend them.

- If the blood sample is very old (more than one week) the cells may look damaged or even ruptured on cytospins of the positive fraction. However, if this happens with fresh blood, you should check your buffers and solutions and make sure that the cells are not left in Ficoll longer than necessary and are resuspended immediately after centrifugation.

References

Bianchi DW (1994) Clinical trials and experience. Ann N Y Acad Sci, 731:92-102

De la Cruz F, Shifrin H, Elias S, Simpson JL, Jackson L, Klinger K, Bianchi D, Kaplan SH, Evans M, Holzgreve W, Gänshirt D (1995) Prenatal diagnosis by use of fetal cells isolated from maternal blood. Am J Obstet Gynecol 173:1354-1355

Elias S, Simpson JL (1994) Prenatal diagnosis of aneuploidy using fetal cells isolated from maternal blood. Ann N Y Acad Sci, 731:80-91

Ferguson-Smith MA, Zheng Y-L, Carter NP (1994) Simultaneous immunophenotyping and FISH on fetal cells from maternal blood. Ann N Y Acad Sci, 731:73-79

Gänshirt-Ahlert D, Börjesson-Stoll R, Burschyk M, Dohr A, Garritsen HSP, Helmer E, Miny P, Velasco M, Walde C, Patterson D, Teng N, Bhat NM, Bieber MM, Holzgreve W (1993) Detection of fetal trisomies 21 and 18 from maternal blood using triple gradient and magnetic cell sorting. AJRI, 30:194-201

Gänshirt D, Börjesson-Stoll R, Burschyk M, Garritsen HSP, Miny P, Neusser M, Smeets F, Velasco M, Walde C, Holzgreve W (1994) Isolation of fetal cells from maternal circulation. In: Zakut H (ed) Proceedings of the 7th International conference on early prenatal diagnosis. Monduzzi Editore, Bologna, Italy, pp 19-26

Park VM, Bravo RR, Price JO, Simpson JL, Elias S (1994) A model system using fetal hemoglobin to distinguish fetal cells enriched from maternal blood. Ann N Y Acad Sci, 731:133-135

Gänshirt D, Garritsen HSP, Holzgreve W (1995a) Fetal cells in maternal blood. Curr Opin Obstet Gynecol, 7:103-108

Gänshirt D, Garritsen HSP, Holzgreve W (1995b) Prenatal diagnosis using fetal cells in the maternal circulation.Fet Mat Med Rev, 7:77-85

Chapter 23

Spectral Karyotyping in Clinical and Tumor Cytogenetics

EVELIN SCHRÖCK, YUVAL GARINI, MICHAEL KÖHLER AND THOMAS RIED

Introduction

Karyotype analysis based on chromosome banding techniques has been the diagnostic "gold standard" in clinical and cancer cytogenetics for more than 20 years.

Screening of patients with mental retardation and/or physical disabilities for chromosomal abnormalities has become a routine procedure in medical genetics and pediatrics. Specific chromosome aberrations were found to cause phenotypic abnormalities, such as trisomy 21 in Down syndrome and a partial deletion of the short arm of chromosome 5 resulting in cri du chat syndrome. Prenatal chromosome analysis is performed for pregnancies with an increased risk for a chromosome abnormality of the fetus. While the diagnosis of a trisomy 21 is readily made, some clinical cases remain unsolved showing either unidentified marker chromosomes or normal karyotypes, that exhibit phenotypic evidence for chromosomal rearrangements. Furthermore, complex chromosomal rearrangements detected in a number of patients afflicted with multiple miscarriages or infertility are often difficult to characterize by traditional banding methods alone.

In tumor cytogenetics, progress has been achieved by revealing consistent chromosomal rearrangements particularly in leukaemias and lymphomas. The detection of numerical aberrations and tumor specific transloca-

Correspondence to Evelin Schröck, National Cancer Institute (NCI/NIH), 49 Convent Drive, Building 49, Room 4C36, BethesdaMD, 20892, USA (*phone* +01-301-402-2008; *fax* +01-301-402-1204; *e-mail* eschrock@nhgri.nih.gov), Yuval Garini, Applied Spectral Imaging, Ltd., P.O. Box 101, Migdal Ha'Emek, 10551, Israel, Michael Köhler, Applied Spectral Imaging, Neurottstr. 12, Edingen-Neckarhausen, 68535, Germany, Thomas Ried, National Cancer Institute (NCI/NIH), 49 Convent Drive, Building 49, Room 4A28, BethesdaMD, 20892, USA tions, like the Philadelphia chromosome [t(9;22)] and secondary aberrations [eg +8, t(3;21)] in chronic myeloid leukaemia (CML), supports diagnostic and therapeutic decisions. In contrast, only a small number of solid tumors reveal specific chromosomal aberrations which could be used as genetic markers. Several sarcomas show translocations resulting in fusion genes which may be of etiologic significance, – like the t(X;18) in Ewing sarcomas and the t(11;22) in synovial sarcoma. Carcinomas, however, show a high number of numerical and structural aberrations. The identification of primary and secondary aberrations and the determination of their biological function is much more difficult due to a number of methodological shortcomings. Problems such as selective cell growth, low mitotic indices and low quality metaphase spreads arise during tumor cell culture and in the process of metaphase preparation. Also, the complexity of the rearrangements often prevents a complete classification.

Molecular cytogenetic (FISH) and microdissection techniques provide helpful tools for the characterization of aberrant chromosomes. However, they cannot be used as initial screening tests, because they do not permit the analysis of the whole genome in a single assay. Recently, two methods were introduced which enable the screening of the chromosome complement in one experiment, namely m-FISH and SKY (Speicher et al. 1996, Schröck et al. 1996).

Here, we describe in detail the technique called spectral karyotyping (SKY), which is based upon combinatorial FISH using chromosome painting probes, optical microscopy, spectral imaging and spectra-based chromosome classification (Garini et al. 1996a). SKY permits the simultaneous visualization of all chromosomes in a metaphase spread in different colors (Figure 1A). A single exposure with the SpectraCube connected to an epifluorescence microscope is sufficient to measure the complete emission spectra at all image points. This spectral information is the basis for chromosome classification (Figure 1B-D). The measurement of the emission spectra is largely independent of differences in fluorescent intensities along the chromosomes. This makes SKY a robust technology. Marker chromosomes and interchromosomal changes like translocations are readily identified. For instance, a translocation der(4)t(4;8) in a patient afflicted with Wolf Hirschhorn syndrome, which was not apparent by conventional G-banding (Hannig et al. 1984), was identified by SKY using metaphase chromosomes with an average of only 500 bands (Schröck et al. 1997). The smallest translocation detected so far was in a range of about 1.5 Mb (Schröck et al. 1996). Intrachromosomal changes like small deletions, duplications and inversions are more difficult to discern by SKY. Therefore, the combination of conventional banding techniques and SKY represents the most comprehensive cytogenetic diagnostic approach (Schröck et al. 1997, Veldman et al. 1997).

The diagnostic potential of this technique can be further increased by using different sets of probe kits, for example probe sets for centromeres, telomeres and gene specific probes. In addition, the availability of an increasing number of region specific DNA probes will allow one to create

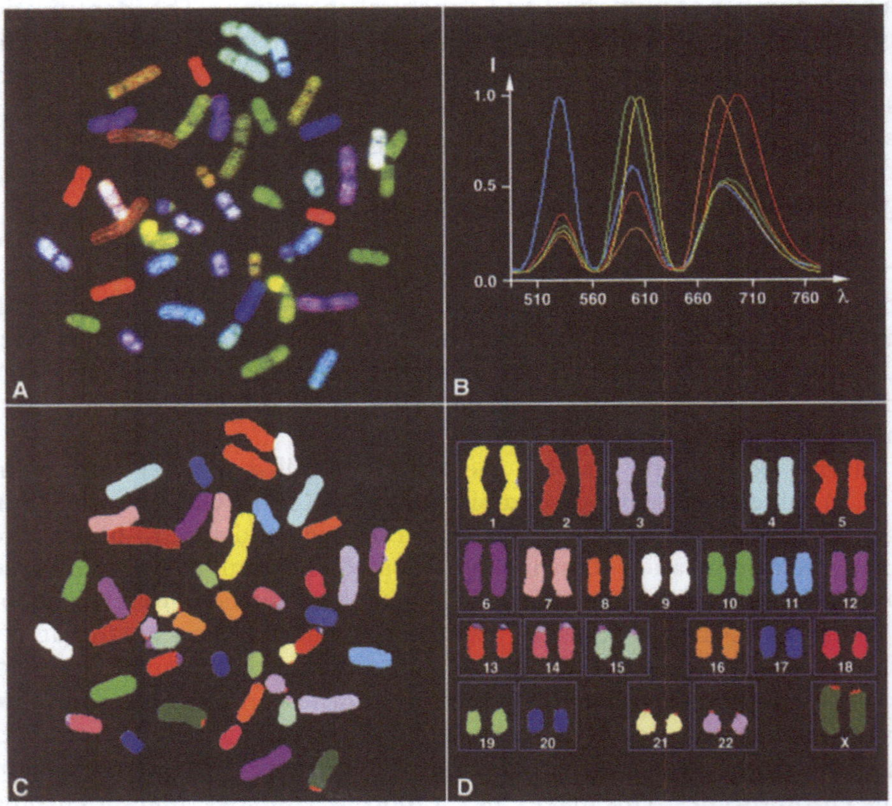

Fig. 1. SKY-analysis of normal metaphase chromosomes. A: SKY-display image. A single image was acquired with the Spectracube connected to an epifluorescence microscope. The hybridization pattern of the chromosome painting probes, differentially labelled with 24 colors, is visualized by assigning an RGB-display. Note that several chromosomes show similar display colors. B: Characteristic emission spectra of the five fluorochromes used for the preparation of the painting probes (blue - rhodamine 110, green - SpectrumOrange, yellow - Texas Red, orange - Cy5, red - Cy5.5). C: Same metaphase as in A after automated spectral classification. The measurement of the spectra for each image point permits one to readily identify chromosomes and chromosome segments based on their unique spectral signature. Chromosomes with similar display colors (compare to A) show different classification colors due to their spectral differences. D: The metaphase chromosomes are arranged as a karyotype.

a multicolor banding pattern along chromosomes. This, in turn, will result in an increase of resolution and sensitivity of SKY to detect intrachromosomal changes like small deletions and inversions.

Outline

SKY methodology

SKY is performed on metaphase chromosomes prepared according to standard procedures (Barch et al. 1997). The SKY kit used for hybridization contains a complete set of chromosome specific painting libraries labelled with different fluorescent dyes and dye combinations for each chromosome. Hybridization and detection procedures follow routine FISH protocols. Image acquisition is performed using the SpectraCube connected to a regular epifluorescence microscope followed by image analysis supported by a comprehensive software package. The single steps of a SKY experiment can be summarized as shown in Figure 2.

Materials

Reagents

- PCR-Nucleotides: 100mM dNTPs Boehringer Mannheim, Indianapolis, IN, USA, (1051440, 1051458, 1051466, 1051482)
- PCR-Primer: Telenius 6MW [5'-CCGACTCGAGNNNNNNATGTGG-3']
- PCR-Polymerase: Native Taq (5U/μl), Perkin Elmer, Foster City, CA, USA, (part no. N801-0046)
- PCR-Buffer: 10X PCR Buffer, Perkin Elmer, Foster City, CA, USA, (part no. N808-0010)
- Spectrum Orange dUTP: Vysis, Downers Grove, IL, USA, (30-803000)
- Texas Red-dUTP: Molecular Probes, Eugene, OR, USA, (C-7631)
- Biotin-16-dUTP: Boehringer Mannheim, Indianapolis, IN, USA, (1093070)
- Rhodamine 110-dUTP: Perkin Elmer, Foster City, CA, (403070C)
- Digoxigenin-11-dUTP: Boehringer Mannheim, Indianapolis, IN, USA, (1558706)
- Fluorolink-Cy5-avidin: Amersham Life Science Inc., Pittsburgh, PA, USA, (PA 45000)

- mouse-anti digoxin: Sigma, St. Louis, MO, USA, (D-8156)
- Fluorolink Cy5.5 sheep anti-mouse-IgG: Amersham Life Science Inc., Pittsburgh, PA, USA (RPQ 0115)
- Cot-1 DNA (human): Life Technologies BRL, Grand Island, NY, USA, (15279-011)
- RNase A: Boehringer Mannheim, Indianapolis, IN, USA, (109169)
- Pepsin: Sigma, St. Louis, MO, USA, (P-6887)
- Salmon Testes DNA: Sigma, St. Louis, MO, USA, (D-7657)

Solutions

- master mix: stock solution (2 x SSC, 20 % dextran sulfate): dissolve 20 g dextran sulfate in 100 ml of 2 x SSC, pH 7.0, autoclave, store aliquots at – 20 °C
- RNase A: stock solution: dissolve 20 mg/ml sterile water, boil for 15 min, cool to room temperature, make aliquots, store at – 20 °C
- Pepsin: stock solution (10%): dissolve 100 mg/ml sterile water, keep on ice, make aliquots, store at – 20 °C
- PBS/$MgCl_2$: add 50 ml 1M $MgCl_2$ to 950 ml 1xPBS, final volume 1l

Equipment

- PCR machine: PTC-100, MJ Research, Inc., Watertown, MA, USA
- Leica DMRXA microscope: Leica Mikroskopie und Systeme GmbH, Postfach 2040, D 35530 Wetzlar, Germany, ++49-(0)6441-292280 (phone), ++49-(0)6441-293399 (FAX)
- 770 U lamphouse for 150 W Xenon lamp: Opti-Quip, Highland Mills, NY, USA
- DAPI-filter TR1, SKY-filter V3.0: ChromaTechnology Corporation, Brattleboro, VT, USA
- SpectraCube and SkyView software: Applied Spectral Imaging Ltd., PO Box 101, Industrial Park, Migdal Ha'Emek 10551, Israel, ++972-6-6547 567 (phone), ++972-6-6547 507 (FAX)

Fig. 2. Steps of a SKY experiment

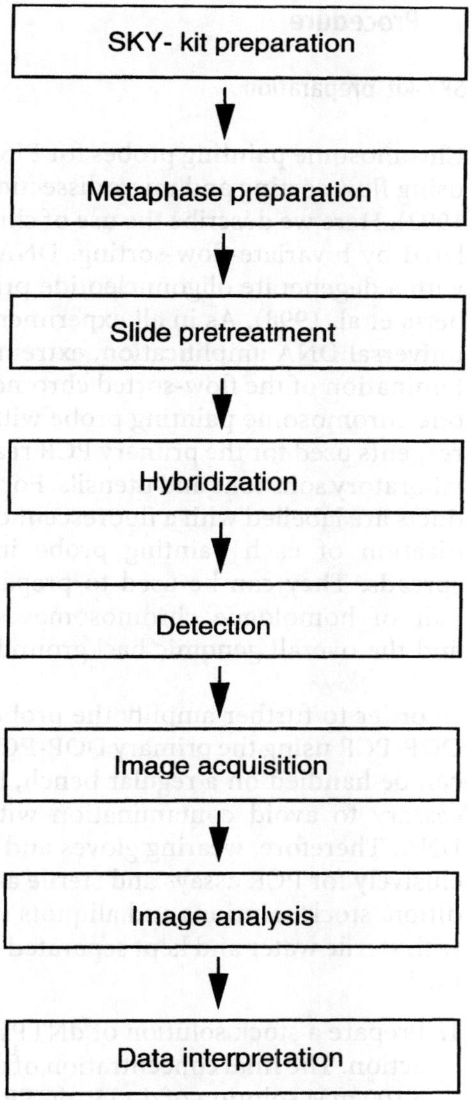

Procedure

SKY-kit preparation

Chromosome flowsorting and primary DOP-PCR

Chromosome painting probes for FISH have been prepared for many years using flow sorting and microdissection (Van Dilla et al. 1986, Meltzer et al. 1992). Here we describe the use of chromosome-specific DNA libraries isolated by bivariate-flow-sorting. DNA-amplification is performed by PCR with a degenerate oligonucleotide primer (DOP) (Telenius et al. 1992, Roberts et al. 1998). As in all experiments that utilize sequence independent universal DNA amplification, extreme care should be taken to avoid contamination of the flow-sorted chromosome samples with genomic DNA or one chromosome painting probe with another. Therefore, equipment and reagents used for the primary PCR reaction should be isolated from regular laboratory solutions and utensils. For quality control, the primary PCR products are labelled with a fluorescent dye via DOP-PCR and tested by hybridization of each painting probe individually onto normal metaphase spreads. They can be used to prepare the hybridization kits if only one pair of homologue chromosomes shows specific hybridization signals and the overall genomic background is low.

Secondary DOP-PCR

In order to further amplify the probe DNA sequences, a second round of DOP-PCR using the primary DOP-PCR products is used. This PCR-reaction can be handled on a regular bench, but again, special precautions are necessary to avoid contamination with genomic or chromosome specific DNA. Therefore, wearing gloves and using a set of pipettes designated exclusively for PCR assays and sterile aerosol barrier tips are required. In addition, stock solutions and aliquots should be handled carefully, prepared with sterile water and kept separated from all other solutions in the laboratory.

1. Prepare a stock solution of dNTPs to be used for the secondary PCR reaction. The final concentration of each nucleotide is 0.2 mM. Use 10 µl of a 100 mM solution of dATP, dCTP, dGTP and dTTP, add 460 µl of sterile water, vortex, aliquot and freeze at -20°C.
2. Pipet a “PCR-master-mix“ for the amplification of all 24 chromosomes, aliquot 98 µl into 24 sterile Eppendorf-tubes and add 2 µl of DNA specific for each chromosome (primary PCR-product). Note: vortex and centrifuge before adding the Taq polymerase.

Table 1.

PCR-master-mix	for 1 reaction (μl)	for 24 chromosomes (μl)
DNA-Primary PCR product	2	add separately
PCR buffer II (10X)	10	250
$MgCl_2$ (25mM)	8	200
dNTP (2mM)	10	250
sterile water	65	1625
Primer (100μM)	4	100
Taq Polymerase (5U/μl)	1	25
Total Volume	100	950

3. Use the following PCR-program:

Table 2.

step	temp. (°C)	minutes
1	94	1
2	56	1
3	72	3 with addition of 1sec/cycle
4	repeat steps 1-3, 29 times	
5	72	10
6	4	until use
7	end	

4. Run 2 μl of the PCR-products on a 1% agarose gel to test the length and amount of amplified DNA (Figure 3.).

The same procedure can be repeated using the secondary PCR product as a template resulting in tertiary DOP-PCR.

Labelling DOP-PCR

Secondary PCR-products are labelled by the incorporation of fluorescently labelled nucleotides (dUTP) via DOP-PCR. Using 5 different fluorochromes up to 31 combinations can be produced resulting in 31 different colors (2^5-1=31). The labelling scheme (Table 1) was designed to achieve optimal color differences between individual chromosomes.

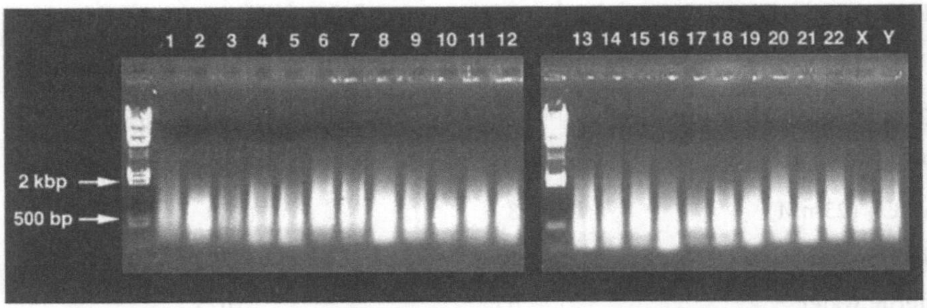

Fig. 3. Gel-electrophoresis pattern of all 24 chromosome painting probes after universal DOP-PCR.

1. Label 57 autoclaved Eppendorf-tubes in accordance to the enclosed labelling scheme (for instance 1B, 1C.... YE).

Table 3. Chromosome labelling scheme

chromosome	A Spectrum Orange-dUTP	B Texas-Red-dUTP	C Biotin-dUTP (Cy5)	D Rhodamine 110-dUTP	E Digoxigenin-dUTP (Cy5.5)
1		x	x	x	
2					x
3			x	x	x
4		x	x		
5	x	x		x	x
6		x			x
7			x	x	
8				x	
9	x			x	x
10			x		x
11	x				
12		x	x	x	x
13	x			x	
14		x			
15	x	x	x		
16	x				x

Table 3. Continous

chromo-some	A Spectrum Orange-dUTP	B Texas-Red-dUTP	C Biotin-dUTP (Cy5)	D Rhodamine 110-dUTP	E Digoxigenin-dUTP (Cy5.5)
17			x		
18	x	x		x	
19	x		x		
20	x		x	x	
21				x	x
22	x	x	x		x
X		x		x	
Y	x		x	x	x

2. Prepare a stock solution of dNTPs to be used for the labelling PCR reaction. The final concentration of dATP, dCTP and dGTP is 0.2 mM and of dTTP 0.15 mM. Use 10 µl of a 100 mM solution of dATP, dCTP, dGTP and 7.5 µl of dTTP, add 462.5 µl of sterile water, vortex, aliquot and freeze at -20°C.
3. Pipet a "PCR-master-mix" for the labelling of all 24 chromosomes in all colors resulting in 57 PCR-reactions. Aliquot 91 µl each into 57 Eppendorf-tubes. Add 5 µl of the respective fluorochrome-dUTP and 4 µl of the chromosome-specific secondary PCR-products. Note: vortex and centrifuge before adding the Taq polymerase.

Table 4.

PCR-master-mix	for 1 reaction (µl)	for 57 reactions (µl)
DNA (400-600ng)	4	add separately
PCR buffer II (10X)	10	600
$MgCl_2$ (25 mM)	8	480
dNTP (2mM)	10	600
sterile water	59	3540
Primer (100µM)	2	120
Taq Polymerase (5U/µl)	2	120
x-dUTP (1mM)	5	add separately
Totale Volume	100	5460

4. Use the following PCR-program:

Table 5.

step	temp. (°C)	minutes
1	94	1
2	56	1
3	72	3 with addition of 1sec/cycle
4	repeat steps 1-3, 29 times	
5	72	10
6	4	until use
7	end	

5. Run 2 µl of the PCR-products on a 1% agarose gel to determine the length and amount of amplified DNA (Figure 4).

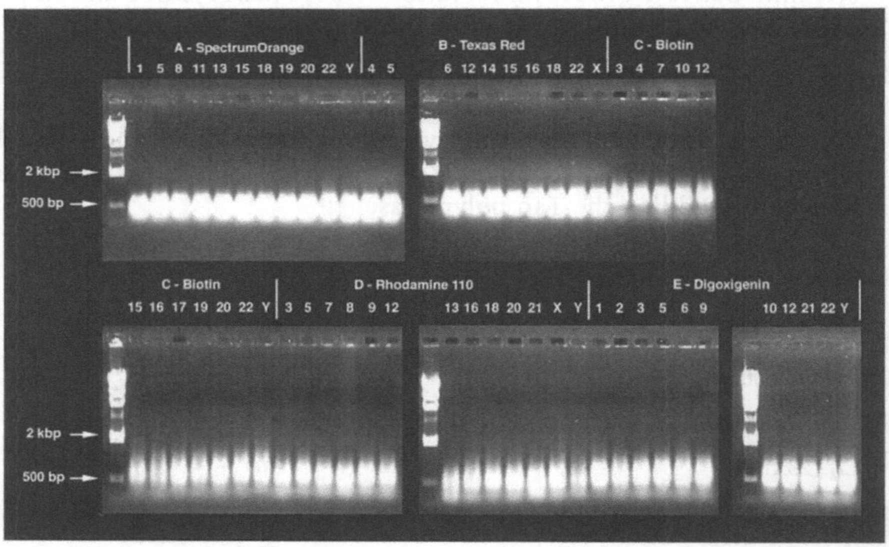

Fig. 4. Gel-electrophoresis pattern of all 24 chromosome painting probes after labelling with SpectrumOrange-dUTP (A), Texas Red-dUTP (B), Biotin-16-dUTP (C), Rhodamine110-dUTP (D), and Digoxigenin-11-dUTP (E). The labelling scheme is shown in Table 1 indicating the 57 different reactions.

6. Prepare a single SKY probe kit and perform a test hybridization as outlined in the section Hybridization. After image acquisition with the SpectraCube, check the following parameters which indicate the quality of the labelling PCR:
   - overall hybridization quality - check painting pattern of all chromosomes and suppression of heterochromatic regions;
   - signal to noise ratio - compare fluorescence intensity along the chromosomes with the intensity measured in background areas; the highest and lowest intensity values within the image are displayed when using the image acquisition software, the difference between those values is supposed to be at least 100 counts;
   - color separation - the difference between chromosomes visualized in red, green and blue must be obvious in the RGB display image (compare Figure 1a);
   - spectra separation - check the spectra of the single dyes and compare them to the reference spectra in the combinatorial table (ctb-file);
   - quality of classification - a 100% classification of a normal metaphase using the automated function in the SkyView software (see section Image Analysis) is the optimal result after a SKY kit preparation.

Metaphase preparation

Metaphase chromosome preparation follows standard procedures. Specific protocols are provided in Chapters 5, 7 and 12 for clinical samples and for tumor specimen in Chapter 9. The cell suspension can be stored in capped tubes at -20°C for several years, but the quality of chromosome-spreading as well as the hybridization efficiency decrease over time. Slides containing metaphase spreads should be dehydrated through an ethanol series of 70%, 90% and 100% for 3 minutes each. They can be stored in an airtight container (plastic box sealed in plastic bag together with drierite) at -20°C or -80°C indefinitely.

Slide pretreatment

To remove remnants of cytoplasm and cellular RNA which can compromise hybridization results, slides are pretreated with RNase and pepsin prior to hybridization. Since the quality of chromosome preparations is variable, the optimal amount of pepsin as well as the length of treatment need to be adjusted empirically for each batch. Note that overpepsinization can impair chromosome morphology, signal intensity and therefore SKY results. If the chromosome preparations show no visible cytoplasm, a pretreatment might not be necessary.

1. Equilibrate slides in 2 x SSC at RT
2. Dilute the RNase stock solution 1:200 in 2xSSC and apply 100 µl to a 24x60 mm^2 coverslip, touch slide to coverslip, incubate at 37°C for 45 minutes.
3. Remove coverslips, transfer slides to a coplin jar and wash 3x5 minutes in 2xSSC at room temperature while shaking.
4. Prepare a 0.01 M HCl solution by adding about 1ml of 1M HCl to 99ml dH_20. Prewarm the solution at 37°C, add 10 – 50 µl pepsin, mix well and adjust pH to 2.0 using 1M HCl.
5. Incubate slides at 37°C in coplin jar for 5 – 10 minutes.
6. Wash slides twice in 1xPBS for 5 minutes each at room temperature, shaking.
7. Wash slides once in 1xPBS/$MgCl_2$ for 5 minutes each at room temperature, shaking.
8. Prepare 1% Formaldehyde solution in 1xPBS/$MgCl_2$, (add 2.7 ml of 37% Formaldehyde to 97.3 ml of 1xPBS/$MgCl_2$) and incubate slides for 10 minutes at room temperature.
9. Wash slides once in 1xPBS for 5 min at room temperature, shaking.
10. Dehydrate slides in 70, 90, 100% Ethanol for 3 minutes each.
11. Air dry slides.

Previously
G-banded slides should be treated as follows:

1. Remove oil with 100% xylene in a coplin jar and wash with 100% methanol.
2. Remove Giemsa stain using 100% ethanol or regular fixative (Methanol:Acetic Acid 3:1).
3. Rehydrate slides through ethanol series of 100, 90, 70% for 3 minutes each.
4. Follow the regular slide pretreatment protocol, skip the pepsin treatment and proceed to step 6.

Hybridization

The products of the labelling PCR are combined which will result in a batch of SKY kits with identical quality. The DNA-mixture is then aliquotted into Eppendorf tubes and precipitated in the presence of cot1-DNA and salmon sperm DNA as outlined below and hybridized onto test slides. Cot1-DNA is enriched for highly repetitive DNA sequences and is used to prevent unspecific binding of repetitive DNA sequences, present in the painting probes, to homologous sequences on all chromosomes.

1. Add to one Eppendorf tube:
   - 4 μl of each chromosome paint probe (400-600 ng each)
   - 50 μl of human cot-1 DNA (1mg/ml)
   - 1 μl salmon sperm DNA (10 mg/ml)
   - Na-acetate (3M), 1/10 of the volume
   - 2.5 -3.0 x total volume of cold 100% ethanol.
2. Vortex, let precipitate at -20 °C overnight or at -80° C for at least 30 minutes.
3. Centrifuge at 13000 rpm at 4 °C for 30 minutes.
4. Remove supernatant carefully and dry the pellet in a speed vac concentrator for 5-10 minutes.
5. Add 5 μl of deionized formamide (pH 7.5), and resuspend pellet at 37 °C for 30 minutes preferably shaking in a thermomixer or in a waterbath, vortex at least twice in between.
6. Add 5 μl of master mix (thaw and vortex Master Mix before use), vortex, spin down briefly.
7. Denature probe DNA at 80 °C for 5 minutes in a waterbath.
8. Allow reannealing of repetitive sequences present in the SKY probes with the cot1-competitor DNA for a minimum of 1 hour at 37 °C.
9. For slide denaturation apply 100 μl of 70% formamide/2xSSC to slides and add 24x60mm^2 coverslip.
10. Denature slides at 75°C on slide warmer for 1.5 min. Denaturation time may vary depending on the nature of specimen (eg destained G-banded chromosomes need shorter denaturation time - approximately 0.5 min).

11. Place slides immediately in freshly prepared ice cold 70% ethanol, followed by 90% ethanol and 100% ethanol for 3 min each.
12. Let slides air dry.
13. After preannealing, add probe DNA to denatured slides, cover with 18 mm^2 coverslips, and seal coverslips with rubber cement.
14. Hybridize at 37°C overnight, protect slides from light.

Detection

After hybridization, biotinylated sequences and DNA labelled with digoxigenin need to be detected using fluorochromes linked to avidin and fluorescently tagged antibodies. The use of directly-linked dUTPs has several advantages. For instance, it is faster and avoids problems related to immunological detection systems. However, the photostability may vary between different fluorochromes applied in a direct or indirect format. We have tested numerous dyes and obtained satisfactory results with regards to incorporation efficiency during PCR and photostability using Rhodamine 110, SpectrumOrange and Texas Red directly linked to dUTPs. With respect to Cy5 and Cy5.5, the best results were achieved if these fluorescent dyes were used in an indirect format, ie via labelling with Biotin-16-dUTP and Digoxigenin-11-dUTP and detection with avidin conjugated to Cy5 and mouse anti dig antibodies labelled with Cy5.5, respectively.

Note: Avoid air drying of slides during the detection procedure.

1. Prepare and prewarm the following solutions:

FA/SSC
- 30 ml 20xSSC
- 120 ml Sterile Water
- 150 ml Formamide
- adjust pH to 7-7.5 by adding 1M HCl
- heat for 30 minutes at 45°C

1 x SSC
- 25 ml 20xSSC
- add H_20, final volume is 500 ml
- heat for 30 minutes at 45°C

4 x SSC/Tween 20
- 100ml 20xSSC
- 400ml H_2O
- 0.5ml Tween 20
- heat for 30 minutes at 45°C

Blocking Solution
- add 0.3g Bovine Serum Albumin to (pre-warmed) 10ml 4xSSC/Tween 20 (final BSA concentration is 3%)
- mix well and dissolve at 37°C

2. Remove rubber cement and coverslips carefully from hybridized slides
3. Wash slides in FA/SSC 3 x 5 min, shaking
4. Wash slides in 1xSSC, 3 x 5 min, shaking
5. Dip slides in 4xSSC/Tween 20, do not let it dry
6. Apply 100 µl of blocking solution to slides. Cover with 24x60 mm^2 coverslips and incubate in moist chamber for about 30 minutes at 37°C.
7. Dip slides in 4xSSC/Tween 20, and apply 100 µl of antibody solution containing avidin Cy5 (diluted 1:200 in 4xSSC/Tween 20/1% BSA) and mouse anti Digoxin (diluted 1:500 in 4xSSC/Tween 20/1% BSA). Add coverslip (24X60 mm^2) and incubate in hybridization chamber for 30 minutes at 37°C

Note: Spin all antibody stock solutions for 3 min at 13,000 rpm, before use

8. Wash slides in 4xSSC/Tween 20, 3 x 5 minutes, shaking
9. Apply 100 µl of antibody solution containing Cy5.5 sheep anti mouse (diluted 1:100 in 4xSSC/Tween20/1% BSA), add coverslip (use 24 X 60 mm^2) and incubate in moist chamber for 30 minutes at 37°C.
10. Wash slides in 4xSSC/Tween 20 at 45°C, 3x5 minutes, shaking
11. Counterstain with DAPI (80ng DAPI per ml 2xSSC; stock solution is 200 µg DAPI per ml sterile water) for 10 min in a lightprotected coplin jar.
12. Wash in sterile H_2O at room temperature, for 5 min, shaking.
13. Dehydrate in an increasing ethanol series of 70, 90 and 100% for 3 min each.
14. Let slides air dry and finally apply 30-35 µl antifade solution, cover with 24x60 mm^2 coverslips, and store slides in the dark at 4°C.

Antifade solution Dissolve 100 mg of 1,4-phenylenediamine in 2 ml 1xPBS, adjust pH to 8.0 with carbonate-bicarbonate-buffer (mix 10 ml of 0.5M sodiumbicarbonate (pH 8.13) and 40 ml of 0.5M sodiumcarbonate (pH 11.32), steril filtrate). Add 1xPBS until a final volume of 10 ml and mix with 90 ml 86% glycerol. Aliquot the antifade solution into uv-safe Eppendorf tubes and store aliquots at − 20°C in the dark.

Image acquisition

The spectral image of the 24 color hybridization and the DAPI image for each metaphase spread are acquired using the SpectraCube (Garini et al. 1996a). DAPI-images are collected with the TR1 filter (Chroma, Inc.) and the SKY-image through a custom designed optical filter set (SKY 3.0, Chroma, Inc.).

Optimal image acquisition requires the proper adjustment of the epifluorescence microscope. The Xenon-lamp needs to be carefully aligned in order to achieve an even illumination throughout the field. To reduce photobleaching, heat protection filters (KG1, BG38) should be fitted into the light pass. Optionally, the BG38 can be removed if the fluorescence intensities in the far red range are relatively low compared to the other fluorescent dyes. The custom-made-triple-filter-cube with narrow excitation, but wide emission bands (Chroma, Inc.) was designed to allow simultaneous excitation for all five fluorochromes. Closing the field diaphragm increases image contrast. Note, that high quality objectives corrected for chromatic aberration are recommended.

When installing the SpectraCube onto a microscope, a set of calibration slides are measured. These are normal slides which were hybridized with single chromosome paint probes labelled with the five fluorescent dyes used for SKY-kit preparation. The spectra measured for those five fluorochromes are stored in a file called "combinatorial table" (ctb-file) and utilized as reference spectra during chromosome classification of the test images.

Image analysis

Using the SkyView software, DAPI- or G-banded images (tiff-format) and SKY-images (raw-files) are analyzed simultaneously. In order to assess the hybridization quality, the SKY-image is visualized first as a RGB display image (Figure 1A). Routinely performed image processing includes automated background subtraction, data connection between DAPI and SKY-images and the separation of the chromosomes. Interactive tools provide the possibility to adjust chromosome segmentation by cutting, editing, joining etc. The reference spectra of the 5 fluorochromes used for kit preparation (Figure 1B) and the PCR labelling scheme (Table 1) are provided in the ctb-file. Based upon this information and the spectra measured for each image point, chromosome classification is performed automatically using a mathematical procedure termed linear combination (Garini et al. 1996b). As a result, a specific pseudocolor is assigned to all image points that have identical spectra. Therefore, chromosomes, chromosomal regions or chromosomal bands showing identical spectra will be displayed in the same pseudocolor (Figure 1C-D).

The quality of the metaphase preparation and the quality of the SKY kits contribute both to the success of the SKY-analysis. A poor hybridization obviously reduces the fidelity of the classification. The SkyView software contains a set of parameters which can be applied to reveal if the test metaphase spread is suitable for classification. Often, SKY-karyotyping of normal metaphase spreads results in a perfect classification by using only the automated karyotyping function. The analysis of highly rearranged metaphases is supported by additional interactive tools provided in the SkyView software. The results of the SKY-classification are displayed in a karyotype fashion and can include DAPI images or conventional banding images of the same metaphase chromosomes, even if they originate from other imaging systems.

Data interpretation

SKY is a molecular cytogenetic technique applied to chromosome preparations obtained from different specimens. Therefore, data interpretation depends not only on the SKY-methodology, but also on the nature of the sample itself. Routine laboratory standards for chromosome analysis (with respect to the number of cells which need to be analyzed, the interpretation of mosaicism or tumor clonality etc.) should be applied for SKY and G-banding alike. Often, the SKY-analysis of only 5 to 10 metaphase spreads allows one to characterize all chromosomal aberrations. In addition, it becomes possible to follow clonal evolution also in cases with complex and multiple rearrangements. The assignment of breakpoints greatly benefits from the combination of G- or DAPI-banding and SKY-analysis. In some instances, – when small markers or subtle translocations are detected –, a subsequent FISH-experiment, using two or three chromosome paint probes, is recommended to confirm the results.

A large amount of data will be produced particularly when analyzing highly rearranged, tumor-metaphase-spreads. The interpretation of these results will be supported by the implementation of database functions into the analysis software.

Results

SKY facilitates the identification of chromosomal aberrations and, therefore, improves and refines conventional cytogenetic diagnostics by solving difficult and questionable cases. Also, clinical cases with normal G-banded karyotypes might sometimes be found by SKY to show a subtle translocation, because the banding pattern of small chromosomal regions can look alike. For instance, one patient was diagnosed with an unbalanced *de novo* translocation (der(18)t(X;18)) by SKY, whereas conventional cytogenetic analysis revealed a normal 46,XY karyotype during prenatal and postnatal diagnoses (Figure5) (Schröck et al. 1997).

In hematological malignancies, the detection of chromosomal aberrations supports diagnostics and therapeutic decisions. Conventional cytogenetics has greatly contributed to the detection of tumor-specific, numerical

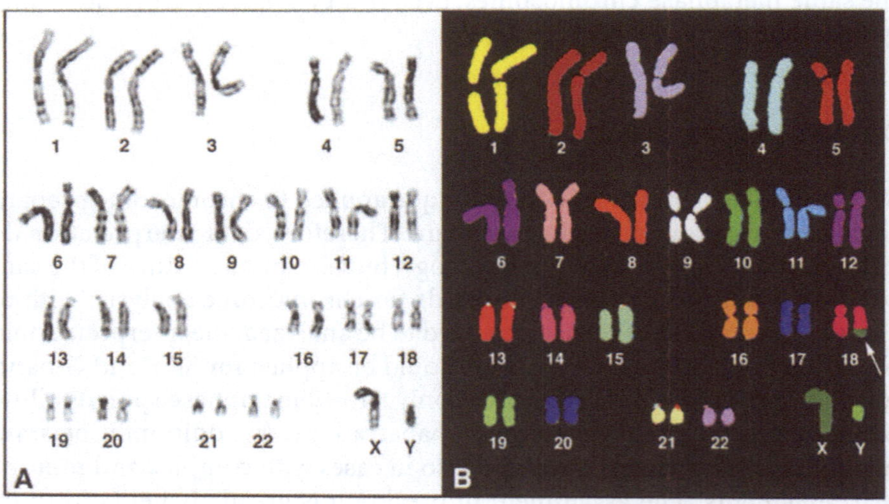

Fig. 5. SKY-analysis performed on previously G-banded metaphase chromosomes prepared from a patient with physical disabilities and mental retardation. A: Metaphase chromosomes after G-banding analysis indicating a normal male karyotype (46,XY). B: The same metaphase was destained and analyzed by SKY revealing an unbalanced translocation der(18)t(X;18). The phenotype of the patient is comparable to an 18q- syndrome (Schröck et al., 1997).

and structural aberrations, eg the t(9;22) in CML, an interstitial deletion of chromosome arm 5q in myelodysplasia and the t(15;17) in acute promyelocytic leukaemia (APL). Nevertheless, the analysis of chromosomal aberrations in leukaemias and lymphomas often remains incomplete. SKY is especially suited in combination with conventional cytogenetics to resolve those unknown rearrangements. In a series of 15 leukaemias, the G-banded diagnosis could be refined by SKY in every case (Veldman et al. 1997). In this study, additional chromosomal material, marker chromosomes and subtle translocations were identified; and apparently normal chromosomes were

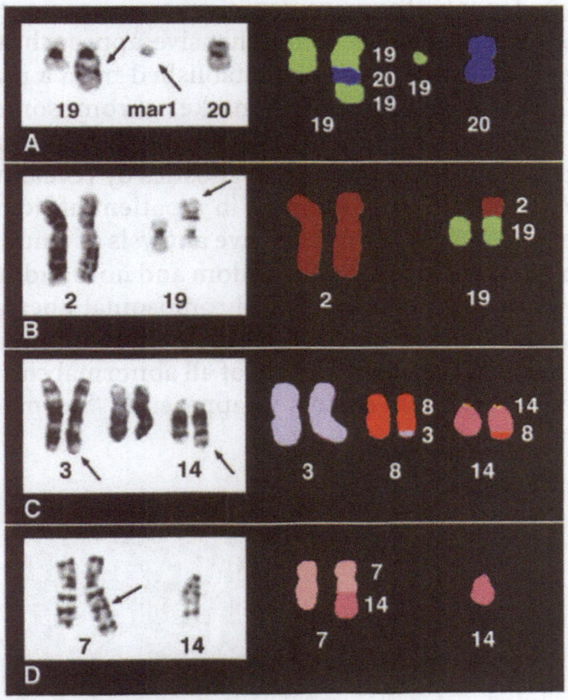

Fig. 6. Examples of chromosome aberrations analyzed by G-banding and SKY in several cases of leukaemias (Veldman et al., 1997). The non-involved homologue chromosomes are shown for comparison. A: The marker chromosome could be identified as being derived from chromosome 19 and an insertion of chromosome 20 material was found on chromosome 19 using SKY. B: The additional material on chromosome 19 was characterized as from chromosome 2. C: A translocation t(3;14) was found by G-banding analysis, however, SKY revealed that the material from chromosome 3 was located on chromosome 8. The terminal band of chromosome 8 was observed on chromosome 14. D: The breakpoint on chromosome 7 was defined by G-banding as 7q36, and the additional material could not be identified. SKY detected that the additional material was derived from chromosome 14. The breakpoint was refined to chromosomal band 7q22.

detected to be involved in translocations. In addition, breakpoints could be refined and complex rearrangements could be completely characterized (Figure 6). Conceivably, the application of SKY in notoriously difficult cases will uncover additional recurrent, tumor-specific, chromosomal aberrations.

In many solid tumors, comprehensive chromosome analysis is frequently not possible. Therefore, cytogenetic data is sparse and does not reflect disease incidence and mortality. Figure 7 shows an example of the analysis of malignant astrocytomas by G-banding and SKY. Despite the highly rearranged karyotype, it was possible to identify all marker chromosomes, detect subtle translocations and resolve the many complex rearrangements. Furthermore, it became possible, using this comprehensive approach, to completely analyze the often-studied Hela cell line established from a malignant cervical adenocarcinoma. Twelve out of 20 marker chromosomes could be resolved for the first time (Macville et al. 1999). In addition, the application of SKY contributed to solid tumor diagnostics by revealing a translocation t(X;18) specific for synovial sarcoma in a patient afflicted with a bone tumor (Cohen et al. 1997). A comprehensive analysis of a number of cases will allow one to differentiate between random and nonrandom abnormalities and define tumor and stage specific chromosomal aberrations (Ried et al. 1997). This goal is now within reach because the combination of G-banding and SKY permits the identification of all abnormal chromosomes and detects additional rearrangements in apparently "normal" chromosomes.

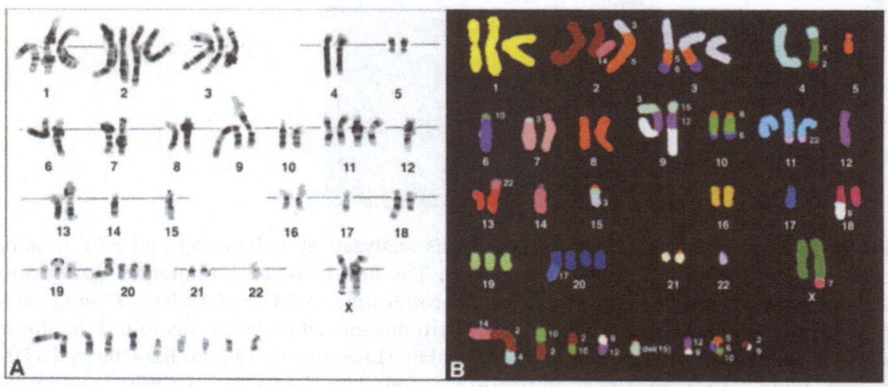

Fig. 7. G-banding and SKY-analysis of metaphase spreads prepared from a glioblastoma multiforme. A: G-banded karyotype indicating multiple rearrangements and numerous marker chromosomes. B: The SKY results from a different metaphase are arranged in accordance to the G-banded karyotype shown in A. Note, that the complex aberrations and the marker chromosomes could be resolved using SKY.

Additional applications of SKY include the analysis of model systems of human cancer, such as experimentally induced tumors arising in knockout or transgenic mice (Liyanage et al. 1996, Barlow et al., Coleman et al. 1997, McCormack et al. 1998). SKY was successfully applied to study chromosomal rearrangements as they occur during the course of human evolution (Schröck et al. 1996). Genotoxic effects of gamma-irradiation were studied by Roschke et al. 1997, indicating that the potential application of SKY in biodosimetry could contribute to the detection of chromosomal aberrations specific for radiation-induced malignancies.

The authors are grateful to Prof. Malcolm A. Ferguson-Smith, Dr. Johannes Wienberg and Patricia O'Brien (Cambridge, UK) for providing chromosome painting probes. We would like to thank Dr. Joan Rankin Shapiro for providing G-banding results and metaphase chromosomes for the SKY-analysis of brain tumors. Dr. Chahira Kozma and Dr. Les Biesecker kindly provided the clinical case. The authors are indebted to Dr. Janet Rowley for contributing metaphase chromosome preparations of leukaemia cases. We would like to acknowledge Stan du Manoir, Tim Veldman, Hesed Padilla-Nash, Merryn Macville, Yi Ning and Marek Liyanage who contributed significantly to the development of SKY. Joel Barnabas is gratefully acknowledged for critically reading the manuscript. SKY was developed under the terms of a Cooperative Research and Development Agreement (CRADA) with Applied Spectral Imaging, Inc. The authors would like to thank Dirk Soenksen (Carlsbad, California) for his generous support and valuable advice.

References

Barch MJ, Knutsen T, Spurbeck JL (eds.) The AGT Cytogenetics Manual, Lippinott-Raven, 1997

Barlow C, Hirotsube S, Paylor R, Liyanage M, Eckhaus M, Collins F, Shiloh Y, Crawley JN, Ried T, Tagle D, Wynshaw-Boris A (1996) *Atm*-deficient mice: a paradigm of ataxia telangiectasia. Cell 86: 159-171

Cohen IJ, Issakov J, Avigad S, Stark B, Meller I, Zaizov R, Bar-Am I (1997) Synovial sarcoma of bone delineated by spectral karyotyping. Lancet 350: 1679-1680

Coleman AE, Schröck E, du Manoir S, Weaver Z, Wienberg J, Ferguson-Smith M, Potter M, Ried T, Janz S (1997) Multicolor spectral karyotyping (SKY) in T(12;15)-positive BALB/c plasmacytomas Cancer Res. 57:4585-4592

Garini Y, Macville M, du Manoir S, Buckwald RA, Lavi M, Katzir N, Wine D, Bar-Am I, Schröck E, Cabib D, Ried T (1996a) Spectral karyotyping. Bioimaging 4:65-72

Garini Y, Katzir N, Cabib D, Buckwald RA, Soenksen DJ, Malik Z (1996b) Fluorescence imaging spectroscopy and microscopy, X. F. Wang and B. Herman, (eds.) (John Wiley and Sons) 137-87-124

Hannig V, Schroer RJ, Martens P, Phelan MC (1984) Chromosome 4p deletion with substitution of unknown chromosomal segment and clinical signs of Wolf syndrome. Proc. Greenwood Genet. Center 3:19-21

Liyanage M, Coleman A, du Manoir S, Veldman T, McCormack S, Dickson RB, Barlow C, Wynshaw-Boris A, Janz S, Wienberg J, Ferguson-Smith MA, Schröck E, Ried T (1996) Multicolour spectral karyotyping of mouse chromosomes. Nature Genet 14:312-315

Macville, M., Schröck, E., Padilla-Nash, H., Ghadimi, B.M., Keck, C., Zimonjic, D., Popescu, N., Ried T. (1999) Comprehensive and definitive molecular cytogenetic characterization of HeLa cells by spectral karyotyping. Cancer Res. 59: 141-50

McCormack SJ, Weaver Z, Deming S, Natarajan G, Torri J, Johnson MD, Liyanage M, Ried T, Dickson RB (1998) Myc/p53 interactions in transgenic mouse mammary development, tumorigenesis and chromosomal instability. Oncogene 16(21):2755-66

Meltzer PS, Guan XY, Burgess A, Trent JM (1992) Rapid generation of region specific probes by chromosome microdissection and their application. Nature Genet 1, 24-28

Ried T, Liyanage M, du Manoir S, Heselmeyer K, Auer G, Macville M, Schröck E. (1997) Tumor cytogenetics revisited: comparative genomic hybridization and spectral karyotyping. J. Mol. Med. 75: 801-814

Roberts I, et al., TIG submitted

Roschke A, Thraves PJ, Kuettel MR, Dritschilo A, Ried T (1997) SKY and CGH analysis of genetic changes involved in radiation-induced neoplastic transformation of human prostate epithelial cells. ASHG abstract #439

Schröck E, du Manoir S, Veldman T, Schoell B, Wienberg J, Ferguson-Smith MA, Ning Y, Ledbetter D, Bar-Am I, Soenksen D, Garini Y, Ried T (1996) Multicolor spectral karyotyping of human chromosomes. Science 273:494-497

Schröck E, Veldman T, Ning Y, Padilla-Nash H, Spurbeck J, Shaffer L, Papenhausen P, Kozma C, Phelan MC, Kjeldsen E, Biesecker L, du Manoir S, Ried T (1997) Spectral karyotyping refines cytogenetic diagnostics of constitutional chromosomal abnormalities. Hum. Genet. 101-255-262

Speicher M, Ballard SG, Ward DC (1996) Karyotyping human chromosomes by combinatorial multi-fluor FISH. Nature Genet 12:368-375

Telenius H, Pelear AH, Tunnacliffe A, Carter NP, Behmel A, Ferguson-Smith MA, Nordenskjöld M, Pfragner R and Ponder BAJ (1992) Cytogenetic analysis by chromosome painting using DOP-PCR amplified flow sorted chromosomes. Genes Chrom Cancer 4, 257-263

Van Dilla MA, Deaven LL, Albroght KL, Allen NA, Aubuchon MR, Bartholdi MF, Brown NC, Campbell EW, Carrano AV, Clark LM, Cram LS, Crawford BD, Fuscoe JC, Gray JW, Hildebrand CE, Jackson PJ, Jett JH, Longmire JL, Lozes CR, Luedemann ML, Martin JC, McNinch JS, Meincke LJ, Mendelsohn ML, Meyne J, Moyzis RK, Munk AC, Perlman J, Peters DC, Silva AJ, Trask BJ (1986) Human chromosome specific DNA libraries: construction and availability. Biotechnology 4:537-552

Veldman, T., Vignon, C., Schröck, E., Rowly, J.D., Ried, T. (1997) Hidden chromosome abnormalities in hematological malignancies detected by multicolour spectral karyotyping. Na. Genet. 15: 406-410.

Chapter 24

Chromosome Analysis by Multiplex-FISH (M-FISH)

MICHAEL R. SPEICHER

Introduction

Traditionally, chromosome staining and karyotyping has made use of simple chemicals that reveal characteristic banding patterns along the chromosome's length. For many years these banding procedures of metaphase chromosomes have been the gold standard for karyotypic analysis. However, metaphase spreads of sufficient quality and quantity are often difficult to prepare, the resolution can be poor, and metaphase spreads from solid tumor tissues frequently have many chromosomal changes that are very difficult to interpret. Thus, it is not surprising, that in recent years the technique of fluorescence *in situ* hybridization (FISH) has grown in popularity. The number of different DNA-probes has grown steadily over the last years and most of them are generally available to the public. These DNA-probes include chromosome-specific painting probes, chromosome specific centromeric probes, unique band-specific probes (eg YACs, BACs, cosmids), and telomere probes.

One limitation of FISH for effective application in clinical diagnosis was the difficulty of choosing the right DNA-probe. Only regions stained by the DNA-probes used can be evaluated. Without prior knowledge about the precise region in question, FISH with a limited number of DNA-probes might be useless. Therefore it was a long-awaited goal of cytogeneticists to be able to distinguish with ease each human chromosome in a cell by some means of specific color labelling. Using a broad palate of paint probes two methods, termed multiplex-FISH (M-FISH/Speicher et al. 1996) and spectral karyotyping (SKY/Schröck et al. 1996), recently realized that goal by showing that they can simultaneously and instantly discern each chromosome.

Michael R. Speicher, Universität München, Institut für Anthropologie und Humangenetik, Goethestr. 31, München, 80336, Germany (*phone* +49-89-5996-622; *fax* +49-89-5996-618; *e-mail* speicher@fish.med.uni-muenchen.de)

This chapter will focus on the M-FISH technique. Advantages of this technique are that both simple and complex chromosomal rearrangements can be detected rapidly and unequivocally. This can be achieved by using either whole-chromosome painting probes or a set of multiple region-specific/unique sequence probes. M-FISH using whole-chromosome painting probes is a powerful screening tool for numerical and structural abnormalities that allows the rapid karyotyping of metaphase spreads. However, some subtle structural changes are difficult to detect (eg cryptic translocations) or can not be detected at all (eg small deletions and duplications, peri- and paracentric inversions). M-FISH with a set of multiple region-specific probes results in a multicolor bar code that increases the resolution of the regions covered by the probe set drastically.

Outline

The flowchart (Figure 1) summarizes briefly the outline of the entire procedure:

Materials

Probe collection - DNA Probes

- Whole chromosome painting probes
  The whole chromosome painting probes used by us were generated either by microdissection or by flow sorting. The microdissected probes were made by Dr. J. Trent (for address see appendix) and generously

Collect probes, amplification by PCR

Probe-Labelling

Preparation of multiplex probe-mix

Probe-mix and chromosome denaturation

Hybridization of multiplex-probe-mix to metaphase chromosomes

Post-hybridization washes and detection of indirectly labelled probes

Microscopy and Image Capturing

Image Analysis

Fig. 1. Outline of the procedure

supplied to us. This laboratory has already generated a large number of microdissected probes (Guan et al. 1993, 1994, 1995, 1996). The flow-sorted whole chromosome painting probes were a generous gift of Dr. Johannes Wienberg (Cambridge University, Department of Pathology, UK). Each probe set has to be amplified by the degenerate oligonucleotide primed (DOP)-PCR following the original protocol as published by Telenius et al. (1992) (see below).

Note: Laboratories wishing to obtain microdissected probes can contact the laboratory of Dr. J. Trent (National Center for Human Genome Research, National Institute of Health, Bethesda, Maryland 20892, USA) for information on the mechanisms established for probe distribution.

- YAC-clones
  A large number of YAC-clones covering more or less the entire human genome has already been identified (Chumakov et al. 1995). A very valuable source for YAC-clones is the CEPH library (information about access to the CEPH-data base and the YAC clones is in the appendix). It is advisable to amplify the YAC-clones by Alu-PCR (Lengauer et al. 1992).

Note: CEPH-YAC clones: Information about the current status of the CEPH-YAC library can be obtained in the internet using the following address: http://www.cephb.fr/bio/ceph_yac.html. It also contains a list of centers that distribute CEPH YAC clones (eg the CEPH YAC Distribution Center in Paris, France; the Whitehead Institute Genome Center in Boston, MA, USA; the Leiden University in The Netherlands).

Reagents for PCR

DOP-PCR and Alu-PCR

- 10xPCR Buffer (same for DOP-PCR and Alu-PCR): 100 mM Tris-HCl, pH 8.4; 500mM KCl; 0.01% (w/v) gelatin. The buffer is stable for several months at -20° C.
- 15mM $MgCl_2$ (Alu-PCR)
- 25mM $MgCl_2$ (DOP-PCR)
- 25 mM dNTP Mix (Alu-PCR): 1:4 dilution of 100 mM dNTPs.
- 5mM dNTP (DOP-PCR): 1:20 dilution of 100 mM dNTPs.

- Oligonucleotide primers:
- Alu-PCR:
- 25mM CL1-primer (5'-TCC CAA AGT GCT GGG ATT ACA G-3')
- 25mM CL2-primer (5'-CTG CAC TCC AGC CTG GG-3').
- DOP-PCR: 100 μM 6MW-primer (5'-CCG ACT CGA GNN NNN NAT GTG G-3')
- Taq Polymerase (Standard concentration: 5 Units/μl)(Taq polymerase can be purchased from different manufactures, no significant differences were noted when Taq's from different vendors were tested.)

Reagents for probe labelling

Probes can be labelled either by Nick translation or DOP-PCR.

Nick translation

- 10 x Nick translation Buffer (0.5 M Tris-HCl pH 8.0, 50 mM $MgCl_2$, 0.5 mg/ml BSA)
- 0.1 M β-mercaptoethanol (0.1 ml of β-mercaptoethanol diluted in 14.4 ml double-distilled water)
- 0.5 mM AGC-Mix (0.5 mM dATP, 0.5 mM dGTP and 0.5 mM dCTP)
- 0.5 mM AGT-Mix (0.5 mM dATP, 0.5 mM dGTP and 0.5 mM dTTP)
- Fluorochromes and haptens:
- 1 mM Biotin-16-dUTP (e.g. Boehringer Mannheim No. 1093 070)
- 1 mM Digoxigenin-11-dUTP (e.g. Boehringer Mannheim No. 1093 3088)
- 1 mM Fluor-X-dCTP (e.g. Amersham No. PA58021)
- 1 mM Cy3-dUTP (e.g. Amersham No. PA53022)
- 1 mM Cy5-dUTP (e.g. Amersham No. PA55022)
- DNase I solution (prepare a stock solution with a concentration of 3mg/ml) dissolve 3 mg DNase I in 0.5 ml 0.3 M NaCl, then add 0.5 ml glycerol, store at -20° C. Before use, dilute 1 μl of this stock solution in 10 ml of ice-cold water
- DNA Polymerase I (eg Boehringer Mannheim, No. 104485, Kornberg fragment, 5 units/μl)

DOP-PCR and Alu-PCR

The reagents needed for probe-labelling with DOP-PCR correspond to the above mentioned reagents for probe-amplification. The same fluorochromes and haptens are used as for the Nick-Translation.

agarose gel

- Seakem ME agarose (FMC Bioproducts, Rockland, ME)
- 50 x TAE (2 M Tris-acetate, pH 8.0; 0.05 M EDTA)
- 1% Ethidium bromide
- DNA gel electrophoresis apparatus.

columns

- Column buffer (10 mM Tris-HCl pH 8.0, 1 mM EDTA, 0.1% SDS)
- Sephadex G-50 (e.g. Pharmacia No. 17-0043-01) (Disperse 30 g of Sephadex G-50 in 300 ml of column buffer and incubate for several hours at 95°C or autoclave. By using the column buffer the spin columns will also contain 0.1% SDS. SDS prevents biotinylated probes from sticking in the column due to the hydrophobic biotin groups).

Reagents for preparation of multiplex probe-mix

ethanol precipitation of DNA probes

- Cot-1 DNA (GibcoBRL/Life Technologies)
- Salmon sperm DNA (The salmon sperm DNA should be sheared or DNAse digested to an average size of approximately 500 bp).
- 3 M NaOAc, pH 5.2
- 70% and 100 % ice cold Ethanol

Reagents for probe-mix and chromosome denaturation

probe denaturation

- Deionized Formamide
- Hybridization buffer (4 X SSC, 20 % dextran sulfate)

slide denaturation

- Formamide (almost any formamide grade is suitable for the slide denaturation; eg Aldrich 18,590-6)
- 20 X SSC (3 M NaCl, 0.3 M sodium citrate, pH 7.0)
- Denaturation solution (70% formamide and 2 x SSC, adjust pH to 7.0).
- 70%, 90% and 100% ice-cold ethanol

Reagents for post-hybridization washes and detection of indirectly labelled probes

- Formamide (almost any formamide grade is suitable for the slide denaturation; eg Aldrich 18,590-6)
- 20 X SSC (3 M NaCl, 0.3 M sodium citrate, pH 7.0)
- BSA (Bovine serum albumine/ fraction V)
- Tween 20
- 4xSSC + 0,2% Tween-20: 200ml 20xSSC pH 7,0 add 1l with bidest + 2ml Tween-20
- Fluorochrome-conjugated reporter binding molecule against biotin and digoxigenin:
- Avidin-Cy3.5 and anti-Dig-Cy7 (both can be ordered from Amersham as special request)

counterstaining and antifade buffer

- 0.2 mg/ml DAPI (4,6-diamidino-2-phenylindole-dihydrochloride)
- Phosphate buffered saline (PBS): 8 g NaCl, 0.2 g KCl, 0.2 g KH_2PO_4, 1 g $Na_2HPO_4 \cdot 2H_2O$, 0.15 g $NaH_2PO_4 \cdot H_2O$. Add distilled water to 1 L and adjust pH to 7.4 with HCl.
- Antifade: Mix 10 ml of p-phenylendiamine solution (100 mg p-phenylendiamine-dihydrochloride in 10 ml of PBS), pH 8.0 and 90 ml of glycerine. Store at -20 C. Alternatively commercially available Antifade (eg Vectashield from Vector) can be used.

Microscopy and image capturing - equipment

Microscope An epifluorescence microscope equipped with the filters as listed in Table 1 is needed. Epifluorescence microscopes with an automated filter wheel should be preferred for two reasons: 1) Filter blocks in a filter wheel can be aligned with such precision that no significant pixel shift occurs. 2) The automated filter wheel allows a very rapid acquisition of all required images in a very user-friendly way. In our lab the newly developed Leica DMRXA-RF8 microscope is used. This microscope is equipped with an eight filter wheel that allows the optimization for excitation-filter, dichroic mirror and emission-filter for eight different fluors. The microscope should have a 100 W Mercury lamp as light source, alternatively, a 75 W Xenon lamp can also be used.

Table 1. This table lists the "first generation" filter set needed for 24-color experiments that was used by us (Speicher et al. 1996). In the meantime a completely new filter generation was developed, details can be obtained from Chroma (address in the appendix).

	DAPI	FITC	Cy3	Cy3.5	Cy5	Cy7
Excitation Filter	Zeiss 365 nm	Omega 455DF70	Omega 546DF10	Ealing 35-3763	Omega 640DF20	Omega 740DF25
Dichroic Beam-splitter	Zeiss 395 nm	Omega 505DRLP02	Omega 560DRLP02	Omega 590DRLP02	Omega 645DRLP02	Omega 777DRLP02
Emission Filter	Zeiss > 397 nm	Omega 530DF30	Ealing 35-3722	Zeiss 630/30	Omega 670DF32	Omega 780EFLP
IR Blocking	Schott BG38	Schott BG38	Schott BG38	Schott BG38	Oriel 58893	Oriel 58895

Camera and computer

A high sensitive charge-coupled device (CCD) camera is needed. This camera should be sensitive in the infrared range and it should be cooled in order to allow longer exposure times. Currently the Sensys-camera (Photometrics; Tucson, AZ) which is cooled to +10°C is used in our lab.

There is already a large number of commercially available software packages for the automated evaluation of M-FISH images from different vendors on the market. It is strongly recommended to test a product carefully before a purchase is made. In our lab we use the Leica-MCK software package that was developed by Dr. Roland Eils at the University of Heidelberg in close collaboration with our lab.

Procedure

Amplification by PCR

Degenerate oligonucleotide-primed PCR /DOP-PCR

- 100 pg to 100 ng DNA
- 5 µl 10x -PCR-buffer
- 4 µl 25mM $MgCl_2$ (Endconc. 2mM $MgCl_2$)
- 2 µl 5mM dNTPs (Endconc. 200 µM of each dNTP)
- 1 µl 6MW-primer 100 µM (Endconc. 2 µM)
- 0.5 µl Taq Polymerase (2.5 units)

- add sterile ddH_2O to final volume of 50 µl

PCR-program

5 min at 93° C, followed by five cycles of 1 min at 94° C, 1.5 min at 30° C, 3 min transition 30-72° C, and 3 min extension at 72° C, followed by 35 cycles of 1 min at 94° C, 1 min at 62° C, and 3 min at 72° C, with an addition of 1 sec/cycle to the extension step and a final extension of 10 min.

Check 5 µl of PCR product on a 1% Agarose gel. The typical amplification product ranges from 100 bp to 2.5 kb, often a distinct band at 400 bp is visible.

If gel looks good, ethanol precipitate DNA, resuspend in 50 µl TE or ddH_2O.

Alu -PCR of YAC DNA:

- 100-150 ng DNA
- 10 µl 10x PCR buffer
- 10 µl 15mM $MgCl_2$ (Endconc. 1.5 mM $MgCl_2$)
- 1 µl 25mM dNTPs (Endconc. 250 µM of each dNTP)
- 1 µl 25mM CL1-primer (Endconc. 250 µM)
- 1 µl 25mM CL2-primer (Endconc. 250 µM)
- 1 µl Taq Polymerase (5 units)
- ad sterile ddH_2O to final volume of 100 µl

PCR-program

3 min at 96° C, followed by 30 cycles of 1 min. at 96° C, 30 sec. at 37° C, 6 min. at 72° C.

Run 10 µl aliquots on a 1.2% agarose gel. The typical amplification product shows a banding pattern with a faint background smear (see Lengauer et al. 1994). If gel shows a good amplification product, ethanol precipitate DNA, resuspend in 44 µl TE or ddH_2O. Store at -20° C or 4° C.

Probe labelling

Labelled probes can be stored for long periods at 20°C without affecting the probe quality. Therefore large probe amounts can be labelled at one time.

Nick translation

The exact Nick translation procedure might vary depending on the probe source used. In general, both microdissected probes and flow sorted probes can be nick-translated in a similar way because in either case the user labels a DOP-PCR amplification product. There are some principles that are important for every Nick translation:

Haptens, such as Biotin or Digoxigenin incorporate generally more efficiently than directly labelled fluors (in our case Fluor-X-dCTP, Cy3-dUTP, Cy5-dUTP). To compensate for these differences the haptens were incubated for 90 minutes at 15°C, the directly labelled fluors for 120 minutes at 15°C. The correct DNase concentration is very important: The DNase concentration depends on a) the DNase stock used; b) the DNA-probe; c) the fluor used for the probe labelling.

ad a) Usually our DNase stock has a concentration of 3 mg/ml for the nick translation when a 1:10.000 dilution of this stock is done. However, different DNase stocks might vary in their activity. Thus, a series of digestions, each with a different DNase solution has to be carried out in order to find the optimal DNase concentration. In general the probe size of whole chromosome painting probes should be in the range of 1kb to 200 bp, the probe size of YAC clones in the range of 600 bp to 200 bp to avoid a strong background. Depending on the DNase stock used the DNase concentrations have to be adjusted.

ad b) Larger DNA fragments should be treated with higher DNase concentrations than smaller probes. This is in particularly true for Alu-PCR products: some YAC clones yield a large number of high molecular weight bands, other YAC clones may yield only a few bands below 1 kb. Thus the correct DNase concentration has to established for each new DNA-probe with some control experiments.

ad c) Different fluors require different DNase concentrations, even if all other parameters are unchanged. In general, the highest DNase concentrations are needed for Biotin and Digoxigenin, Fluorescein and Cy3 need somewhat lower DNase concentrations, Cy5 needs the lowest DNase concentration of all fluors used.

Labelling by DOP-PCR

Labelling is done in a 25 μl volume, example for Biotin labelling:

- 100 ng DNA
- 2.5 μl 10x -PCR-buffer
- 2 μl 25mM $MgCl_2$ (Endconc. 2mM $MgCl_2$)
- 1 μl 5mM dAGCs (Endconc. 200 μM of each dAGC)

- 0.75 μl 5mM dT (Endconc. 150 μM)
- 1.5 μl 1mM Biotin-dUTP (Endconc. 50 μM)
- 0.5 μl 6MW-primer 100 μM (Endconc. 2 μM)
- 0.25 μl Taq Polymerase (2.5 units)
- add sterile ddH_2O to final volume of 25 μl

PCR-program

3 min at 94° C, followed by 35 cycles of 1 min at 94° C, 1 min at 56° C, and 4 min at 72° C, and a final extension of 20 min.

Check 5 μl of PCR product on a 1% Agarose gel. The typical amplification product ranges from 100 bp to 2.5 kb, often a distinct band at 400 bp is visible.

If gel looks good, ethanol precipitate DNA, resuspend in 50 μl TE or ddH_2O. The directly labelled nucleotides need a higher concentration of 100 μM, the concentration of the dT should be correspondingly reduced to 100 μM. In case that one of the nucleotides is not linked to dUTP (eg our Fluoresceine is linked to dCTP) change the dAGC- (for Fluoresceine-dCTP to dAGT) and dT-mixture (to dC) correspondingly.

Check of probe size

Check of probe size after nick-translation or DOP-PCR labelling: A 10th volume of the reaction mix should be used to check the probe size on a 1% Agarose gel. Optimal probe length is in the range of 200 to 800 bp.

Enzyme inactivation

After nick translation inactivate the enzymes by adding 1.5 μl of 0.5 M EDTA (15 mM final concentration), 0.5 μl 10% SDS (0.1 % final concentration) and heat for 15 min at 68° C. Store probes at -20° C.

Preparation of multiplex probe-mix

Probe precipitation

A scheme for probe precipitation of whole chromosome painting probes is shown in Table 2. Prepare your probe mix in this order:

1. Add all probes as listed in Table 2 in a 1.5 ml Eppendorf tube.
2. Add Cot 1-DNA, Salmon DNA, 1/10 volume of 3 M NaOAc and 2 volumes of ice cold Ethanol.

Table 2. Hybridization Scheme: Date _____/_____/____

WCP probe	DNA µl	Evaluation	WCP probe	DNA µl	Evaluation
1-Flu	7		14-Bio	3	
2-Dig	18		14-Cy5	3.5	
3-Cy3	8.5		14-Dig	4	
4-Bio	11		15-Flu	4	
5-Cy3	14.5		15-Bio	2	
5-Dig	11		15-Dig	3	
6-Flu	6.5		16-Cy3	10	
6-Bio	4		16-Cy5	9	
7-Flu	3		17-Flu	2.5	
7-Cy3	3		17-Dig	4	
7-Bio	2		18-Bio	5	
8-Cy5	6.5		18-Dig	9	
8-Dig	4		19-Cy3	2.5	
9-Flu	6.5		19-Bio	2	
9-Cy3	8		19-Cy5	4	
9-Cy5	7		20-Flu	2	
10-Cy5	14		20-Cy5	4	
11-Bio	4		20-Dig	2	
11-Cy5	7.5		21-Flu	5	
12-Cy3	4.5		21-Bio	4.5	
12-Bio	2		21-Cy5	6	
12-Dig	4		22-Flu	8	
13-Flu	2		22-Cy5	9.5	
13-Cy3	3		X-Cy3	3	
13-Dig	3		X-Bio	2	

Table 3. Continued

WCP probe	DNA µl	Evaluation	WCP probe	DNA µl	Evaluation
			Y-Flu	2.5	
			Y-Cy3	2.5	
σ: 283 µl					
Cot:	70 µl		3M NaOAc:	37.3 µl	
Salmon:	20 µl		ETOH:	822 µl	
σ:	372.5 µl				

WCP: Whole chromosome painting probe

Flu: Fluoresceine

3. Precipitate probe mixture at -20° C over night. This is the most efficient ethanol precipitation with a minimal loss of DNA. An ethanol precipitation at -80° C for 30 min gave consistently poorer results. Therefore an overnight precipitation is strongly recommended.
4. Spin probe mixture at 13.000 rpm for 30 min.
5. Discant supernatant and air dry probe. It is very important that the probe is not dried too much. Therefore instead of using a speed vac, we prefer to let the probe air dry. Check probe frequently and add 6 µl of formamide as soon as probe is dry enough. If the probe set is too dry, it will be difficult to resuspend the probe.
6. Transfer tube to a heat block or water bath at 37°C. Let probe mixture dissolve completely, until no pellet is visible. Usually we keep the probe mixture in formamide only for at least one hour at 37°C. If probe is dissolved add 6 µl of hybridization buffer.

Probe-mix and chromosome denaturation

1. Denature probe mix at 75°C for 5 minutes, transfer tube back to 37°C and let probe preanneal for at least one hour. Longer preannealing times are also possible. During this step proceed with chromosome denaturation.
2. Prewarm denaturation solution in a Coplin jar in a water bath to 70°C.

3. Label hybridization field on the slide by scratching the slide with a diamond pen.
4. Put slides into the denaturation solution and incubate for about 2 minutes.
   The denaturation time can vary from slide to slide, it is usually in the range of 1 min 45 seconds to 2 min 30 seconds.
5. Transfer the slides to ethanol series on ice, incubate for 3 minutes each (70%, 90%, 100%).
6. Air-dry slides.
7. Add the denatured hybridization mixture to the denatured chromosome preparation.
8. Put an 18 mm^2 coverslip on the hybridization mixture droplet and seal the edges with rubber cement.
9. Incubate the slides at 37°C for at least two nights. Slides can be incubated longer, however, no increase in signal intensity is observed after two days of incubation.

Post-hybridization washes and detection of indirectly labelled probes

1. Wash 5 min with formamide/2xSSC (1:1, v:v,) pH 7.0 with 1N HCl, prewarmed at 45°C, shaking
2. Wash 5 min with 0.1xSSC prewarmed at 60°C, shaking
3. Incubate slides short in 4xSSC/Tween (few seconds)
4. Blocking: 3% BSA in 4xSSC/Tween, drop 1ml of each slide and incubate 20-30 min at 37°C
5. Remove blocking solution in 4xSSC/Tween (short, only few seconds)
6. Cy3.5-Avidin (1:300) and anti-Dig Cy7 (1:200) diluted in 4xSSC/Tween plus 1% BSA
7. Cy3.5 and Cy7-working solution per slide, coverslip, incubate 45 min at 37°C in a moist chamber, in the dark
8. Washing 3 x 5 min in 4xSSC/Tween prewarmed at 45°C, shaking
9. Counterstaining with DAPI:

DAPI-staining solution: 10ml 4xSSC/Tween + 0.5µl DAPI-Stock solution (2mg/ml)
1ml per slide, incubate for 3 Minutes at room temperature in the dark
Rinse with water, air dry slides, embed with antifading

Results

Typical results for a 24 color experiment with whole chromosome painting probes are presented in Figures 2 and 3.

Troubleshooting

Unsuccessful experiments have usually one or a combination of three reasons: poor probe labelling, poor hybridization, poor metaphase quality.

- Poor probe labelling
  There are some important principles for the probe labelling: Whenever work with a new probe set is started do some test labelling procedures either with Nick translation or DOP-PCR to establish the optimal parameters. The protocols given here are guidelines, they can vary for different probe sets. Directly labelled nucleotides are difficult to evaluate on an agarose gel. The non-incorperated Fluorescein- and Cy3-nucleotides yield strong bands at about 300-400 bp, these bands are often so intense that the DNA smear of the probe is difficult to visualize and to assess. Cy5 emits in the same wavelengths as Ethidium bromide, this results in a quenching of Cy5 so that this fluor appears very weak on the gel. Thus, the agarose gel has a very limited value for the evaluation of the incorporation of directly labelled-nucleotides. Instead of running a gel, we prefer a test-hybridization with newly labelled probes. For an M-FISH experiment use only probes that passed these test-hybridizations.
- Poor hybridization
  There are probably many reasons why a hybridization might have a poor outcome. Some steps that are often easily overlooked: After ethanol precipitation be careful that probes do not get too dry, because this might result in poor probe-dissolving. After resuspending probe in formamide incubate mixture for several hours at 37°C to ensure that the entire DNA will dissolve. The probe concentrations listed in Table 3 are also just guidelines. Every probe set will have some probes that hybridize poorly,

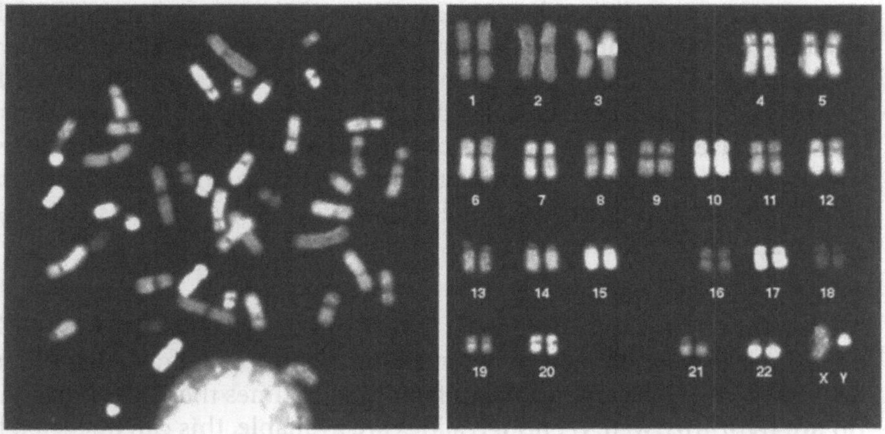

Fig. 2. Normal male metaphase spread after hybridization with a 24 chromosome-specific DNA-probe cocktail. The left column of the figure shows the unprocessed fluorescence source images and the right column the segmentation masks computed for each fluor. The fluorescence banding pattern obtained after DAPI staining was used for chromosome identification. a) DAPI: The DAPI source image was inverted in order to produce a G-band like pattern; b) FITC; c) Cy3; d) Cy3.5; e) Cy5; f) Cy7

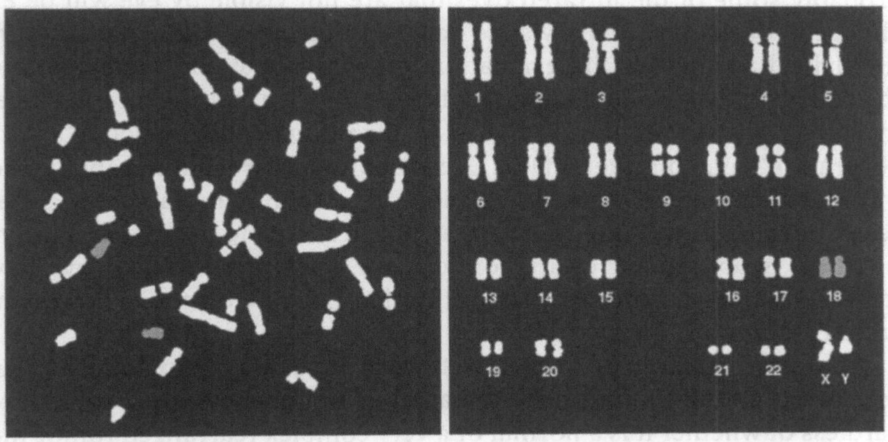

Fig. 3. a) Metaphase spread of Figure 2 as a pseudocolored image. b) Final karyotype generated on the basis of the boolean spectral signature. Note that the heterochromatic block of chromosome 14 has a different color than the q-arm due to some non-specific staining (for details see text).

this can be compensated by increasing the probe concentrations of the respective probes. Only a series of tests will yield the exact probe-concentrations for this complex probe set. The hybridization can often be improved by a pretreatment of slides with RNase and subsequent Pepsin or Proteinase K digestion. Let probe mixture hybridize for at least two nights, this results in a better signal intensity than a one night hybridization. Hybridizations can be done even longer, however, this does not yield significantly better results.

- Poor metaphase quality
  The quality of metaphase spreads is very important. Successful M-FISH experiments were already done on slides that were several years old, so age seems not to be an important factor. Sometimes there are slides that do not hybridize well. If enough slides are available, this can be tested by hybridizing a single painting probe. It is our experience, that conditions that allow a good hybridization of a single painting probe should also yield good results for an M-FISH experiment.

It can easily by anticipated that some parameters for the M-FISH procedure will change in the near future. Most likely some of the fluors will be changed, probably some of the infrared dyes that are not visible by eye will be exchanged for fluors in the visible range. While the labelling-procedure, the hybridization and the evaluation will stay about the same, some of the filters will have to be adapted to the new fluors. Thus, the filter set listed in Table 1 should not be considered as the ultimate, but rather as a "first generation" set. This filter set was already optimized by Chroma and further improvements are very likely. In addition, several manufacturers are already offering or will offer in the near future complete packages including filter sets, probe kits and evaluation software. Thus, most users will not have to worry about an optimal filter set design but rely on the vendor from whom a system is purchased.

The analysis of M-FISH hybridizations is done by a computer algorithm that needs about 3 minutes for the analysis of one metaphase spread, regardless of whether it is a normal or a very complex rearranged tumor metaphase spread. Although the procedure is very reliable and reproducible the accuracy of the results depends on the quality of the hybridization and the metaphase spreads. It should be clear that a poor hybridization might lead to inaccurate probe assignments. Therefore a good evaluation-program should provide built-in quality controls such that the user can check the reliability of the hybridization results.

A prerequisite for both, the M-FISH and the SKY technique, is the availability of DNA-probes, of five different fluors, the correct filter sets, and image analysis software. A (temporary) disadvantage is the limited availability of these items. However, it can be expected that entire kits for multicolor applications will be made commercially available from different companies in the near future.

References

Chumakov IM, et al. (1995) A YAC contig map of the human genome. Nature 377, 175-297

Guan XY, Trent JM and Meltzer PS 1993 Generation of band-specific painting probes from a single microdissected chromosome Hum. Mol. Genet. 2 1117-1121

Guan XY, Meltzer PS and Trent JM 1994 Rapid generation of whole chromosome painting probes (WCPs) by chromosome microdissection Genomics 22 101-107

Guan XY, Meltzer PS, Burgess A and Trent JM 1995 Complete coverage of chromosome 6 by chromosome microdissection: Generation of 14 band region-specific probes Hum. Genet. 95 637-640

Guan XY, Zhang H, Bittner M, Jiang Y, Meltzer P and Trent J 1996 Chromosome arm painting probes Nature Genet. 12 10-11

Lengauer, C. et al. Metaphase and Interphase Cytogenetics with Alu-PCR-amplified Yeast Artificial Chromosome Clones containing the BCR Gene and the Protooncogenes c-raf-1, c-fms, and c-erbB-2. Cancer Res. 52,2590-2596 (1992)

Schrock E, du Manoir S, Veldman T, Schoell B, Wienberg J, Ferguson-Smith MA, Ning Y, Ledbetter DH, Bar-AM I, Soenksen D, Garini Y and Ried T 1996 Multicolor spectral karyotyping of human chromosomes Science 273 494-497

Speicher MR, Ballard SG and Ward DC 1996 Karyotyping human chromosomes by combinatorial multi-fluor FISH Nature Genet 12368-375

Speicher, M.R., Ballard, S.G. & Ward, D.C. Computer image analysis of combinatorial multi-fluor FISH. Bioimaging 4,52-64 (1996)

Speicher MR and Ward DC 1996 The coloring of cytogenetics Nature Medicine 2, 1046-1048

Telenius H, Pelmear AH, Tunnacliffe A, Carter NP, Behmel A, Ferguson-Smith MA, Nordenskjöld M, Pfragner R and Ponder BAJ (1992) Cytogenetic analysis by chromosome painting using DOP-PCR amplified flow-sorted chromosomes. Genes Chrom Cancer 4:257-263

Subject Index

MIX
Papier aus verantwortungsvollen Quellen
Paper from responsible sources
FSC® C105338

If you have any concerns about our products,
you can contact us on
ProductSafety@springernature.com

In case Publisher is established outside the EU,
the EU authorized representative is:
Springer Nature Customer Service Center GmbH
Europaplatz 3, 69115 Heidelberg, Germany

Printed by Libri Plureos GmbH
in Hamburg, Germany